HANDBOOK *of* OPTOMECHANICAL ENGINEERING

HANDBOOK *of*
OPTOMECHANICAL
ENGINEERING

Editor-in-Chief
Anees Ahmad

CRC Press
Boca Raton   New York   London   Tokyo

Acquiring Editor:	Tim Pletscher
Assistant Managing Editor:	Paul Gottehrer
Marketing Manager:	Susie Carlisle
Direct Marketing Manager:	Becky McEldowney
Cover Design:	Dawn Boyd
Prepress:	Kevin Luong

Library of Congress Cataloging-in-Publication Data

Handbook of optomechanical engineering / edited by Anees Ahmad.
    p. cm.
    Includes bibliographical references and index.
    ISBN 0-8493-0133-5 (alk. paper)
    1. Electrooptics--Handbooks, manuals, etc. 2. Mechanical engineering--Handbooks, manuals, etc. I. Ahmad, Anees.
  TA1750.H36 1997
  681'.4--dc20
                                           96-41742
                                               CIP

No claim to original U.S. Government works
International Standard Book Number 0-8493-0133-5
Library of Congress Card Number 96-41742
Printed in the United States of America 1 2 3 4 5 6 7 8 9 0
Printed on acid-free paper

DEDICATION

This book is affectionely dedicated to my wife Rukhsana, and my two daughters Iram and Rabia for their support, understanding and encouragement during the writing of this handbook. I would also like to thank my colleagues and peers in the optics world, who have shared their practical knowledge with me, and provided valuable help, suggestions and advice for making this handbook more useful and informative.

Foreword

Optomechanical engineering plays an increasingly important role in our everyday life. The success of new imaging entertainment systems, next generation computers, laser systems, biomedical instruments, space and aircraft telescopes/instruments, high bandwidth optical communications, and many other emerging industries depend on high quality optomechanical engineering for their performance and competitiveness in the world markets. Optomechanical engineering is a major driving force behind the information society of tomorrow. The material covered in this book forms the foundation of new multi-billion dollar industries over the entire globe.

An optomechanical engineer uses multidisciplinary skills: optics, thermal, structures, materials science, precision mechanics, mechanisms, productions/process engineering, control electronics, focal plane engineering and image visualization to design, specify, fabricate, integrate, align, test, and verify an optomechanical product. The need for these multi-disciplinary engineering skills is great.

The U.S. National Academy of Sciences, through its committee on optical sciences and engineering (COSE) has recently recognized the importance of optics to global economic and defense security. Unlike electrical engineering and mechanical engineering, there are no comprehensive textbooks on optomechanical engineering. There are only two or three universities that teach optomechanical engineering classes and these are within other programs. With the publication of this book there is, for the first time, a much needed handbook of optomechanical engineering available in English.

This handbook will be of very significant value to the working engineer who needs to acquire knowledge of optomechanical systems in a timely manner to satisfy the customer's need. The book is excellent and long overdue.

James B. Breckinridge
Jet Propulsion Laboratory
California Institute of Technology
SPIE Past President
June, 1996

PREFACE

This is the first handbook on the subject of optomechanical engineering, which has gained a very significant importance in the optics world during the past fifteen years. The discipline of optomechanical engineering plays a vital role in optimizing the cost, performance and reliability of all optical systems ranging from common consumer items such as cameras, CD players and binoculars to sophisticated space-based telescopes, navigation, guidance and surveillance systems for military applications; and specialized cameras for remote sensing, medical imaging and environmental monitoring. In spite of a critical role of good optomechanical engineering practices in the successful operation of all photonic systems, very few books and other sources of knowledge and information are available on this subject. This problem is compounded by the fact that just a couple of the universities offer formal courses in optomechanical engineering to the undergraduate and graduate students enrolled in numerous optical science and engineering programs throughout the country. Therefore, most of the today's practicing optical and optomechanical engineers involved in the research, design and development of sophisticated photonic systems have learned their optomechanical engineering skills from on the job training and through a trial and error approach.

Currently, only one book devoted to the subject of optomechanical system design is available. A few of the other books covering broad subjects of optics have only one chapter each covering only a few topics encompassed by the field of optomechanical engineering. Other sources of information on this subject include some SPIE conference proceedings, one Milestone series and short course notes. As a result of this limited availability of literature on such a critical subject, my peers and colleagues decided to collaborate in compiling this first and unique *Handbook of Optomechanical Engineering*. This self-contained book, for the first time, covers all major aspects of optomechanical engineering from conceptual design to fabrication and integration of complex optical systems. This handbook contains practical and valuable information for the practicing optical and optomechanical engineers and scientists involved in the design, development and integration of modern optical systems for commercial, space and military applications.

The handbook consists of ten chapters containing numerous charts, tables, figures and pictures. Each chapter is authored by a world-renowned expert in that area. This unique collaboration of experts makes this handbook a comprehensive source of the latest information and knowledge in the important field of optomechanical engineering. It is almost an impossible task to include design information and guidelines, which would cover each and every imaginable optical system and application. Therefore, to keep this book concise and useful for a broad spectrum of optical engineers and scientists, we chose to cover only those practical topics in depth, which are applicable to the design and fabrication of about 90% of the optical systems for all types of applications. Chapter 1 of this handbook covers some fundamentals of geometric optics for the benefit of those mechanical engineers who may not have taken a formal optics course. This chapter also covers ANSI and ISO drafting standards, the subject of dimensional tolerances and their effect on fabrication cost of optical components. Some important environmental effects, which can affect the performance of optical systems during storage and operation are also addressed in this chapter. Some important optomechanical design principles, including structural design, athermalization, and vibration control are covered in Chapter 2. Chapter 3 contains applications, tables of physical properties, and material selection criteria for a large variety of materials used for optical and structural components of reflective and refractive optical systems. Chapter 4 includes discussion on the important subjects of dimensional stability, materials and fabrication methods for lightweight metal mirrors. The topics relating to lightweight design of open-back and sandwich-type mirrors are addressed in detail in Chapter 5.

Chapter 6 includes comprehensive design guidelines and numerous designs of mirrors, lenses, prisms and optical windows mounts. It addresses the adverse effects of metal to glass contact in these mounts, and gives very useful equations to calculate axial and radial stresses in optical elements. Chapter 7 covers the design and applications of linear, tilt and rotary adjustment mechanisms in sophisticated optical systems requiring micron-level assembly and alignment accuracies. This is the first time that practical design aspects of adjustment mechanisms have been covered in detail in any book. Chapters 8 and 9 also make this handbook unique, because it is the first time that topics of structural and thermal finite element analysis for optical systems have been covered in any book on optics. Chapter 8 provides useful information about practical finite element modeling and design optimization techniques for the optics, mounts, metering structures and adhesives. Thermal and thermoelastic analysis of optics is covered in Chapter 9, including the analytical methods for modeling temperature gradients, coefficient of thermal expansion variations and the resulting adverse effects on the performance of optical systems subjected to large temperature variations. The last chapter of the book addresses the selection and specification of optimum fabrication and assembly methods for optical and structural components made from diversified materials such as metals and high performance composites. The replication, heat treatment, stabilization, finishing, coating , plating and electroforming methods for optical systems are covered in great detail in this chapter.

A large number of people and organizations have provided help in the preparation of this book by providing advice, photos, figures and other technical information. I would like to express my special thanks to Dr. John Caulfield, who initiated and encouraged me to undertake the task of writing this handbook. I am also indebted to the staff of the *Center for Applied Optics* for providing word processing and drafting help, and to many people at CRC Press, who made the decision to publish this first ever *Handbook of Optomechanical Engineering*, and showed great understanding and patience during numerous delays and missed deadlines.

Anees Ahmad
Editor

THE EDITOR

Anees Ahmad is a Senior Research Scientist at the Center for Applied Optics, and an Associate Research Professor in Optical Science and Engineering Program at the University of Alabama in Huntsville, Huntsville, AL. Dr. Ahmad has over 17 years of professional experience in precision optomechanical design and optical engineering. Prior to joining the Center for Applied Optics, he worked at Perkin-Elmer Corporation in Wilton, CT and Martin Marietta in Orlando, FL. There, he was involved in research and development of advanced microlithographic systems, and state of the art photonic systems for navigation, surveillance and remote sensing applications. His current research interests include optomechanical design and fabrication of ultra-lightweight instruments for biomedical, space and military applications. He is also involved in the development of knowledge-based expert systems, and integrated computer-aided optical system design and development. He has several publications and patents in various areas of optomechanical engineering. He is a member of SPIE and OSA. He received his Ph.D. in mechanical engineering from the University of Houston, Houston, Texas in 1979.

Contributors

Anees Ahmad
Center for Applied Optics
The University of Alabama in
   Huntsville
Huntsville, Alabama

Darell Engelhaupt
Center for Applied Optics
The University of Alabama in
   Huntsville
Huntsville, Alabama

Victor Genberg
Eastman Kodak Co.
Rochester, New York

Roger A. Paquin
Advanced Materials Consultant
Oro Valley, Arizona

Robert E. Parks
Optical Perspectives Group, LLC
Tucson, Arizona

Daniel Vukobratovich
NOAO
Tucson, Arizona

Ronald R. Willey
LexaLite Scientific Center
Charlesvoix, Michigan

Paul R. Yoder, Jr.
Consultant in Optical Engineering
Norwalk, Connecticut

Contents

1

Optical Fundamentals

Ronald R. Willey and Robert E. Parks

1.1 Introduction

This handbook consists of chapters written by authors of considerable experience in the practical application of **optomechanics** and covers a broad range of related subjects. In this chapter, we introduce general background information, techniques, and concepts which may be useful to the practitioners of optomechanics. These topics include definitions, fundamentals of geometrical optics, drawing standards, tolerancing concepts, and environmental effects.

What Is Optomechanics?

We will draw upon the definition of optomechanical design from Dan Vukobratovich who is a contributing author in this handbook. We will define **optomechanics** as the science, engineering, and/or art of maintaining the proper shapes and positions of the functional elements of an optical system so that the system performance requirements are satisfied.

Vukobratovich also points out in his chapters why optomechanics is different from conventional mechanical engineering in that the emphasis is on strain or deformation rather than stress.

The recent difficulties with the Hubble Space Telescope (HST) have been attributed to an optomechanical error in the relative positioning of two mirror surfaces in the null corrector used to test the primary mirror. Great pains were undoubtedly taken in the design and construction of the HST which was a major achievement in the field of optomechanics. Aside from its well-publicized difficulty, it still is a good example of the application of optomechanics. The shape and

0-8493-0133-5/97/$0.00+$.50

the position of the functional elements such as the primary and secondary mirrors must be maintained very precisely in order for the instrument to obtain results not previously achieved. It is hoped that the repairs made to the HST in 1993 will allow most of the original scientific goals to be attained.

Another example of optomechanics at the other extreme are the eyeglasses that many of us wear. The frame must hold the lenses such that their principal points and astigmatic correction are in the right position and orientation with respect to the user's eyes to within appropriate tolerances. The frame must also interface to the user's head in a comfortable and reliable way, and the whole system must perform and survive in an appropriate variety of environments.

Role of Optomechanical Design and Its Significance

The optical design of an optical instrument is often less than half of the design work. The mechanical design of the elements and their support and positioning also are at least as critical. If we look at the typical astronomical telescope consisting of two or three mirrors, there are many more parts in the mechanical structure that position the mirrors and attempt to do so without distorting the functional surfaces. It can be seen that the mechanical design or the **optomechanics** plays a **major** role in any optical instrument development, particularly where the optics and mechanics interface.

It can be seen from the above examples of the HST and eyeglasses that optomechanics is significant to our lives over the whole gamut from the mundane to the sublime. An out-of-tolerance condition in our glasses can cause us headaches and other pains. A similar condition in the HST caused the scientific community emotional and fiscal pain. On the other hand, proper eyeglasses enhance our individual abilities and comfort, and a proper HST can expand our knowledge of the universe. The reader's own imagination can provide a myriad of other examples of the role and significance of optomechanical design. These might include adjustable rearview mirrors in cars or very sophisticated military optical systems.

The optical and mechanical designers of instruments have by far the greatest influence on the ultimate cost and performance of an instrument. All others, including the manufacturing operations, cannot have as much influence as the designers to change the potential satisfaction of the user and profitability of the producer. Therefore, once the instrument's function is satisfied, economics is of great significance in optomechanical design. Figures 1.1 and 1.2 illustrate these effects and show some of the typical steps in the process of developing an optical instrument.

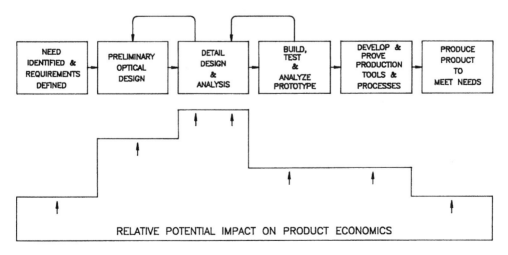

FIGURE 1.1 Overall process to develop and produce a new optical product.

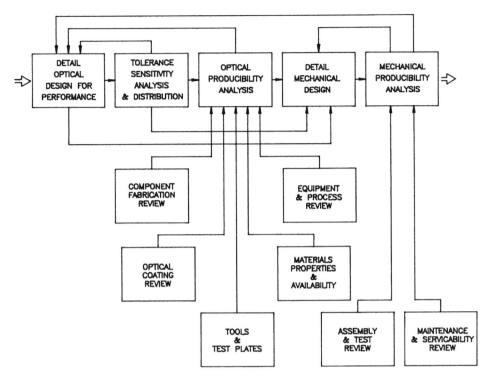

FIGURE 1.2 Detail design and analysis process of an optical product.

1.2 Geometric Optics Fundamentals

Basic Terminology

In this section, we will discuss some of the more pertinent optical terms and definitions which are often needed by the optomechanical design engineer. These terms will be highlighted in **bold** where they first occur in the discussion below. There are many good texts such as Smith's[1] for more detail if needed. We will provide some tools and concepts here which we think are useful for the designer.

Graphical Tools in Geometrical Optics and System Layouts

Image Position and Magnification

In many instances, an assembly of lenses such as a 35-mm camera lens, a telescope, a magnifier, etc. can be treated as a **black box** where only three parameters are known about the lens system. These three parameters are the **effective focal length** (EFL) and the positions in the box of the **front and rear principal points** (P_1 and P_2). Figure 1.3 illustrates such a black box lens system. We will assume throughout this discussion that the optical system is rotationally symmetric about the **optical axis** which passes through the two principal points and that there is air or a vacuum on both sides of the lens system. With this information, we can find the position and size or **magnification** of the image of any object which the lens system can image. This is the first major area of concern in an optical system which the optomechanical designer can easily work out. The second area has to do with how much light can get through the system as a function of the angle relative to the optical axis. We will address these issues subsequently.

Everything that we will deal with here is referred to as **first-order optics.** Departures from the answers which first-order calculations give are **aberrations** or deviations from these answers. These are higher-order effects which lens designers attempt to reduce to practical values in their detail

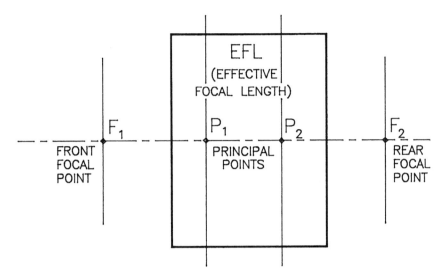

FIGURE 1.3   The three properties which define a "Black Box" lens: effective focal length (EFL) and the two principal points (P_1 and P_2).

design processes. We will not discuss aberrations in detail here since they are not something that the optomechanical design process is expected to improve upon. However, we will briefly introduce the subject. The famous scientist Hamilton viewed optical aberrations as being of three types. The first aberration is composed of the effects which cause the image of a sharp point (stigma) object not to be a sharp point. This he referred to as **astigmatism.** The second Hamiltonian aberration is that the image of a flat plane perpendicular to the optical axis is not on a flat plane but a curved surface. This he called **field curvature.** The third is that the mapping from the object plane to the image plane has **distortion.** This would make a rectangle look like a pincushion or a barrel. Optical designers today tend to divide astigmatism into several parts called **spherical aberration, coma, astigmatism,** and **longitudinal and lateral chromatic aberration.** This latter definition is much less inclusive than Hamilton's astigmatism. These details are mostly of only academic interest to the optomechanical design engineer.

The term "first-order" in optics can be described as simplified equations which are derived by using only the first terms of the series expansions of the sine of an angle and the coordinates of a spherical lens surface as shown in Equations 1 and 2.

$$\sin a = a - a^3/3! + a^5/5! + \dots \tag{1}$$

$$z = y^2/2R + y^4/8R^3 + y^6/16R^5 + \dots \tag{2}$$

In the equation for the sphere, the y is the **zonal radius** from the intersection of the optical axis with the surface of the lens, or its **vertex.** The R is the radius of curvature of the lens surface. The z is the **sagittal** distance or height in the direction of the optical axis from the plane containing the vertex and which is perpendicular to the optical axis. If we use only the first term to the right of the equal sign in each case, we have first-order optics. If we include the next term in each case, we have the basis of third-order optics and the associated aberrations. Seidel worked out the third-order relations whereby all of the above-mentioned aberrations can be calculated. The next terms give fifth-order aberrations etc. With the availability of computing power in modern times, we find only first-order and rigorous calculations to be useful; the higher-order aberrations are of mostly academic value. For our present purposes, no calculations are required, rather only graphical constructions based on first-order principles are needed.

With reference to Figure 1.3, we define a plane containing a principal point and perpendicular to the optical axis as a **principal plane.** The convention is that light passes in a positive direction through a lens if it moves from left to right. The first principal plane is the one which the light reaches first and its principal point is P_1. Similarly, the second principal plane and point P_2 is where light reaches after passing the first. When light is dealt with before it reaches the first principal plane it is said to be in **object space** because the object to be imaged is in that space. After the light has passed through the lens and departed from the second principal plane it would then be in **image space.** The first **focal point** (F_1) is at a distance EFL to the left of P_1 and the second focal point (F_2) is EFL to the right of P_2 on the optical axis. The **focal planes** are planes which contain the focal points and are perpendicular to the optical axis. With only these data, we can construct the position and size of images formed by any centered optical system. Although we will illustrate this with simple examples, almost all image-forming systems can be reduced to an effective focal length and its two principal points and thereby treated by this same technique.

The principles needed for this construction are simple and are illustrated in Figure 1.4. First, when a ray parallel to the optical axis in object space intersects the first principal plane at a given height, it will exit the second principal plane at the same height and pass through the second focal point F_2 in image space. Similarly, a ray parallel to the optical axis in image space must first pass through the first focal point F_1. Second, any ray passing through the first principal point will exit from the second principal point in the same direction (parallel to the first ray). These are really the **nodal points** which correspond to the principal points as long as the lens is bounded on both sides by the same medium (usually air or vacuum).

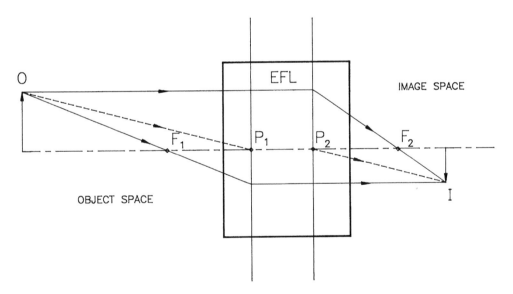

FIGURE 1.4 Construction of an image point from a general object point to find the size and position of an image.

In Figure 1.4, the image point I of an object point O is found by using the above rules on the ray from the point O parallel to the axis and the ray from O through F_1. The rays through the nodal points are also shown (dashed) as an alternative or check. These principles can be applied to any object point. Note that a point on one principal plane is imaged as a unit magnification on the other principal plane.

There are cases where the principal planes are "crossed" such that the first is to the right of the second. In such cases, the rays are still traced from the object point to the first principal plane and emanate from the second at the same height. The principal planes may also be outside of the physical lens system. This is particularly true of some telephoto lenses where, by definition, the lens system is shorter than its effective focal length. We, therefore, have the simple tools we need to find the position and size of the image of any object.

For an afocal telescope as a whole, the above rules have no clear meaning since the focal length is infinite. However, an afocal system can be dealt with when broken into front and rear sections of finite focal length whose focal planes are coincident.

Amount of Light Through a Lens System

The amount of light which can get through a lens system at a given angle from the optical axis is determined by the pupils, apertures, and vignetting. We will neglect the effects of the transmittance of the lenses and reflectance of the mirrors in the system and only address the relative difference of some angle off-axis from the on-axis light. The **entrance pupil** of a lens is the aperture viewed from object space which can pass light to the image space. The **exit pupil** is the aperture viewed in image space which passes light from object space. Both of these pupils are the images of the same **aperture stop** as viewed from object and image space. In a photographic lens, the aperture stop is typically an iris diaphragm which is of adjustable aperture for light brightness control. The **F-number** of a lens is the effective focal length divided by the entrance pupil diameter. The **numerical aperture** is another statement of the same quantity where it is 1/(2*F-number) when objects are at infinity and in air or vacuum. The greater the numerical aperture, the more light will pass through the lens.

We will now use a simple lens to illustrate pupils and the use of the previous principles to find an exit pupil. Figure 1.5 shows a meniscus lens with the aperture stop well in front of the lens and gives the principal and focal points (known as **cardinal points**). In this case, the aperture stop is also the entrance pupil because there are no other lenses in front of it. We want to find by construction the size and position of the exit pupil, which is the image of this stop. We use the point at the top of the stop as an object point to find where it is seen in image space. The ray through that point and the front focal point F_1 is extended until it intersects the first principal plane, and then a line parallel to the optical axis is drawn from that height in the second principal plane and on to the second focal plane. Note that this line extends from minus to plus infinity and we will find that it intersects the next ray on the left of the lens. The next ray is the ray through the stop parallel to the axis in object space which passes through the second focus F_2 in image space. When the image space ray through F_2 is extended backward to the left, it intersects the first ray at the image of the top of the stop in image space. By symmetry, this gives the position and size of the exit pupil. Even though it is to the left of the lens, it is still in image space because it is composed of light that has passed through the lens. Therefore, this exit pupil is larger than the entrance pupil and is farther to the left, but note that the numerical aperture or F-number is still correct.

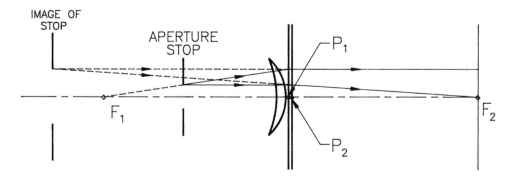

FIGURE 1.5 Finding the exit pupil of a system by constructing the image of the aperture stop.

As seen in Figure 1.6, we now trace rays, which are parallel to a line from the object to the front node (P_1), from the top and bottom of the exit pupil to the first principal plane. Then, from the same height on the second principal plane as those intersections on the first principal plane, we draw rays to the image point (determined by the line from the second node [P_2] to the focal plane which is parallel to the line from the object to the first node). This shows where this off-axis beam will be on the lens. Note that a ray (dashed) through the nodes of the lens to this off-axis image does not pass through the stop; it is **vignetted.** This is not a problem because the nodal ray is just used to find image position from an input angle and vice versa. If we go to a greater angle off-axis as in Figure 1.7, the aperture of the lens itself limits the rays which can pass through it; this is then a **limiting aperture.** The interaction of this limiting aperture and the exit pupil creates a vignetting pattern also shown in Figure 1.7. The only light which can pass is where the two apertures overlap, in this case about 50% of the on-axis value. More complex lenses will typically have front and rear limiting apertures which interact with the exit pupil to give vignetting. When these block all of the light coming to the focal plane, we reach the absolute limit of the **field of view** (FOV). Vignetting is sometimes used to block highly aberrant rays so that the image is sharper even though it is correspondingly dimmer off-axis. In many photographic or television systems, the film or detector (CCD) sizes are usually the limiters of the FOV rather than the vignetting. In visual instruments like binoculars or telescopes, there may be a physical aperture at the final or intermediate focal plane which limits the FOV and is called a **field stop.**

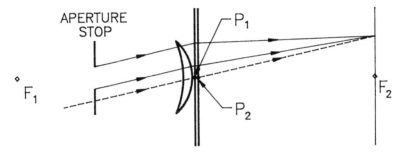

FIGURE 1.6 Construction of the ray paths from an off-axis object through the aperture stop or pupil to the image plane.

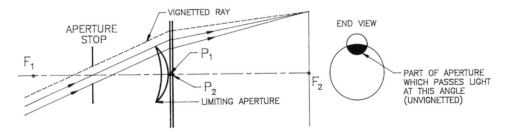

FIGURE 1.7 Off-axis beam which is partially obstructed by a limiting aperture showing its vignetting effect and limiting the ultimate field of view.

The display of the vignetting of the apertures can be easily constructed. In Figure 1.8 we project the limiting aperture onto the plane of the exit pupil. This is done by a first ray from the image point through the center of the limiting aperture to the exit pupil plane. This defines the center of the projected circle. The second ray is traced from the image point through the top or bottom of the limiting aperture to the exit pupil plane. This defines a point on the circumference of the circle whose center was found above. The common overlapping area pattern of this circle and the circle of the exit pupil (or image of the aperture stop) give the vignetting of the system for this image point as illustrated to the left in Figure 1.8. This same result can be obtained by projecting the exit pupil onto the limiting aperture plane. The whole procedure could also have been equally

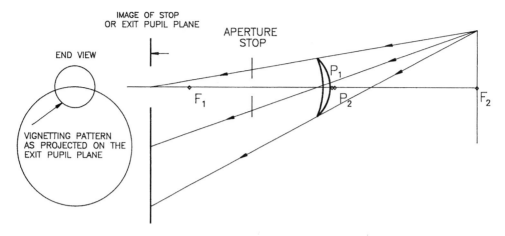

FIGURE 1.8 Construction of the vignetting pattern at a given field angle (image position) in image space.

well applied in object space instead of image space. In this latter case, the vignetted pupil is viewed from the object point.

Additional CAD Techniques for Design and Ray Tracing

It is now practical to trace rays that are rigorously correct using computer-aided design (CAD) tools. This has been true for over four decades with respect to ray tracing and lens design where equations are used to calculate ray paths in three-dimensional space. In fact, optical design may have been the most active CAD process over the majority of that period. However, today's drafting CAD tools allow additional ray tracing to be done with relative ease in a graphical setting. This would not be a replacement for optical design in the usual sense, but it allows the optomechanical designer to find the exact path of a specific ray when needed. The data required is the same prescription which the lens designer produces. This includes: the surface position in three dimensions, its radius, the index of refraction on each side of the surface, and the incoming ray coordinates. The relation which we will execute graphically is **Snell's law** as given in Equation 3.

$$n \sin i = n' \sin r \qquad\qquad (3)$$

The index of refraction on the side where the ray is incident is **n.** The angle of the incident ray to the surface normal at the point where the ray intersects the surface is **i.** After refraction, the ray is in a medium of index **n′** and makes an angle **r** with the surface normal. The refracted ray lies in the plane which is defined by the incident ray and the surface normal.

The steps to trace the refracted ray as seen in Figure 1.9 are as follows:

Define (draw) the refracting curved surface in a plane which contains its center of curvature and the incident ray.

Draw the incident ray to the point where it intersects the refracting surface and draw the surface normal at that point. This is a line through that point and the center of the circle.

Draw two circles about the intersection point of the incident ray and the refracting surface whose radii are the same multiple of n and n′. It might be convenient to use 1.000″ for an index of 1.000 and 1.517″ for an index of 1.517 or possibly 2× these values.

Through the point where the incident ray intersects the n-circle, draw a line parallel to the surface normal which intersects the n′-circle. This could be a copy command offsetting from the incident ray intercept with the refracting surface to the ray intersection with the n-circle.

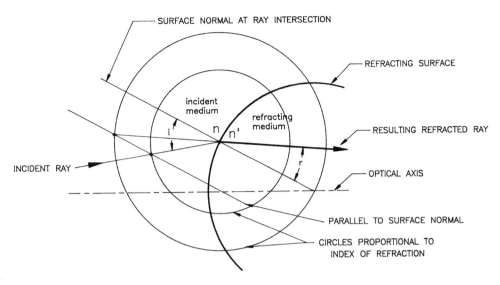

FIGURE 1.9 Rigorous ray tracing by construction of a refracted ray.

From the intersection of this parallel line with the n′-circle, trace a line which passes through the point where the incident ray intersects the refracting surface. This line, when extended beyond that point, is the refracted ray which we wanted to construct.

Trim away any excess lines and the process is done for that refraction of the ray.

It can be shown that this construction satisfies Equation 3. Because of the rigor of modern CAD drawing systems, this should provide real rays that are as accurate as the system can draw.

If exact ray trace is available, the ray can be drawn by connecting intercept points on surfaces. This may be more precise than the above-described graphical means.

If we need to find the path of a reflected ray, the process is even simpler on CAD. The "MIRROR" function of CAD will provide the reflected ray by mirroring the incident ray about the surface normal at the point of intersection.

Rays could be propagated through entire systems this way if needed. The most likely use might be only occasional, such as to check the path of rays near the edge of a lens mounting cell.

All of the tools and concepts described above are general and can be applied with or without CAD, but the availability of CAD has made them that much easier and more useful.

1.3 Drawings of Optical Components and Systems

Units of Measurement

The optical industry has probably been the first in the U.S. to be most familiar with the metric system. Most optical shops work in inches or millimeters with almost equal facility. Some recent U.S. Government military contracts even require the drawings to be in metric units, millimeters, etc., which is a change from the practice of the past. The choice of the units of measurement will probably continue for some time to vary with the customers and the contract specifications. The purpose of the new standards described below is to make specifications more uniform and to reduce ambiguity.

ISO and ANSI Drafting Standards

There are far more international (ISO) and national (ANSI) voluntary mechanical and optical technical drawing standards available than most people are aware of. In this section, we will describe most of these standards, explain why the tendency now is toward using the ISO standards, and where copies of these standards can be obtained.

First a word about why these standards are not better known. At least in the U.S., the majority of optical drawings were either in-house drawings for commercial products made on site and proprietary standards were used, or the drawings were done for military optical systems and were done to military standards. Thus there was little need for optical firms to use voluntary national or international standards. Now, however, there are few commercial optics being made in the U.S. and orders for military optics have decreased dramatically. Instead, commercial systems are being designed in the U.S., Europe, and Japan for manufacture elsewhere and the value of standards that are understood in all parts of the world is increasingly valuable.

Another point to keep in mind about optomechanical drawing standards is that they explain how to indicate on a drawing what features and dimensions for the features are desired on the finished product. The standards, in general, do not tell what values should be used for various desired features. In this sense, this section is merely a lead in to the next section on tolerances, where the actual numerical values to be used on the drawing will be discussed.

It should be pointed out, however, that the ISO standards do offer some guidance on suggested values for certain features, a guidance that is not often found in ANSI standards. For one, virtually every feature that should be considered in the drawing is listed so there is a reminder to think about whether this item needs to be considered or not. In addition, the ISO optical drawing standard contains a section that lists default tolerances. If, for example, chamfers are not called out on a drawing, this section governs the widths of chamfers on the finished components.

Before describing the available standards, we would like to re-emphasize why, when there is a choice of using either ANSI or ISO standards, we would recommend using the ISO ones. More and more, we are dealing in a global economy and the use of methods that make international commerce easier will lead to greater productivity. The ISO standards have been written more recently than most of the ANSI standards, are much more thorough in their treatment of features to be indicated on drawings, and they also deal with a broader range of possible features than the existing national standards.

Another reason to adopt these standards is that they are being built into optical design software. Several of the major lens design code suppliers have added or are adding the ISO indications to their optical drawing software packages. In addition, these same suppliers are working with the electronic data transfer standards people doing IGES/PDES and the more comprehensive STEP work for transfer of CAD/CAM data files. In other words, by working in the ISO environment, drawings will be forward compatible into the newer methods of electronic data transfer.

There is one final reason for using the ISO approach. There has been a concerted effort to make optical drawings virtually noteless, a big change from the past U.S. practice. Most of the ISO indications on drawings use alphanumeric symbols which stand for certain features or parameters of the features. Once the code for the features and parameters is learned, and whatever may be the language of either the designer or the manufacturer, the drawings can be interpreted by workers having almost any language background without the need for translation. An optical drawing created in Japan or Russia using the ISO notation should be readily usable in the U.S., for example, without any need for translation and vice versa. Figure 1.10 is an illustration of the application of the ISO standards to the drawing of a lens. This example was provided by Sinclair Optics and was automatically generated by their optical design software package.

Mechanical Drawing Standards

The principal U.S. national standard covering mechanical drawing practice is ANSI Y14.5M, *Dimensioning and Tolerancing*. This standard explains how to represent on drawings concepts such as maintaining parallelism between two surfaces or that a hole is located a certain distance from a pair of right angle edges. In addition to defining the symbols needed to express these ideas, the standard gives examples of what one should expect in terms of a finished part when a given set of symbols and dimensions are put on a drawing. Two other related standards used in conjunction

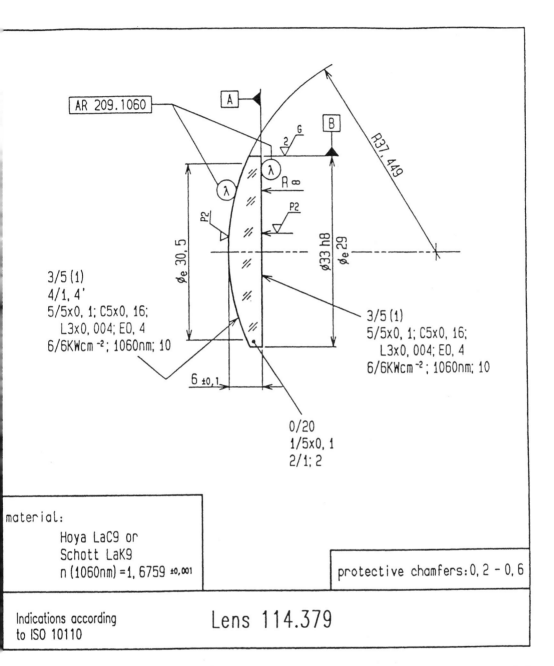

AR 209.1060

3/5 (1)
4/1, 4'
5/5x0, 1; C5x0, 16;
  L3x0, 004; E0, 4
6/6KWcm⁻²; 1060nm; 10

3/5 (1)
5/5x0, 1; C5x0, 16;
  L3x0, 004; E0, 4
6/6KWcm⁻²; 1060nm; 10

6 ±0,1

0/20
1/5x0, 1
2/1; 2

material:

Hoya LaC9 or
Schott LaK9
n (1060nm) =1, 6759 ±0,001

protective chamfers: 0, 2 - 0, 6

Indications according
to ISO 10110

Lens 114.379

FIGURE 1.10 Example of a lens drawing to the ISO standards. Note the limited notes and dependence on any particular language (i.e., English).

with ANSI Y14.5M are ANSI Y14.36, *Surface Texture Symbols*, and ANSI B46.1, *Surface Texture*, dealing with the finish of machined parts.

We should mention two books that are tutorials on ANSI Y14.5M. These are *Geo-metrics II*,[2] written by the vice chairman of the ANSI Y14.5 committee when the latest version of the national standard was published, and *Design Dimensioning and Tolerancing*,[3] which is written like a textbook and is a good place to start if one knows little about ANSI Y14.5M.

The ISO standard analogous to ANSI Y14.5M is ISO 1101–1983, *Technical Drawings — Geo-metrical Tolerancing — Tolerancing of Forms, Orientation, Location and Run-Out Generalities, Definitions, Symbols, Indications on Drawings*. In fact, there are so few differences between these two documents that the differences are easily listed in the Preface of ANSI Y14.5 and the members of both committees are working to remove even those.

A companion to ISO 1101 is ISO/TR 5460–1985, *Geometrical Tolerancing — Verification Principles and Methods — Guidelines*. For every feature designation in ISO 1101, ISO/TR 5460 shows in schematic form, kinematically correct methods for verifying that each feature meets the specified criteria. For every kind of feature control call out, this standard shows several examples of how to fixture the part to a surface plate and how to set up an indicator to verify that the feature is in tolerance. The effect is to make clear in a functional way what each feature control call out means in terms of the instruments used to inspect the feature.

The above-mentioned ISO drawing standards are bound together along with 58 other standards related to mechanical drawings and metrology in a 600-plus-page book called *ISO Standards Handbook 33, Applied Metrology — Limits, Fits and Surface Properties*. This, along with an earlier publication, *ISO Standards Handbook 12, Technical Drawings*, is a convenient and relatively inexpensive way of obtaining all the ISO standards on mechanical drawings published through 1988. These handbooks as well as more recent ISO standards are available through ANSI,[4] the U.S. sales agent for ISO.

Before leaving the mechanical drawing standards, we should mention that the official system of units for ISO standards is the SI system (the International System of Units). While ISO mechanical drawing standards allow for the use of English units on drawings if they are identified on the title block, English units are neither extensively used in world trade nor in optics, in general. Thus, it is suggested that the reader have a copy of ISO 1000–1981, *SI Units and Recommendations for the Use of Their Multiples and of Certain Other Units*. This standard is available along with 14 others dealing with definitions of physical quantities and units in *ISO Standards Handbook 2, Units of Measure*.

Optical Drawing Standards

There are a handful of U.S. voluntary standards relating to optical drawings and optomechanics. These are listed in Table 1.1 along with the issuing organization and page length of the standard. The only one of these standards that has had much if any practical impact on domestic optics is ASME Y14.18M-86, *Optical Parts*, and that only in its original incarnation as MIL-STD-34, *Preparation of Drawings for Optical Elements*. The ANSI standards on the list are all the product of the photographic industry and were written about the time when all camera manufacturing ceased in the U.S.

TABLE 1.1

	Ref.
ASME Y14.18M-86 Optical Parts, 37 p	5
ANSI PH3.617-80 Appearance Imperfections, Test for, 22 p	
ASTM F1048-87 Surface Roughness by TIS, 6 p	6
SAE AMS 2521B-89 Antireflection Coating for Glass, 8 p	7
ASTM F1128-88 Abrasion Resistance of Coatings Using Salt Impingement, 3 p	
ASTM D4541-85 Pull Strength Using Adhesion Testers, 7 p	
ASTM F768-82 Measurement of Specular Reflectance and Transmittance	
ANSI PH3.616-90 AR Coatings for Photographic Lenses	
ANSI PH3.713-85 Environmental Testing of Photographic Equipment, 15 p	
ASTM D2851-86 Liquid Optical Adhesive, 4 p	

On the international front, ISO Technical Committee 172, Optics and Optical Instruments, was founded in 1979 and has been writing optical standards ever since. Over 150 standards are either being worked on or are now published. The two standards that are of greatest interest are ISO 10110, *Indications in Optical Drawings*, and ISO 9211, *Optical Coatings*. ISO 10110 is similar to ASME Y14.18M but has 13 parts and is over 100 pages long in draft form. There is no U.S. equivalent of ISO 9211.

We will briefly go over the contents of the two standards as they bear directly on optomechanical drawings. Part 1 covers the mechanical aspects of optical drawings that are specific to optics and not already covered in one of the ISO mechanical drawing standards. Parts 2 to 4 cover material-related parameters such as stress birefringence, bubbles and inclusions, and inhomogeneity. Part 5 concerns figure measurement and differentiates between a figure measured visually with a test plate and that measured with a phase measuring interferometer. Part 6 deals with centering errors and allows either an entirely mechanical method of tolerancing or an optomechanical one.

Part 7 is the equivalent of what we now call the scratch and dig or surface beauty specification and appears to be a more workable method than present schemes. Part 8 concerns ground and polished surface texture and is unique to this standard. Part 9 tells how to indicate that a surface will be coated, but not what the specifications of the coating are. The latter are covered in ISO 9211. Part 10 tells how to describe the parameters of an optical element in tabular form and is the foundation of the effort to be able to transfer data about optical elements electronically.

Part 11 is a table of default tolerances on optical parameters so that if a particular parameter is not specified, it should then be made to the tolerances given in this table. Part 12 defines how to describe an aspheric surface and the method has been coordinated with the major vendors of lens design software so the definitions are consistent. Finally, Part 13 tells how to specify a laser power damage threshold on an optical component, again a parameter that goes far beyond any other existing standard.

The four-part ISO 9211, *Optical Coatings* standard exceeds any existing standard in its thoroughness and detail. Part 1 covers definitions of coating terminology and the definition of ten coating types by function. It also has an extensive table of types of coating imperfections including diagrammatic illustrations.

Part 2 concerns the optical properties of coatings and outlines the properties of the coating that need to be specified to be a complete description. It also shows graphical formats for specifying the transmission or reflection properties of coatings. Several example illustrations of coating specifications are given.

Part 3 is about the environmental durability of coatings as a function of intended use. There are five categories of use ranging from completely sealed within an instrument to surfaces exposed to severe outdoor conditions and unsupervised cleaning. There is a list of 14 different environmental tests for coated surfaces ranging from abrasion to mold growth. Part 4 defines environmental test methods specific for coatings. These are abrasion and solubility tests. The abrasion test includes specifications for the cheesecloth and the eraser for the tests.

It should be mentioned that there are many other ISO optical instrument standards that involve various aspects of optomechanics. For example, there are subcommittees working on standards for binoculars and riflescopes, microscopes, geodetic instruments, medical instruments such as tonometers and endoscopes, and all types of lasers and instruments using lasers. The laser subcommittee has been one of the most active groups and is working on such things as standardizing the diameters and thicknesses of optical elements used in lasers.

Information on these international standards is available from several sources. As has already been indicated, published ISO standards, Handbooks, and catalogs of standards are available from the "Foreign Order Department" at ANSI.[4] For draft versions of ISO standards and lists of drafts being worked on, contact the Optics and Electro-Optics Standards Council (OEOSC),[8] the secretariat for U.S. participation in ISO optical standards writing activities and administrator for the domestic ANSI OP Committee on Optics and Electro-optical instruments. There is a growing list of military, domestic, foreign, and international optical standards on the OEOSC Internet Homepage at http://www.optstd.org as well as calendars of standards meetings and other publically accessible optical standards information. SPIE[9] and OSA[9] have optical standards information on their Homepages as well. Finally, the Optical Society of America has published a handbook[10] designed to be a companion to the ISO 10110 optical drawing standard and an aid to its interpretation.

1.4 Dimensional Tolerances and Error Budgets

The focus of this section is concerned with the practical aspects of tolerancing the designs of optical instruments which are intended for production in large or even small quantities. Certain performance is required of an instrument in all applications. The design and tolerancing aspects of the process have a major effect on the life cycle cost and efficiency of the system. We will discuss what factors make up the cost of a lens and the effects of tolerances and other factors on that cost. This, incidentally, will result in a lens cost estimation formula. We will describe the interactions of lenses and lens cells/mounts from the tolerance viewpoint. We then explain the principles whereby the system tolerances can be determined to minimize the cost of a system, which meets the performance requirements.

The assignment of tolerances to various dimensions and parameters of an optical system is a **critical** element in determining the resulting performance and cost of the system. Much of the tolerancing of systems in practice has been done by art and experience rather than by scientific calculation. Here, we attempt to make the **engineering** principles as simple and clear as possible so that these may be applied in a straightforward manner. We use the term "engineering" to imply that practical approximations based on empirical data are used to reduce the problem to practical terms that can be handled in the real world. In the production of an optical system, random errors in parameters occur. These cause the results to be statistically predictable, but not exactly calculable. Therefore, the use of reasonable engineering approximations is appropriate and justifiable.

It is sometimes possible to tolerance a system such that each of the components is fabricated to an accuracy which will ensure that the instruments will be adequately precise and aligned to give the required performance by simple assembly with no alignment or adjustment. This may be the case with certain diamond-turned (a high accuracy process) optical assemblies of high precision, or with systems having low performance requirements with respect to the process capability. The other extreme is where almost every parameter of a system is loosely toleranced but can be adjusted, with proper skill and labor, to allow the system to deliver the desired performance. However, neither of these approaches is usually the least-cost method to meet the performance requirements. We next discuss philosophical principles and practical ways of approaching the least-cost solutions and give an illustration of the application of the techniques.

In 1982 we started a series of investigations[11-13] of how to achieve the least-cost tolerancing of an optical system. Since that time, Wiese[14] has compiled a very useful collection of papers covering many aspects of tolerancing. This volume includes a paper by Plummer[15] which played a role in our earlier work and a paper by Adams[16] which judiciously utilized some of our findings. Fischer[17] did a more recent survey based on Plummer's work, which we compare with some of our updated cost vs. tolerance data below. Parks[18,19] and Smith[20] have papers in Wiese's collection that have many practical and helpful discussions on the subject. The referenced works form a good general background for this section, but we will reiterate the salient points below for the convenience of the reader. We will also cite other specific references as they apply. It is the authors' experience and opinion that a great deal of resources have been wasted in the past due to poor tolerancing "art". A rigorous and all-encompassing treatment of all but the simplest system can be **very** complex. It is our aim to move the practice of tolerancing from the art stage to the engineering stage with as much simplification as is reasonably justifiable. Warren Smith[20] has made significant contributions to move the status of the practice in this direction. We are attempting to move another step in this direction by adding the real influence of cost into the tolerancing process.

Effect of Tolerances on Cost

Base Costs

We will first review the concepts of base costs from our previous work, and then present the results of later work for the estimation of the base costs directly from the data on a drawing and/or the specifications of a given component.

Let us take the example of fabricating a single lens. For fabricating a biconvex lens of glass, we would typically have to go through the following steps:

- Generate (or mill) a radius on the two sides.
- Mount the lens on a spindle.
- Grind and polish the first side of the lens.
- Remount the other side of the lens on a spindle and grind and polish the second side.
- Edge the lens.

There are obviously a few other minor steps such as obtaining the materials and grinding and polishing tools, dismounting, cleaning, etc. We have said nothing to this point about adding other specifications such as diameter, radii, thickness, and tolerances. Even without these parameters, there is a minimum cost in time, materials, and equipment necessary to make the biconvex glass lens. This is what we define as the base cost. As we get more specific about the lens and add more restrictive tolerances to the parameters, more care, time, and equipment will probably be required to make the lens to the new specifications. Therefore, the cost will increase with increasingly stringent requirements/tolerances/specifications.

Our previous work[11-13] went into some detail on the relationships of the increase above the base cost for the changing tolerance values. However, neither we nor Plummer[15] quantified those base costs. We will quantify the base costs here in order to enhance the usefulness of the technique and the accuracy of the results. It was not made clear in the earlier work that the base costs for a given cost vs. tolerance case are not the same for the grinding and polishing as they are for the centering and edging. As we shall show later, the centering and edging operation is not influenced by the grinding and polishing costs or tolerances and vice versa. Not incorporating this concept can lead to some error in the application of our earlier work and Adams'[16] extension of it. There is a different base for the increase if tolerance is to be applied for the two classes of operation. Figures 1.11 and 1.12 show the effect of tolerances of lens diameter and eccentricity on the costs as a percentage of base cost for the centering and edging operations. This applies to the centering and edging base cost (CE). Figures 1.13 through 1.18 show the tolerance costs as a percentage of the base grinding and polishing costs (GP) for tolerances of radius of curvature, irregularity, diameter-to-thickness ratio, center thickness, scratch and dig, and the glass stain characteristics. It was also interesting to us to recognize that the milling or generating costs are not particularly affected by tolerances with today's equipment and they are not part of the base cost affected by tolerance factors. The milling costs are, therefore, part of the base cost, but only as a function of material and dimensions and not tolerances.

Before we discuss the cost vs. tolerances in detail in the next section, we will show the development of the base cost formula. The present view and cost-estimating scheme are the authors' best effort to date resulting from their experience and discussions based on the work and experience of Stephen Cupka, Manager of Estimating, and Reinhard Seipp, Assistant Manager of Optical Manufacturing at Opto Mechanik (OMI). These people also draw upon their experience from one or more other shops. Let us call the total base cost to make a lens MT, which is in either time or money, which differ only by some multiplicative factor. If we call the milling or generating cost MG, we can represent the total base cost MT as the sum of milling, grind and polish, and centering and edging costs as given in Equation 4.

$$MT = MG + GP + CE \qquad (4)$$

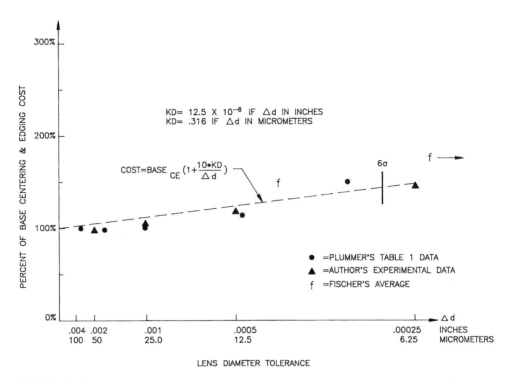

FIGURE 1.11 Relative cost vs. reciprocal tolerance according to various authors concerning lens diameter.

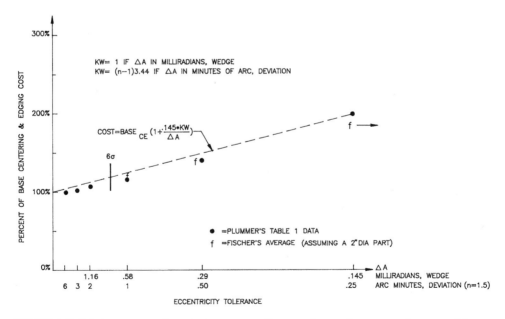

FIGURE 1.12 Relative cost vs. reciprocal tolerance according to various authors concerning eccentricity.

Milling/Generating Costs. We find a reasonable fit with experience for the milling cost to be given by Equation 5, where LM is the number per lot to be milled and d is the diameter in inches. The milling of both sides of the lens is included here.

$$MG = 4 + 90/LM + 0.1 \times d^2 \qquad (5)$$

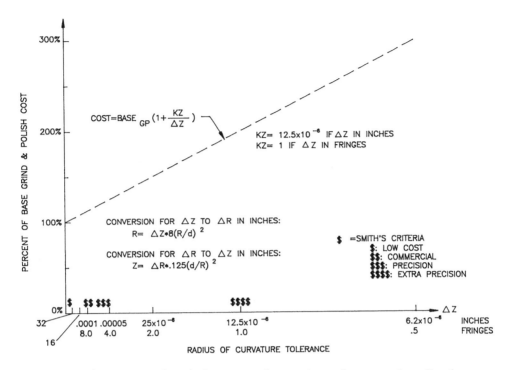

FIGURE 1.13 Relative cost vs. reciprocal tolerance according to various authors concerning radius of curvature (when expressed as sagittal error).

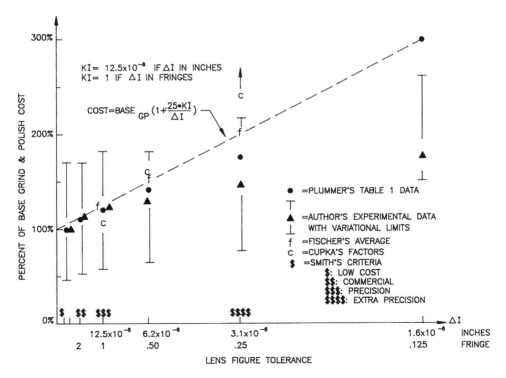

FIGURE 1.14 Relative cost vs. reciprocal tolerance according to various authors concerning lens figure irregularity.

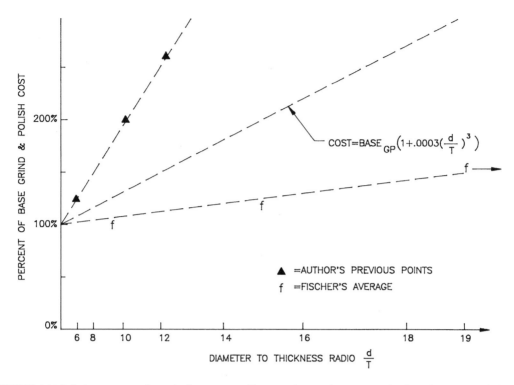

FIGURE 1.15 Relative cost vs. reciprocal tolerance according to various authors concerning lens diameter-to-thickness ratio.

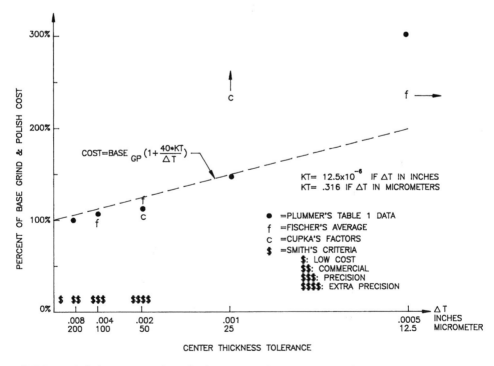

FIGURE 1.16 Relative cost vs. reciprocal tolerance according to various authors concerning lens center or axial thickness.

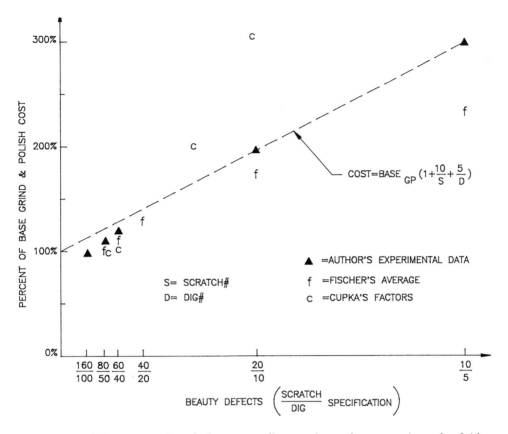

FIGURE 1.17 Relative cost vs. reciprocal tolerance according to various authors concerning surface finish or scratch and dig.

This implies that there is some base cost to mill the optic plus some setup cost divided by the number of parts to be run from that setup plus a factor due to lens size. Since the material to be removed from a molded blank is usually about the same thickness independent of blank diameter, the generating time is only a function of a blank's area (d^2).

Centering and Edging Costs. We will deal with centering and edging (CE) costs first because they are simpler. The cost is a function of the number in the lot LC to be centered in one setup, the diameter d, the number of chamfers C, and the number of flats F (planes perpendicular to the lens axis). Equation 6 represents our collective best estimate.

$$CE = \left(2 + d + C + F\right)/3 + \left(30 + 10 \times C + 15 \times F\right)/LC \qquad (6)$$

This accounts for a setup cost for the diameter, chamfers, and flats plus the edging of each lens.

Grinding and Polishing Costs. The GP cost is a very strong function of the number of lenses which can be ground and/or polished at one time on a block. If the radius is short, the number NS which can be blocked for that side is determined by the radius R and the lens diameter d. If the radius is long, the number which can be blocked is determined by the maximum block diameter G and the lens diameter d. The precise calculation of this number can be performed when flats, chamfers, and center thicknesses are properly accounted for. For simplicity and practicality, we will use a conservative approximation without showing its derivation here. If the factor R/d is smaller than 0.87, only a single lens can be polished at one time. At least three per block can be

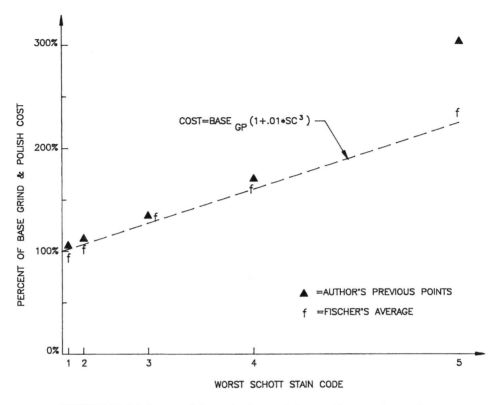

FIGURE 1.18 Relative cost of glass stain characteristics according to various authors.

processed if the ratio is greater. If the radius is short, the number per block NS will be given by integer value in Equation 7.

$$NS = INT\left[2.9 \times \left(R/d\right)^2\right] \qquad (7)$$

Here R is the radius and d is the diameter of the side in question, and N1 would be NS for side 1 and N2 would be NS for side 2. If the radius of the side is long, the number per block will depend on the diameter G of the largest block which can be used as shown in Equation 8.

$$NS = INT\left[0.64 \times \left(G/d\right)^2\right] \qquad (8)$$

Whichever NS is smaller for a given side (for R or G) will be the number which can be ground and polished per block for that side. Our experience is that less than the maximum number will often be used when the radius is long. This is not of great cost consequence, however, because the change in cost for a few parts, when NS is large, as given in Equation 9 is small relative to other costs. Our collective experience with the GP per lens and per side in a block of NS lenses is given in Equation 9.

$$GP = 7 + 14/NS \qquad (9)$$

It is also usually appropriate to consider a yield-related factor Y due to scrappage etc. This factor Y multiplied by the number of lenses to be delivered is the number of lenses to be started to give the required yield. The factor Y is then actually an inverse yield. When this factor is applied and both sides of lenses are considered, the total GP base costs are given in Equation 10:

$$GP = Y \times \left[14 + 28 \times \left(1/N1 + 1/N2 \right) \right] \qquad (10)$$

where N1 and N2 are the numbers of lenses per block for side 1 and side 2. We now have all of the components of the base cost (in relative units) for a lens based on a collective empirical history and in a fairly workable form. We only need to apply the effects of tolerances and other influences such as material properties and diameter-to-thickness ratio to these base costs in the next section and we can then predict the cost to produce a given lens to its specifications. The derivative of that cost for a change in any tolerance or parameter will be used to determine the distribution of tolerances which will minimize the system cost and ensure the desired yield.

Effect of Tolerances and Other Factors on Costs

We have updated and included here the cost vs. tolerance data presented in our previous work[13] in Figures 1.11 through 1.18. In these figures, the dots are Plummer's work,[15] the triangles are our previous experimental and/or estimated values, the c's are factors from Cupka's earlier estimates, and the f's are the averages from Fischer's survey.[17] The lines drawn are the functions which we currently choose to use as the best estimates from experience and the functional equations are indicated on each figure. In the figures, where appropriate, we have included information from Smith.[20] He has defined tolerances that are "low cost" which are plotted as "$". His "commercial" tolerances are shown as "$$", "precision" tolerances as "$$$", and "extraprecise" tolerances are shown as "$$$$". These are in general agreement with the other data. For a more extensive discussion of the previous work, the reader is encouraged to review the references.

Figure 1.11 shows the cost effect of lens diameter tolerances which is not a strong factor up to the limit of the capabilities of the edging process. The various authors' data are in good agreement. We have shown the tolerance scale in inches and micrometers on this and some of the other figures for the readers' convenience. Figure 1.12 shows the cost effect of the lens centering tolerances. The centering tolerance is sometimes expressed as light deviation, wedge, or total indicator runout (TIR) at the edge of a lens or window. These different versions of the requirements can be reconciled in the following way. The wedge angle in radians is the same as the TIR divided by the diameter being measured. The light deviation depends on the wedge times the refractive index minus one and converted to arcminutes as needed. The KW-factors in Figure 1.12 account for this. Plummer's and Fischer's data are in good agreement and are well represented by the approximation which we use.

Figure 1.13 deals indirectly with the cost of radius of curvature tolerances. The work of Thorburn[22] reported in Wiese's collection made it clear that we should refine our approach for the cost effect of radius of curvature. We had previously just used the percentage of the radius as a measure of stringency of the radius tolerance. But it becomes apparent as a result of simple analysis that the change in sagitta over a surface is much more meaningful. This is what a spherometer or an interferometer can measure. One can derive the change in sagitta with a change of radius as approximated by Equation 11.

$$\Delta Z / \Delta R = 0.125 \times \left(d / R \right)^2 \qquad (11)$$

This equation has been used to plot Figure 1.13. The graph is plotted as cost vs. the reciprocal of the delta sagitta ΔZ which is a function of delta radius ΔR, R, and d. The ΔR and R alone are not enough to estimate the real cost of the tolerance; the surface diameter must also be taken into account. It might be more appropriate to refer to the sagitta tolerance of a surface than the radius tolerance, but shops are used to seeing the radius tolerance spelled out on a drawing. Equations 12 and 13 can be used to convert either way between ΔR and ΔZ (typically in inches).

$$\Delta R = \Delta Z \times 8 \times \left(R/d\right)^2 \tag{12}$$

$$\Delta Z = \Delta R \times 0.125 \times \left(d/R\right)^2 \tag{13}$$

The chosen cost function is based on Thorburn's[22] statement that a tight tolerance on sagitta is 0.000005 in. and 0.0001 in. is loose.

Figure 1.14 shows the impact of surface figure irregularity tolerances. We described the earlier experimental data in Reference 12. We have added the points from Fischer and Cupka with a slight reformulation and labeling. The data from the various sources seem sufficiently consistent and are reasonably represented by the chosen function.

Figure 1.15 shows the effects of the diameter-to-thickness ratio of a lens. This has to do with the flexibility of the part when trying to hold a good surface figure and sometimes the temperature effects when working with thin negative lenses. Our simplistic treatments of this to date are probably not as adequate as we would like. This is probably worthy of further study. Fischer[17] reported a much less severe effect than our previous work.[13] In our previous work, we estimated that the effect was inversely proportional to the flexibility of a disk which goes as the cube of the diameter-to-thickness ratio d/T. We will now accept (conditionally) Fischer's data collection as possibly more representative and we have fit our function to a compromise between our old data and that of Fischer, where both are seen in Figure 1.15. There is no doubt that plano–plano windows can be worked very thin by contacting them to rigid plates which lend the effect of their own d/T ratio. In the case of lenses, this is not the practice, however. Structural shape, thermal effects, etc. need to be studied in more detail. We will have to use this function until we or someone else can refine it further.

Figure 1.16 deals with the cost of center thickness tolerances. A major cost factor here is that once the thickness goes under the tolerance limit, the lens is lost and must be replaced. We believe that this is what is reflected in the radical change in cost in Plummer's data as a 0.0005-in. tolerance is approached. At some point, a given tolerance would not allow a scratch or pit to be ground or polished out and the part would be lost due to the combination of the thickness and scratch and dig specifications. Plummer's and Fischer's data are in reasonable agreement, and Cupka only divided the costs into above and below a 0.002-in.-thickness tolerance.

Figure 1.17 shows the effects of scratch and dig tolerances. The various data cluster reasonably near the very simple function that we choose to employ.

Figure 1.18 shows the simple function of cost vs. glass stain code that we fit to our previous data and Fischer's data. We might comment that we feel that even the 5-code glasses can be worked on a fixed price basis now, but at a significant added cost.

The "polishability" factor should also be applied as a cost impact. In general, pyrex takes more time to grind and polish than BK7 and fused silica takes longer than pyrex, etc. In an attempt to include this factor, we have collected the estimated time factor for a variety of materials as compared to BK7. Some suggest that germanium has the same polishing time as BK7, and others think it takes significantly longer. This will, of course, vary from shop to shop and the procedures used. The authors believe that, typically, germanium lenses will polish to specifications in about the same time as BK7, but that the typical polishing specification may be 160/100 for Ge and 80/50 for BK7. Therefore, the Ge surface will take longer to meet all of the same specifications as a piece of BK7. The numbers in Table 1.2 reflect the authors' best estimates based on a variety of inputs, but each shop needs to examine which factors to use in its own case. The grind and polish time will then be multiplied by this material polishability factor P.

TABLE 1.2 Polishability Factor of Various Materials

	%
BK7	100
SF56	120
Pyrex	125
Germanium	130
Fused silica	140
Zerodur	150
ZnS, ZnSe	160
FK2, BaF2, Amtir	170
LaKN9, LaFN21	200
Electroless Ni	250
CaF2, LiF	275
MgF2, Si	300
Electrolytic Ni	350
Ruby	700
Sapphire	800

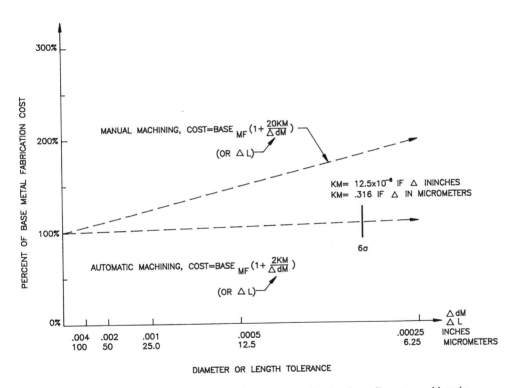

FIGURE 1.19 Relative cost vs. reciprocal tolerance concerning lens bore diameters and lengths.

Once the lenses are fabricated, they are typically mounted in metal cells. We have to coordinate the tolerancing of the metal and the glass to get the desired results. Similar cost vs. tolerance curves can be developed. Figure 1.19 shows the cost vs. lens cell diameter tolerance for both manual and automatic machining. The same curve applies to the length tolerances between bores as in spacers. In an automatic setup, the parts repeat within the capability of the machine with little change in the cost vs. tolerance. The manual operations are more and more labor intensive as the tolerances increase. Figure 1.20 shows the cost vs. tolerances of bore concentricity and length run-out such as tilt in a spacer. There is no difference between manual and automatic here. However, the big difference is whether both bores are cut without removing the part from its holder (chuck). If so, then the concentricity will be limited only by the accuracy of the machine. If the part must be rechucked for the second bore, much more time is consumed to hold a tight tolerance in the rechucking or mandrel-type operation. This points out the strong motivation to design the cell for single chucking as much as is practical.

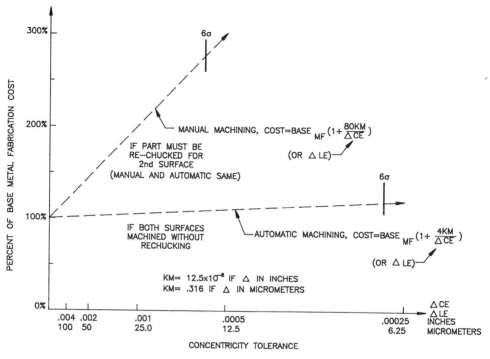

FIGURE 1.20 Relative cost vs. reciprocal tolerance concerning lens bore concentricity (ΔCE) and tilted and length run-out (ΔLE).

We will use the results of Figures 1.19 and 1.20 when we distribute the tolerances in the instrument design. The base cost for the machining operation will be multiplied by the metal tolerance factors in this process. For simplicity, we will ignore the setup costs and use a base machine fabrication cost of six (6) cost units per bore or spacer cut in the same denomination as those used for the lenses above. For an automatic setup, we will use a number of four (4) units. This aspect could be made much more complex and refined. This approximation is adequate to allow us to properly distribute the tolerances in the later sections, but cannot be used to estimate the total machining costs. These cost vs. tolerance and cost vs. other parameter functions are the major factors influencing the optical component fabrication cost. Now that we have them reasonably well characterized, we can apply them to the base costs specified earlier to estimate the total component cost. We can also find (to a sufficient approximation) the change in total component cost with a change in any of the parameters or tolerances. This, coupled with the sensitivity of the system performance characteristics to each of the parameter and tolerance changes, will then allow us to distribute the tolerances in such a way as to minimize the system cost while maintaining the required successful product yield. This is the major goal of the cost optimization process.

Total Lens Cost Estimation

When we combine the base costs given in Equations 5, 6, and 10 with the tolerance and other factors of Figures 1.11 through 1.18 and Table 1.2, we can estimate the total lens cost as given in Equation 14. This obviously is not scientifically rigorous, but is only a practical estimate for engineering or business purposes. The detail factors will vary from shop to shop and time to time.

It is the authors' belief that this type of estimation formula is more accurate and consistent than any other method now available.

TOTAL LENS COST	$= MT =$
(GENERATING)	$+ 4 + 90/LM + 0.1 \times d^2$
(PART SETUP)	$+ Y \times 14$
(SIDE 1, G&P)	$+ (P \times Y \times 14/N1)$
	$\times \{(1 + .25 \times KI/\Delta I1)$
	$\times [1 + (R1/d)^2 \times 8 \times KZ/\Delta R1 + .0003 \times (d/T)^3]$
	$+ 40 \times KT/\Delta T \times (1 + 10/S1 + 5/D1) \times (1 + .01 \times SC^3)\}$
(SIDE 2, G&P)	$+ (P \times Y \times 14/N2)$
	$\times \{(1 + .25 \times KI/\Delta I2)$
	$\times [1 + (R2/d)^2 \times 8 \times KZ/\Delta R2 + .0003 \times (d/T)^3]$
	$+ 40 \times KT/\Delta T \times (1 + 10/S2 + 5/D2) \times (1 + .01 \times SC^3)\}$
(CENTERING)	$+ CE \times (1 + 10 \times KD/\Delta d + .145 \times KW/\Delta A)$ (14)

Although Equation 14 is extensive, it is not particularly complex. Table 1.3 defines the parameters of this equation.

TABLE 1.3

MT	Total lens cost estimate per piece (in relative units)
MG	Milling/generating cost from Equation 5
LM	Number of parts milled in one lot setup
Y	Yield factor of parts started/parts acceptable
P	Polishabilty factor from Table 1.2
N1	Number of parts/block for R1 side from Equation 7 or 8
N2	Number of parts/block for R2 side from Equation 7 or 8
R1	Radius of side 1
R2	Radius of side 2
G	Diameter of largest block to be used for grind and polishing
d	Diameter of lens
T	Thickness of lens
S1	Scratch number spec for side 1
S2	Scratch number spec for side 2
D1	Dig number spec for side 1
D2	Dig number spec for side 2
SC	Worst stain class of glass type Ks and deltas as in Figures 1.11 to 1.18
CE	Centering and edging base cost per Equation 6
C	Number of chamfers on the lens
F	Number of flats on the lens
LC	Number of lenses centered and edged in one lot setup

The generating cost and a certain portion of the setup costs are virtually independent of the tolerances of the part. The centering and edging operation and tolerances are virtually independent of the grinding and polishing operations and tolerances. The operations and tolerances of the two radii of the lens are in most cases independent. This is all reflected in Equation 14, and some logic has been applied to whether the tolerance factors are added or multiplied and to what they are applied. The irregularity cost factor is multiplied with the base cost of grinding and polishing each side. This irregularity cost is further multiplied by the radius tolerance cost factor plus the flexure (d/T) difficulty factor which makes the figure irregularity harder to achieve. The decision to add the d/T influence to the radius tolerance influence was made on the basis that they do not significantly affect each other, but they both affect the cost of meeting the irregularity specifications. The stain class interacts with and affects the scratch and dig requirements and they both increase the difficulty of holding the thickness tolerance, but they do not affect the irregularity degree of difficulty. The edging operation is affected by the diameter tolerance and the wedge or deviation tolerance. We will discuss the unique interaction of the diameter and centering tolerances in the next section. This formula then reduces the estimation of the fabrication cost of most lenses to a clerical task of entering the parameters from the lens drawing into a simple computer program. This could also, in principle, be worked into a CAD program to allow the designer to see immediately the cost impact of the design and tolerances. The key factor for tolerancing, however, is that we can find the partial derivatives of Equation 14 with respect to each of the reciprocal tolerances for use in finding the minimum cost tolerance distribution for a system as discussed below.

Interactions of Lenses and Mounts

A lens system typically consists of lenses in metal cells. The cells have bores that closely fit the lens diameters with "seats" or rings of contact between one or both of the radii and a metal locating surface in the direction of the optical axis. The relationship of one lens to another will be determined by the spacing dimensions and tolerances of the mount and the concentricity and tilts between the locating surfaces. The mounting cell interfaces must be toleranced to be compatible with the glass tolerances. In this section, we will address how to handle this task. There are several interesting references[21-23] dealing with tilts, decentrations, and rolls (by whatever names), but none seem to have addressed them in an analytic form appropriate to our cost minimization goals.

Lens Centering. Figure 1.21 shows a lens in a cell bore where only centering factors are considered, not tilts. It is easy to evaluate the effect or sensitivity of decentering a lens from the intended optical axis in most lens design software. This decentering in a system is the sum of several factors. The decentering of the optical axis of the lens with respect to the outside diameter of the lens is what the optical shop works on. The centering of the mounting bore with respect to the ideal axis is what the machine shop works on. There needs to be some fit clearance, f, for the assembler to insert the lens into the cell. The tolerances of the lens diameter d and the bore diameter dM, Δd, and ΔdM give rise to more potential clearance. These clearances will allow the lens to move to extreme positions in the cell and cause more decentering. Equation 15 expresses the possible total decentering td as a function of the lens decentering dc and the factors mentioned above.

$$td = dc + \left(f + \Delta d + \Delta dM\right)\big/2 \qquad (15)$$

The costs vs. tolerances have been defined and quantified above. We can see that the least cost distribution of tolerances for a lens in a bore with respect to total decentering will dictate a certain ratio between the dc, Δd, and ΔdM. We will then reduce the tolerancing of the "set" to a function of only Δd since the dc and ΔdM can be taken as the dependent variables. The dc of the lens is measured by the wedge ΔA in Figure 1.21 (here we work in milliradians) or in the arcminutes deviation that it causes which we relate by the factor KW in Figure 1.12. We can show that dc can be expressed in terms of ΔA as in Equation 16. (Some RSS possibilities will be ignored here.)

$$dc = \left[R1 \times R2\big/\left(R2 - R1\right)\right]\big/\left(1000 \times KW\right) \times \Delta A \qquad (16)$$

ΔA in terms of Δd is found to be

$$\Delta A = SQR\left(.145 \times KW\big/10 \times KD\right) \times \Delta d \qquad (17)$$

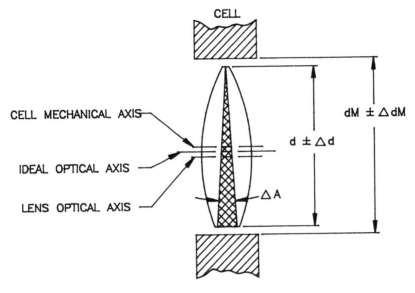

FIGURE 1.21 Decentering factors of a lens in a cell.

If we call the base machining cost MF, then ΔdM in terms of Δd is given by Equation 18.

$$\Delta dM = SQR\left(MF \times 20 \times KM/CE \times 10 \times KD\right) \times \Delta d \tag{18}$$

The minimum fit clearance factor f has to be determined by the assembly plans for the cell and whatever allowances are made for differential thermal expansion. At nominal temperature, it will allow a shift in an otherwise perfectly fitting cell of f/2. This decentering will have to come right off the top of the total decentering budget for this lens/bore set leaving the residual budget to be divided among Δd, ΔdM, and ΔA. We express the result in Equation 19.

$$td - f/2 = \Big[1/2 + SQR\big[(MF \times 20 \times KM)/(CE \times 10 \times KD) \big]$$
$$+ SQR\big[.145/(10 \times KD \times KW) \big] \times ABS\big\{ R1 \times R2/\big[(R2-R1) \times 1000 \big] \big\} \Big] \times \Delta d \tag{19}$$

This factor multiplied by Δd is to be used in the tolerance allocation process with the decentering sensitivity to determine Δd. The Δd can then be used to assign ΔA and ΔdM (which are dependent on Δd) in a secondary operation using Equations 17 and 18. The diameter of the cell bore dM is also determined by this process as expressed in Equation 20.

$$dM = d + f + \Delta d + \Delta dM \tag{20}$$

Although Equation 19 might not appear to simplify anything, it is a relatively straightforward application of the cost and geometric factors which allows us to properly spread the cost and tolerances in the lens and its cell.

Lens Tilt and Roll. An otherwise perfect lens might be tilted with respect to the system's ideal optical axis because the metal locating surface of the cell is tilted by an angle ΔAT, as shown in Figure 1.22a. This can be simply dealt with using the cost vs. tilt curve in Figure 1.12 and the sensitivity of the system to tilt of the whole lens. This assumes that the tilt is not otherwise limited

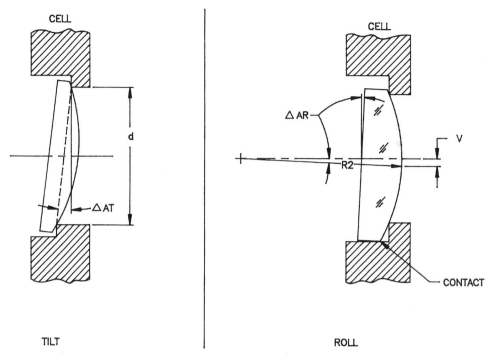

FIGURE 1.22 Tilt and roll factors of a lens in a cell.

by such factors as a retainer on the other side of the lens or the fit of the cylindrical lens diameter into the cylindrical cell bore which prohibits that much tilt. The perfect lens might also "roll" in an oversized bore as shown in Figure 1.22b. This shows that the left-hand surface tilts while the right-hand surface is correctly located against the "perfect" cell. The lens will roll about the center of curvature of the right-hand surface R2. The left-hand surface, which we have shown as plano for clarity, will tilt through an angle of ΔAR which is approximately V/R2 radians. We can show that:

$$V = \left(f + \Delta d + \Delta dM\right)\big/2 \qquad (21)$$

We know ΔdM in terms of Δd from Equation 18. This allows us to express ΔAR1 of the left surface R1 as a function of Δd in Equation 22.

$$\Delta AR1 =$$

$$\left(f + \left\{1 + \text{SQR}\left[\left(MF \times 20 \times KM\right)\big/\left(CE \times 10 \times KD\right)\right]\right\}\right)\big/ \text{ABS}\left(2 \times R2\right) \times \Delta d \qquad (22)$$

This and the system performance sensitivity to a tilt of surface R1 will allow us to allocate the tolerance budget for a tilt of R1. However, note that controlling this requires the control of Δd which is already determined by the decentering requirements! Generally, one or the other will be the more demanding on Δd. It would appear that we should find which is the more stringent and use it to determine the tolerance allocations. The other parameter would still make some contribution to the error budget, but not be independently determined. As in the tilt case, the ability of the element to roll to the full extent indicated in Equation 22 may otherwise be restrained by (but might also be caused by) a retainer ring, etc. Since this can be the case, it may be appropriate for the designer to use some judgment in the application of tilts and rolls after looking at the sensitivities and the mounting designs to decide how these equations may be applied. This unfortunately seems to bring us back a bit from the engineering toward the "art" in tolerancing. Having now dealt with the necessary elements of cost vs. tolerances and other parameters and the applicable geometric simplifications, we will next proceed with how we can apply the foregoing equation to distribute the tolerances for the maximum performance and cost benefit.

Allocation of Tolerances

The term **Six Sigma** (6σ) has come to be referenced extensively in recent years with respect to quality and tolerancing functions. Six Sigma is a tool and philosophy which can improve yield and reduce rework, and thereby improve costs and customer satisfaction. There are two major elements to it. First is to measure the routine capabilities of the processes used to achieve a result. This might include the part-to-part repeatability of a lens diameter. If there are things that can be done to reduce the variations, these will improve the capabilities and therefore yields. Statistical process control (SPC) is a part of this approach. The second major element is to make the tolerances as large as possible and still meet the system performance requirements.

We have indicated in Figure 1.11 where the 6σ point would be for the lens diameter. That is, a tolerance less stringent than ±0.0003 in. would give 6σ results or better. Anything more stringent (to the right in Figure 1.11) would be beyond the capability of that process to provide 6σ results. Figures 1.12, 1.19, and 1.20 also show the 6σ points for the processes of lens centering and metal machining.

Over the last decade, we have been heavily involved in finding methods to best distribute the tolerances in a way as to minimize the cost. This is essentially consistent with 6σ philosophy. One aspect has been to keep the designs as simple as possible and still meet the requirements. This is a key element of a 6σ design. The less complex the design, the lower will be the cost, risk, rework, etc.

Let us digress from the subject to describe our design philosophy. We first try the simplest design we think has a chance of working. If testing shows that it must be more complex, we change it. If we had made it more complex to start with, we would never know if it could be simpler and less costly. This is often a necessary part of the development process.

6σ graphically demonstrates the fact that it is important to identify the weak link in the capability chain and work to improve it. Even if the rest of the links in the process are better than 6σ, the results will still only be as good as the weakest link.

6σ is valuable in focusing the attention on tools and philosophies which can be used to continually improve the processes and designs. Both manufacturing and design organizations have been generally doing many of the right things from a 6σ point of view, but now have a better focus and understanding of how to measure and execute the process. The effects of 6σ will be evolutionary in the future.

Assigning Tolerances for Minimum Costs — An Example

We described the general principles of distributing the tolerances for a minimum cost in our first work[11] on this subject. Adams[16] made some significant additions to our work which we shall apply below. If there is more than one performance criterion that must enter into the tolerancing process, the solution to the equations is somewhat involved, but it is feasible. However, many systems, including the one which we will use as an example, have one performance criterion which dominates all of the others as it relates to the tolerances. That is, if the tolerances are chosen to meet that dominant performance requirement, then all other requirements will also be met. This reduces the computation considerably and makes it easier to visualize. For the balance of this discussion, we will use the single requirement case with the understanding that it can be extended to multiple criteria by the methods of the previous papers as needed.

Figure 1.23 shows the multifocal length tracking telescope (MFLTT) that we will use as an example of the tolerancing process. It has a catadioptric telescope section of 300-mm aperture and about 2000-mm focal length with a 25% central obstruction due to the secondary mirror. The telescope image is then collimated by a focus lens set. The afocal beam is then imaged by one of three imaging lenses to the final focal plane. These lenses are alternately positioned in the beam to give system effective focal lengths of 1000, 2000, and 4000 mm. Before the final focal plane, there is an auto-iris system of variable neutral density filters and a reticle projection unit (AIR). There is also a 500-mm system which is partially separated from the others to allow a larger field of view. The 500-mm system is folded into the same optical path as the others by a movable prism. There are sealing windows in front of the telescope and the 500-mm lens. In this complex telescope system example, the most stringent requirement of the system is the on-axis MTF at 30 lp/mm. When this is satisfied, the off-axis MTF at 30 lp/mm, the on- and off-axis MTF at 10 lp/mm, and the boresight, etc. requirements will all also be satisfied without additional tolerance requirements.

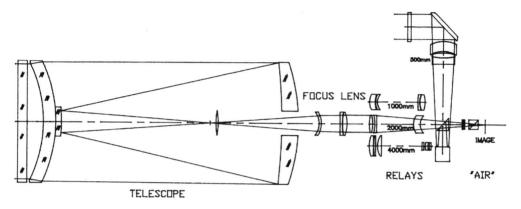

FIGURE 1.23 Example system: Multifocal length tracking telescope.

To be consistent with our previous report,[11] we will designate this performance requirement by E which represents the maximum permissible error in MTF from ideal for the system. We will actually convert this E to units of RMS wavefront error (RMSWE) for simplicity. The total E will eventually be partitioned among each of the tolerances which affect it. (To make a tractable example for this discussion, we will partition the total E among the various sections of the system.) The partial E will then be allocated to the parameter tolerances within one section based on the cost minimizing technique. This "divide and conquer" approach is needed here, plus any justifiable simplifications, in general, to reduce the overwhelming magnitude of the problems that may have multitudes of component tolerances to be determined. In the final analysis, it is best to "tolerance" the whole optical train from object to image in one operation. This will truly allocate the tolerances to achieve the required performance at the minimum cost. The simplifying partitioning will cause some deviation from the ideal result unless the estimate used in the partitioning was exactly correct. In the example used here, it would be best to tolerance the 4000-mm system from end to end, but the data would be too cumbersome to make a good illustration in this discussion.

Simplifying Approximations. The MTF of a system is often the best performance measure to use because it most directly relates in many cases to the performance of an overall system when it is used. It is, however, not generally possible to measure the MTF effect of each component lens of a system in the production process. The characteristics that are readily measured on a lens were discussed above, such as irregularity, radius, centration, etc. We chose to work here with the effects of each tolerance on RMSWE, because it can be reasonably related to the system MTF. We estimated the reduction in MTF per wave of RMSWE at 30 lp/mm for the 2000-mm effective focal length, f/8 system by introducing errors into the system and evaluating it for MTF and RMSWE. With parameter deviations, we produced defocus, spherical aberration, coma, and astigmatism. Defocus was introduced by evaluating the system at different focal planes from the best focus. Spherical aberration was introduced by varying the y^4 aspheric coefficient from the nominal. Coma was

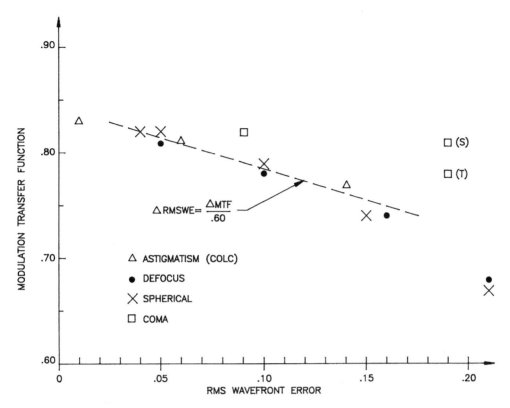

FIGURE 1.24 Diffraction MTF at 30 lp/mm vs. wavefront error for a 2000-mm, F/8 lens with various aberrations.

evaluated in an "equivalent" (f/8, 2000 mm) parabolic mirror system with the stop at the focal plane so that astigmatism was zero. The system was evaluated off-axis to introduce coma.

Last, astigmatism was introduced in an "equivalent" (f/8, 2000 mm) Ritchy-Chretien telescope where coma and spherical aberration were zero. The system was evaluated off-axis to introduce astigmatism. The results appear in Figure 1.24. All of the data form a reasonably consistent pattern except the coma. We do not presently understand this anomaly which may be worthy of a separate study. However, since the effects of coma are less severe than the others, we will ignore them and use the conservative numbers indicated by the others. Therefore we will use ΔRMSWE = ΔMTF/0.60 as the amount of reduction in MTF that will be accompanied by a corresponding RMSWE. This will allow us to work with the effects of tolerances on the RMSWE which we will assume are quasilinear in the regions where we are applying them. This may be a conservative estimate, but we would like to err on that side. Another approximation that we will draw upon comes from Smith[20] where

$$\mathrm{RMS} = \left(\mathrm{Peak\text{-}to\text{-}Valley}\right)\big/3.5 \tag{23}$$

approximates the RMSWE expected from most types of error. It would seem that sharp departures over a small portion of the wavefront would violate this rule, but those are not usually encountered. We conducted a small investigation of our own by comparing the RMS and P-V data on many interferograms from a ZYGO interferometer. We discovered that the factor in Equation 23 might be more like 7 than 3.5 when small irregularities, such as those on spherical surfaces, are examined. For this example, however, we will use Smith's value. In the example system, the apertures were selected at the first-order stage to yield the required MTF when the diffraction effects of the obscuration plus one-quarter wave of design and fabrication errors were taken into account. This is not much error to spread across the many elements from the object to the focal plane. One benefit is the fact that certain compensating alignments can be made at assembly since such systems are not made in very large quantities. We will use the approximation of Equation 23 to establish a preliminary total error budget of 0.071 RMSWE (1/4 wave P–V) from all sources in laboratory tests. In the final application, obviously, atmospheric and other effects might influence the results further.

Error Budgets.   Next we need to decide how to distribute this 0.071 RMSWE among the many facets and tolerances of the system. Smith[20] describes how to work with the root sum of the squares (RSS) to combine error effects. McLaughlin[25] shows that RSS tends to be too pessimistic and Smith himself concludes that it may err on the conservative side. McLaughlin shows that the total system error will tend to be 0.42 times the RSS prediction if the fabrication errors have a Gaussian distribution which is truncated at the 2σ level. Although there is a major move at this time in industry to apply 6σ tolerancing as mentioned above, there is still work to be done along the lines of Adams'[16] contributions to properly incorporate it. For the present case, we have used 2σ in this case where individual adjustment and testing are required. We will therefore use McLaughlin's 0.42 factor for the fabrication errors. To simplify the example, we will partition the 4000-mm path of the system. In looking at Figure 1.23, we count 32 surfaces through the 4000-mm optical path. The authors chose to emphasize the sensitivity effects of mirrors by counting them twice to give 34 as the surface count. Of this 34, 8 are in the telescope, 8 in the focus system, 12 in the 4000-mm relay, and 6 in the AIR. The other paths are less complex. This one will be the critical path and set the pace for the telescope, focus system, and AIR tolerances.

We will allocate the budget to the four sections of the 4000-mm path (telescope, focus, relay, AIR) in proportion to the square root of the number of surfaces in the section divided by the total number of surfaces. This is an engineering estimate of the relative influence of each section. The division of the system into these sections is also logical because each section can be tested independently for RMSWE in production. Figure 1.25 shows the error budget broken down this way. The top level requirement was determined above to be effectively 0.071 RMSWE. We know from the design stage that the design has used up 0.030 RMSWE. Another analysis indicates that the effects of alignment focus errors and the laboratory environment should be on the order of 0.009 RMSWE. This leaves 0.0637 RMSWE to RSS with the other two (three) parts of that level of the budget to give 0.071 RMSWE. From McLaughlin's information and the assumption of Gaussian errors, we then divide the fabrication budget of 0.0637 by 0.42 to give 0.1517 RMSWE, which can be distributed over the four sections of the 4000-mm system. The bottom four boxes of the budget in Figure 1.25 show how these work out when the above argument is applied. We will work through the simplest section, the focus lens with 0.0736 RMSWE budget, as an example of the procedure for tolerance distribution to give the minimum cost while meeting the performance requirements (to within some statistical uncertainty).

Derivatives of Costs with Respect to Tolerances.   In the assignment of tolerances for a minimum cost, it is necessary to have the partial derivatives of the total cost with respect to the reciprocal of each tolerance. These are basically derived from Equation 14 and the metal cell tolerance costs in Figure 1.20. Since we showed that the metal cell diameter and lens-centering tolerances can be made dependent on the glass diameter tolerance, we only have five types of tolerances to allocate in the framework which we have been discussing. These are irregularity, radius, thickness, diameter, and tilt of a lens due to the errors in the cell. Roll is shown in Figure 1.22b and can be derived from Δd. Tilt is just the cell run-out parallel to the optical axis (ΔLE of Figure 1.20) divided by the diameter d of the lens. We will call the partial derivatives of the total system cost with respect to the reciprocal of these tolerances $I, $R, $T, $d, and $L. The $L applies to both the metal cell bore diameters and axial run-out. The first three derivatives will be functions of the base grind and polishing costs with a common factor that we will call BP, which is defined in Equation 24.

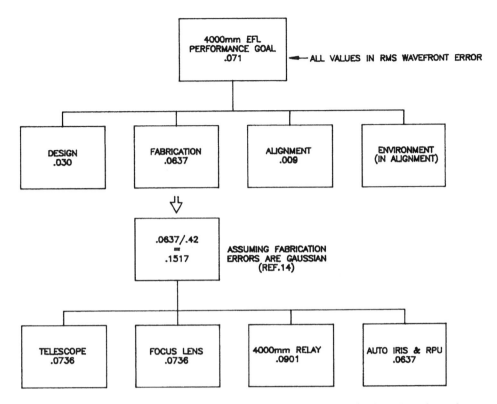

FIGURE 1.25 Error budget allocation for the 4000-mm channel of the multifocal length tracking telescope.

$$BP = P \times Y \times 14/N\#$$ (24)

The N# is the N for the given surface number just as we will use ∆I#, ∆R#, etc. for those values associated with that surface number. The $d has the base centering cost CE as a factor, while $L has as a factor the base machining cost MF. There are additionally three "fudge" factors FI, FR, and FT associated with $I, $R, and $T which can be taken as unity for a simplifying assumption or calculated in each case as we will explain below. Equations 25 through 29 give these partial derivatives which will be needed to allocate tolerances.

$$\$I\# = BP \times FI \times 0.25 \times KI$$ (25)

$$\$R\# = BP \times FR \times 8 \times KZ$$ (26)

$$\$T = BP \times FT \times 40 \times KT$$ (27)

$$\$d = CE \times 10 \times KD$$ (28)

$$\$L = MF \times 80 \times KM$$ (29)

The factors FI, FR, and FT come from taking the partial derivatives of Equation 14 with respect to the reciprocal of the tolerances. They are due to the modifying effects of other parameters and are given in Equations 30, 31, and 32.

$$FI = \left[1 + 8 \times KZ \times \left(R\#/d\right)^2 \big/ \left(\Delta R\#\right) + 0.0003 \times \left(d/T\right)^2\right] \qquad (30)$$

$$FR = \left[1 + 0.25 \times KI/\left(\Delta I\#\right)\right] \qquad (31)$$

$$FT = \left(1 + 10/S\# + 5/D\#\right) \times \left(1 + 0.01 \times SC^3\right) \qquad (32)$$

There is a problem with Equations 30 and 31, however. The values of $\Delta R\#$ and $\Delta I\#$ are not known until after the tolerancing process. We can include the effects of FT in Equation 27 because all of its parameters are known at the start. We chose to include the effects of FI and FR after the tolerances are calculated by dividing the resulting tolerances for irregularity and radius by the cube roots of FI and FR, respectively. The reasons for this will be more apparent from Equation 35 below. The other alternative would be substituting FI and FR back in an iterative procedure. We now need only to address the simple allocation equations and procedures and we can finish the task of finding the least cost tolerances.

Tolerance Allocation Process.   We showed previously[11] that the total error E was the sum of all of the contributions from each error source which contributions are the products of the tolerance value t_i times the sensitivity S_i of the performance to variations of that parameter, as seen in Equation 33.

$$E = \sum_{i=1}^{n} \left(S_i \times t_i\right) \qquad (33)$$

(It may be found in the future that the application of 6σ to this process is simply accomplished by a proper adjustment of E, but we must "press on" without this understanding at this time.)

We have shown that the tolerances could be distributed for minimum cost by applying Equation 34 to each tolerance in turn. The A_i's are the coefficients of the derivatives of cost given in Equations 25 through 29. In Equation 34, the sum of all the square roots of the products of the cost coefficients A_k and the sensitivities S_k is divided into the total error budget E to get a constant which is multiplied by a function of the cost and sensitivity of each tolerance.

$$t_i = \sqrt{\frac{A_i}{|S_i|}} \; \frac{E}{\sum_{k=i}^{n} \sqrt{A_k |S_k|}} \qquad (34)$$

Adams[16] pointed out that Equation 34 gave the solution which represented the tolerances all going to the worst-case limit condition, and this is obviously too severe. He showed that the RSS condition would be satisfied by the tolerance distribution if Equation 35 was applied instead.

$$t_i = \sqrt[3]{\frac{A_i}{S_i^2}} \; \frac{E}{\sqrt{\sum_{k=1}^{n} \sqrt[3]{\left(A_k S_k\right)^2}}} \qquad (35)$$

The allocation is then very straightforward in the single E case. It is only necessary to develop a table of A_i and S_i for all appropriate i and process the data in accordance with Equation 35.

Table 1.4 gives these values for the example case, and the resulting t_i's which we seek are in Table 1.5. The example case has a total of four lenses which we illustrate in Figure 1.26. The field lens is close to the focal plane and has negligible sensitivity. We therefore remove it from consideration since we can assign minimum cost tolerances to it without affecting the rest of the task. This then reduces the example to a three-element lens where the surfaces and dimensions are numbered as in Figure 1.26. The necessary data processing is very conveniently set up and done using a spreadsheet program for data entry and all of the necessary calculations. Table 1.4 starts with the parameters determined from the "lens drawing" or design parameters. The number per block for each side is determined from Equations 7 and/or 8. The sensitivities of the system performance to each parameter are determined from the lens design program. The base costs and cost vs. the reciprocal tolerance derivative coefficients are calculated from Equations 25 to 29. The constant multiplier in Equation 35 is calculated. The individual factors from the individual sensitivities and cost derivatives are used to compute the tolerances for each of the toleranced parameters.

The adjustments to the I and R tolerances are made for FI and FR as mentioned in the previous section. Table 1.5 contains the resulting tolerances which needed to be determined. It is the set of tolerances for each of the toleranced parameters which will give the least cost solution and meet the performance with some statistical "RSS" certainty. The assumption per Adams[16] is that the errors will be distributed about the norm in a Gaussian manner and the tolerance limits will be 2σ.

However, a significant problem appears in Table 1.5. Many of the tolerances are well beyond what can be achieved in normal practice; they are off the chart in Figures 1.11 to 1.20! This is a disappointing result for the designer, but hopefully not the end of the road. Finding the problem at the design stage is not nearly as frustrating or expensive as finding it at the production stage. The lower part of Table 1.5 shows the application of the above formulas to compute the base cost for each of the lenses and the total costs when the tolerances are included. The total costs are about 15 times that of the simple base lens if they could even be made. "Off the chart" implies in most cases that it cannot be made or at least it would be much more expensive than the linear chart data would predict. We may have, therefore, identified here an impractical design. The designer then has the challenge of finding a design and/or an approach which will be less sensitive. The addition of more lens elements is not at all out of the question if they can reduce the sensitivity significantly. If the added cost of one or more elements reduces the tolerance costs sufficiently and all the lenses can be built, the total cost will be less than the first design. It might be possible to cement a doublet to get rid of a sensitive air space. It might be practical to make a centering adjustable and/or add other assembly tricks.[26] The designer can now evaluate the impact of design changes on cost by using the tools put forth in this discussion. The "bottom line" in Table 1.5 can tell him if he has improved the situation or not. What we see here is an example of the processes in Figure 1.2 where the tolerance sensitivity analysis and distribution feed into the producibility analysis which sends us back to the detail optical design for further work.

To solve this particular problem mentioned above, we first tested the ability to cement the achromats to reduce the critical air space sensitivity. It was found that the differential expansion of the glasses caused the cemented lenses to break at the extreme temperatures required for the system. The final solution was to design the lens cells so that the centering of each lens could be adjusted at assembly. This allowed the centering requirements on the lenses and cells and lens-to-cell fits to be significantly relaxed. Although the adjustment process was skill and labor intensive, an otherwise impossible result was achieved. The end result of this subsystem and the other subsystems in the telescope was essentially a diffraction limited performance.

A complex system such as the complete 4000-mm example system will have several times as many columns as Table 1.4, but the process is the same and relatively straightforward to apply. The most difficult aspect can be obtaining the sensitivities of the performance criterion to parameter variations. Existing lens design programs can do this with greater or less facility, but all are readily modifiable to generate the data needed. This data generation is computer intensive and time consuming, but probably unavoidable. The tolerancing program described above takes only a few seconds on a personal computer to calculate the tolerances for the six-surface case. It should

TABLE 1.4   Tolerance Data and Computations — Example Focus Lens Set

	Surface #					
	1	2	3	4	5	6
Radius	3.54	−6.504	−6.346	59.203	2.319	1.941
Diameter	2.35	2.35	2.31	2.31	2.04	2.04
Thickness	0.472	0.045	0.237	1.969	0.237	10.63
Scratch	80	80	80	80	80	80
Dig spec	50	50	50	50	50	50
Refractive index	1.5		1.6		1.5	
Stain ·	5.2	5.2	5.12	5.12	2	2
Polishability	3	3	1	1	1	1
Chamfers	1	1	1	1	1	1
Flats	0	0	0	0	0	1
Number/milling lot	7		7		7	
Centering lot size	7		7		7	
Yield	1.5	1.5	1.3	1.3	1-3	1.3
# Per block	6.58065	7	7	7	3	1
KI (fringes)	1	1	1	1	1	1
KZ (sag, in.)	1.25E-05	1.25E-05	1.25E-05	1.25E-05	1.25E-05	1.25E-05
KT (thk, in.)	1.25E-05	1.25E-05	1.25E-05	1.25E-05	1.25E-05	1.25E-05
KD (dia, in.)	1.25E-05	1.25E-05	1.25E-05	1.25E-05	1.25E-05	1.25E-05
KM (met dia, in.)	1.25E-05	1.25E-05	1.25E-05	1.25E-05	1.25E-05	1.25E-05
KW (mrads wedge)	1	1	1	1	1	1
SI dRMS/fringe	0.03	0.03	0.06	0.05	0.03	0.04
SZ dRMS/in.	375	750	176.5	250	400	117.6
ST dRMS/in.	4	8	2	3	2	0.1
Std dRMS/in.	80		90		25	
SLE dRMS/radian	225		400		25	
f, clearance	0	0	0	0	0	0
MF (MFG base)	6	6	6	6	6	6
BP (Equation 21)	9.573522	9	2.6	2.6	6.066667	18.2
CE (Equation 3)	7.497619		7.484286		7.394286	
A$I (Equation 22)	2.393381	2.25	0.65	0.65	1.516667	4.55
A$Z (Equation 23)	0.000957	0.0009	0.00026	0.00026	0.000607	0.00182
A$T (Equation 24)	0.004787	0.0045	0.0013	0.0013	0.003033	0.0091
A$d (Equation 25)	0.000937		0.000936		0.000924	
A$L (Equation 26)	0.006		0.006		0.006	
[A(K)*S(K)]^2/3 I	0.172736	0.165767	0.114991	0.101829	0.127437	0.321136
[A(K)*S(K)]^2/3 Z	0.505111	0.769479	0.128164	0.161648	0.389033	0.357792
[A(K)*S(K)]^2/3 T	0.071561	0.109015	0.018904	0.024772	0.033258	0.009388
[A(K)*S(K)]^2/3 d	0.177792		0.192088		0.081117	
[A(K)*S(K)]^2/3 L	0.691039		1.02581		0.175497	
Constant E/SQR (sum) =		0.018141				
[A(I)/S^2]^1/3 I	13.85082	13.56855	5.65104	6.381321	11.89711	14.16387
[A(I)/S^2]^1/3 Z	0.001896	0.00117	0.00203	0.001609	0.00156	0.005089
[A(I)/S^2]^1/3 T	0.0669	0.041287	0.068772	0.052484	0.091213	0.969055
[A(I)/S^2]^13 d	0.005274		0.004873		0.011398	
[A(I)/S^2]^1/3 L	0.003697		0.002534		0.016765	
Delta I (fringes)	0.251274	0.246153	0.102518	0.115766	0.215831	0.256953
Delta Z (in.)	3.44E-05	2.12E-05	3.68E-05	2.92E-05	2.83E-05	9.23E-05
Delta AT (tilt, rd)	6.71E-05		4.6E-05		0.000304	

Note: E = 0.0736 RMSWE.

be approximately linear with the number of surfaces as long as only one performance criterion is
to be considered. Multiple criteria would be more cumbersome to evaluate as we showed in our
first work.[11] However, Adams[16] shows that it is most likely that one criterion is all that is needed,
and more than two is highly unlikely.

TABLE 1.5 Resulting Tolerances and Costs

Tolerances	Surface					
	1	2	3	4	5	6
Delta R (in.)	0.000786	0.001644	0.003355	0.225109	0.000378	0.000839
Delta I'(fringes)	0.279124	0.32987	0.113803	0.13038	0.245058	0.268062
Delta T (in.)	0.001214	0.000749	0.001248	0.000952	0.001655	0.01758
Delta AR (roll, rd)	1.81E-05	9.88E-06	8.05E-06	8.63E-07	4.65E-05	5.56E-05
Delta td (in.)	9.57E-05		8.84E-05		0.000207	
Delta LE (in.)	0.000158		0.000106		0.00062	
Delta d (in.)	5.67E-05		4.51E-05		9.49E-05	
Delta dM (in.)	7.17E-05		5.71E-05		0.000121	
Delta A (mrad)	0.001931		0.001535		0.003231	
Delta A ('dev)	0.003322		0.003168		0.005558	
Metal diam = dm	2.350128		2.310102		2.040216	

Base Cost Computation Base = MG + GP + CE

LENS #	1–2	3–4	5–6
Milling = MG	17.40939	17.39075	17.2733
Grind and polish = GP	39.57352	23.4	42.46667
Center and edge = CE	7.497619	7.484286	7.394286
Base cost total =	64.48053	48.27504	67.13426

Computation of Total Lens Cost = MT

Lens #	1–2		3–4		5–6	
Milling/generate	17.40939		17.39075		17.2733	
Part setup	21		18.2		18.2	
Grind and polish	35.68302	716.8573	15.47956	13.71988	21.21353	39.65565
Centering	586.8878		735.275		348.9406	
Total lens cost	1377.837		800.0652		445.2831	

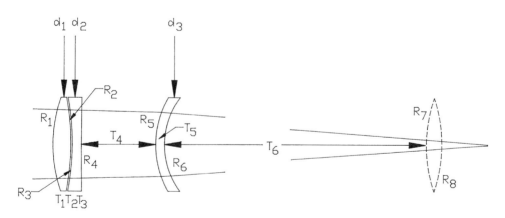

FIGURE 1.26 Focus lens set of telescope used as an example.

Tolerancing Summary

We had previously[11-13] shown the principles of how to assign tolerances to achieve a minimum production cost and we mentioned the possible application of the results to estimating total lens production cost. In this discussion, the previous data and principles have been refined and some of the results provided by Adams,[16] Smith,[20] Parks,[18] and Fischer[17] have been incorporated. The lens cost estimating formulas mentioned in the earlier work[13] have been developed into useful tools. A new analysis is presented of the interdependency of the lens and cell diameter tolerances as a result of the cost vs. tolerance knowledge. And finally, a minimum cost tolerancing procedure has been reduced to practice in a form which is straightforward and accurate enough for practical engineering application. It is practical to estimate the production cost of most lenses by entering the drawing data into a spreadsheet program. This is essentially an "expert system" estimator which can be applied by a person with very little training or experience to get as good or better cost estimates for most lenses than an expert. It is also now practical to distribute lens tolerances using a spreadsheet program such that the production costs are minimized and to find the impact of design changes on lens costs. Both of these tools have advanced what was an "art" to an engineering discipline but not quite a "science" because of the necessary simplifying assumptions. However, these tools are more accurate than what has been the common practice to date, and they are more accurate than the cost data as they typically are measured at this time. The application of the estimating program can reduce the business overhead cost of a production operation, and it can point to the cost drivers of any particular lens such as: setup, milling, polishing, or centering. The application of the tolerancing program can decrease the production cost of systems from what they typically have been in lens production, assembly, and testing. If the tolerances are unnecessarily tight, the lens production cost is wasteful. If the tolerances are not tight enough to give a good yield of deliverable parts, the assembly, rework, and testing costs are wasteful. This tool addresses the economic problem at the point of the most potential impact on the life cycle costs as shown in Figure 1.1, the detail design and analysis phase.

1.5 Environmental Effects

Laboratory optical instruments may operate in a relatively benign environment, but most others are subject to environmental effects which can be a major consideration in the design of an instrument. Astronomical instruments must operate satisfactorily over a broad range of nighttime temperatures, or daytime in the case of solar telescopes. Personal cameras and binoculars may experience a broad range of weather conditions. Military instruments probably have the most widespread and severe environmental exposure to temperatures, moisture, shock, vibration, dust, chemicals, etc. It is very relevant and important that all of the environmental requirements be properly dealt with in the optomechanical design and development process. Subsequent chapters will cover many of these aspects in more detail, but we will discuss a few things here and give some guidance with respect to available standards and specifications.

The U.S. government through its military procurement activities over the past century or more has evolved a very extensive set of specifications and test methods to ensure that the optical instruments (and anything else they buy) will perform as required in the expected environments. Most of the current U.S. optical industry has worked to these requirements as exemplified by MIL-STD-810.[27] This standard might be a worthwhile document to consult to check whether you have considered all of the possible effects that might be important to a given instrument under development and how to test it. For that same reason, we will also describe below the new international standards that exist or are in development.

Survivability under Temperature, Vibration, and Shock Loads

The quality of the image and pointing direction are two of the major performance factors of most optical instruments. These might also be referred to as resolution or modulation transfer function (MTF) and boresight. With a 35-mm camera, the MTF is the key factor and pointing is usually not an issue. In binoculars, one becomes concerned also with boresight so that the views through each eye are not uncomfortably divergent. Tilt and equality of magnification also are important in binoculars. A surveying instrument or military aiming sight places a great deal of emphasis on the repeatability of the line of sight or boresight.

The designer needs to be sure that the expected temperature changes, vibrations, and shocks will not damage or disable the instrument, but also will not degrade the image quality and boresight to unacceptable levels. For example, glass lenses in aluminum housings are at risk of becoming too loose at high temperatures and being "squeezed to breakage" at cold temperatures. This is, of course, due to the difference in the thermal coefficients of expansion (TCE). In the bonding of optical components, the TCE differences can cause major difficulties of distortion and glass fracture. Paul Yoder's chapter and others in this handbook touch on how to deal with some of these problems. Other materials can be used, but at penalties of weight, cost, and sometimes performance. In many instruments, thermal distortions of the shape of optical surfaces (particularly mirrors) can degrade the image quality severely. We have touched on some examples of how these factors are dealt with in the section on tolerancing, and later chapters will discuss many design techniques in some detail. The optomechanical designers' challenge is to develop an instrument which will survive and perform in all of the required environments.

Humidity, Corrosion, Contamination

Designing an instrument which will not degrade due to humidity, corrosion, and contamination poses another class of challenges. Whenever practical, instruments are sealed with an internal atmosphere such as dry nitrogen. This prevents internal condensation of moisture at low temperatures and optical coating degradations due to humidity. It also will keep out dust, contamination, and corrosive agents. A large astronomical telescope usually cannot be sealed. In such a case, the mirrors tend to become dusty and the coatings tend to degrade. The scattering of light by the contamination reduces the contrast (MTF) and can be disastrous when the telescope is used to look at faint objects while it is illuminated by bright objects. Handling contamination in large telescopes and designing to minimize its effects is a major specialty.

In sealed instruments, corrosion is only an issue for external surfaces and interfaces. The most difficult environment is usually salt fog. Unprotected metals such as aluminum and steel will deteriorate rapidly. Plating and/or painting is usually required. Unless specific treatments have been well tested before, it is highly recommended that samples be extensively tested for durability. Subsequent chapters contain some suggestions with respect to corrosion and contamination.

Environmental Testing Standards

To date there are two ISO standards dealing with the effect of environmental conditions on the performance of optical instruments, ISO 9022 — *Environmental Test Methods*, and ISO/CD 10109 — *Environmental Requirements*. ISO 9022 has been issued as a standard while ISO/CD 10109 is still being written. It is complete in outline form but the performance criteria for many instruments are still being defined.

ISO 9022 defines terms relating to environmental tests for optical instruments and for instruments that contain optical assemblies and components. In addition, it specifies the essential steps for conducting an environmental test and defines some 20 types of tests along with various subcategories of these tests.

ISO 10109 specifies the environmental requirements to be met regarding the reliability of particular optical instruments when exposed to various applicable environmental influences. It also defines the geographical and technological areas of applicability of the instruments. The standard does **not** apply to specifications for **packaging** for transportation and storage.

ISO 9022 — Environmental Test Methods

The 20 parts of ISO 9022 are listed in Table 1.6, along with an indication of how many subdivisions there are of the basic test. We do not list all the subdivisions but it is informative to see an example in Table 1.7 where the subdivisions of Parts 2 and 3 are listed.

TABLE 1.6   ISO 9022 — Environmental Test Methods

Part 1	Definitions, extent of testing
Part 2	Cold, heat, humidity — 7
Part 3	Mechanical stresses — 8
Part 4	Salt mist — 1
Part 5	Combined cold, low air pressure — 2
Part 6	Dust — 1
Part 7	Drip, rain — 3
Part 8	High pressure, low pressure, immersion — 3
Part 9	Solar radiation — 1
Part 10	Combined sinusoidal vibration, dry heat or cold — 2
Part 11	Mold growth — 1
Part 12	Contamination — 4
Part 13	Combined shock, bump or free fall, dry heat or cold — 6
Part 14	Dew, hoarfrost, and ice — 3
Part 15	Combined random vibration wide band — 2
Part 16	Combined bounce or steady state — 4 — acceleration, in dry heat or cold
Part 17	Combined contamination, solar radiation — 2
Part 18	Combined damp heat and low internal — 3 — pressure
Part 19	Temperature cycles combined with sinusoidal or random vibration — 3
Part 20	Humid atmosphere containing sulfur dioxide or hydrogen sulfide

TABLE 1.7

Part 2	Cold, Heat, Humidity
10	Cold
11	Dry heat
12	Damp heat
13	Condensed water
14	Slow temperature change
15	Rapid temperature change
16	Damp heat, cyclic
Part 3	**Mechanical Stresses**
30	Shock
31	Bump
32	Drop and topple
33	Free fall
34	Bounce
35	Steady-state acceleration
36	Sinusoidal vibration
37	Random vibration (wide band)

The beginning part of ISO 9022 defines basic terms relating to conducting these tests. For example, an **environmental test** is defined as a laboratory simulation of (usually severe) climatic, mechanical, and chemical influences likely to occur during transport, storage, and operation on a test specimen in order to quickly determine changes in the behavior of the specimen due to the influences. The act of subjecting the specimen to these influences is called **conditioning.**

Conditioning is considered to be the sum of external influences acting on the specimen during the test including the conditioning method (or particular environmental test), the degree of severity of the test, and the internal influences due to the state of operation such as motion and/or temperature change. Also defined are three **states of operation:** state 0 — in a storage or transportation container; state 1 — unprotected, ready for use, but not turned on; and state 2 — unprotected, turned on, and operating.

In order to evaluate what has happened during a test or conditioning, there are three types of tests or examinations. The first is simply a **visual examination** to see if, for example, some part of the specimen became loose during conditioning. The second is a **functional** test to see if the device still functions after conditioning. Finally there is the **measurement,** an objective determination of a physical quantity by comparison with a specified quantity.

A **test sequence** is defined as given in Table 1.8. While this table looks trivial, it does define precisely what is meant by a test sequence and that there are some important matters that must be noted before applying the conditioning so that changes may be recognized after the conditioning. In order to specify what test(s) are to be performed on a particular instrument, a one-line environmental test code is used for each type of test required. This test code is illustrated in Figure 1.27.

TABLE 1.8 Definitions — Test Sequence

1   Preconditioning — prepare specimen for testing
2   Initial test — state of device prior to testing
3   Conditioning — apply conditioning method at degree of severity and state of operation
4   Intermediate test — does it function in state 2
5   Recovery — bring back to ambient conditions
6   Final test — state of device after testing
7   Evaluation — determination if specifications met

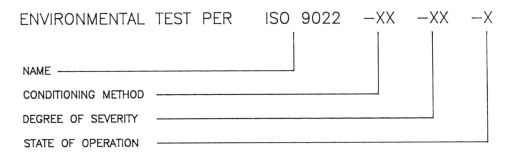

FIGURE 1.27 Definition of an environmental test code which can be communicated in a one line call-out under ISO 9022.

ISO 10109 — Environmental Requirements

Whereas ISO 9022 outlines the details and degrees of severity of nearly 100 conditioning methods, ISO 10109 is concerned with which of these methods and with what degree of severity they should be applied to a particular optical instrument designed for a particular type of use or for use in a particular climatic region.

ISO 10109 is an 11-part standard that specifies the types of testing needed to establish the suitability of an optical instrument for its intended conditions of use. The first part deals with definitions while the remaining ten parts are instrument-type specific sets of environmental requirements.

Part 1 starts by defining five **climatic zones** listed in Table 1.9. Zone 5 is a normal, protected environment in which most analytical instruments would be expected to operate. On the other hand, many types of optical instruments must survive rather severe climates and virtually all instruments must be unaffected by air shipment, much of the reason for zone 4, high altitudes. The severe climates can be thought of as Arctic (zone 1), Maritime (zone 3), and all other outdoor regions (zone 2).

The standard then goes on to define three instrument types: field instruments, instruments used in weather-protected locations, and these same instruments with the additional provision that they must be sterilizable. Also defined are ten groups of instruments along with subclasses as shown in Table 1.10. These begin with ophthalmic optics and cover the field to electro-optical systems.

TABLE 1.9   Definitions — Climatic zones

Zone 1	Not weather protected, cold and extremely cold
Zone 2	Not weather protected, global locations
Zone 3	Not weather protected, maritime locations
Zone 4	High altitudes to 30,000 m
Zone 5	Weather-protected, laboratory environment

TABLE 1.10 Definitions — Groups of Instruments

Group 1	Ophthalmic optics — 3 types
Group 2	Photographic instruments — 3 types
Group 3	Telescopes — 4 types
Group 4	Microscopes — 5 types
Group 5	Medical devices — 3 types
Group 6	Metrology instruments — 4 types
Group 7	Military instruments — 3 types
Group 8	Geodetic instruments — 3 types
Group 9	Photogrammetric instruments — 3 types
Group 10	Electro-optical systems — 3 types

Once this standard is issued, it will specifically define how well each of these instrument types must hold up under various environmental influences depending on the intended type of use the instrument will be subjected to in normal operation.

Summary of Environmental Effects

It can be seen that the optomechanical designer is usually required to give extensive thought during the design process to what environmental conditions will be encountered, and how to maintain the instrument performance under these circumstances. The available standards provide a good checklist to avoid overlooking a pertinent condition. They also provide guidance as to how the instrument might be tested to prove its performance.

REFERENCES

1. Smith, W. J. 1966. *Modern Optical Engineering.* McGraw-Hill, New York.
2. Foster, L. W. 1986. *Geo-metrics II.* Addison-Wesley, Reading, MA.
3. Wilson, B. A. 1992. *Design Dimensioning and Tolerancing.* Goodhart-Willcox, South Holland, IL. See also Gooldy, G. 1995. *Geometric Dimensioning & Tolerancing*, Prentice-Hall, Englewood Cliffs, NJ.
4. ANSI, 11 West 42nd St., New York, NY 11036, (212) 642-4900.
5. ASME, 345 East 47th St., New York, NY 10017, (212) 705-7000.
6. ASTM, 100 Barr Harbor Dr., West Conshohocken, PA 19428-2959, (610) 832-9585.
7. SAE, 400 Commonwealth Ave., Warrendale, PA 15096, (412) 776-4841.
8. OEOSC, 7913 B Fingerboard Rd., Frederick, MD 21704, (301) 607-4249.
9. SPIE's Homepage is at http://www.spie.org and OSA's at http://www.osa.org.
10. Optical Society of America, 2010 Massachusetts Ave., NW, Washington, D.C. 20036, (202) 223-8130.
11. Willey, R. R., George, R., Odell, J., and Nelson, W. 1982. Minimized cost through optimized tolerance distribution in optical assemblies. In *Optical Systems Engineering III, Proc. SPIE*, 389, Taylor, W. H., ed., p. 12. SPIE, Bellingham, WA.
12. Willey, R. R. 1983. Economics in optical design, analysis, and production. In *Optical System Design, Analysis, and Production, SPIE*, 399, Rodgers, P. J. and Fischer, R. E., eds., p. 371. SPIE, Bellingham, WA.

13. Willey, R. R. 1984. The impact of tight tolerances and other factors on the cost of optical components. In *Optical Systems Engineering IV, Proc. SPIE*, 518, Yoder, P. R., ed., p. 106. SPIE, Bellingham, WA.

14. Wiese, G. W., ed. 1991. *Optical Tolerancing*. SPIE, Bellingham, WA.

15. Plummer, J. L. 1979. Tolerancing for economies in mass production of optics. In *Contemporary Optical Systems & Component Specifications, Proc. SPIE*, 181, Fischer, R. E., ed., p. 90. SPIE, Bellingham, WA.

16. Adams, G. 1988. Selection of tolerances. In *Simulation and Modeling of Optical Systems, Proc. SPIE*, 892, Fischer, R. E. and O'Shea, D. C., eds., p. 173. SPIE, Bellingham, WA.

17. Fischer, R. E. 1990. Optimization of lens designer to manufacturer communications. In *International Lens Design Conference, Proc. SPIE*, 1354, Lawrence, G. N., ed., p. 506. SPIE, Bellingham, WA.

18. Parks, R. E. 1980. Optical component specifications. In *International Lens Design Conference, Proc. SPIE*, 237, Fischer, R. E., ed., p. 455. SPIE, Bellingham, WA.

19. Parks, R. E. 1983. Optical specifications and tolerances for large optics. In *Optical Specifications: Components & Systems, Proc. SPIE*, 406, Smith, W. J. and Fischer, R. E., eds., p. 98. SPIE, Bellingham, WA.

20. Smith, W. J. 1985. Fundamentals of establishing an optical tolerance budget. In *Geometrical Optics, Proc. SPIE*, 531, Fischer, R. E., Price, W. H., and Smith, W. J., eds., p. 196. SPIE, Bellingham, WA.

21. Willey, R. R. and Durham, M. E. 1992. Maximizing production yield and performance in optical instruments through effective design and tolerancing. In *Optomechanical Design, SPIE*, CR43, Yoder, P. R., ed., pp. 76–108. SPIE, Bellingham, WA.

22. Thorburn, E. K. 1980. Concepts and misconcepts in the design and fabrication of optical assemblies. In *Optomechanical Systems Design, Proc. SPIE*, 250, Bayar, M., ed., p. 2. SPIE, Bellingham, WA.

23. Beyeler, B. H. and Tiziani, H. J. 1966. Die optische Ubertragungsfunktion von dezentrierten optischen Systemen, *Optik*, 44(3), 317 (included in Ref. 4).

24. Kojima, T. 1979. Yield estimation for mass produced lenses based on computer simulation of the centering tolerance. In *Optical Systems Engineering, Proc. SPIE*, 193, Yoder, P. R., ed., p. 141. SPIE, Bellingham, WA.

25. McLaughlin, P. O. 1991. A primer on tolerance analysis, *Sinclair Optics Design Notes*, 2(3).

26. Willey, R. R. 1989. Optical design for manufacture. In *Recent Trends in Optical Systems Design II, SPIE*, 1049, Fischer, R. E. and Juergens, R. C., eds., p. 96. SPIE, Bellingham, WA.

27. MIL-STD-810, Environmental Test Methods.

2

Optomechanical Design Principles

Daniel Vukobratovich

2.1 Introduction

Optical engineering is defined as the control of light. Light is controlled by interaction with surfaces, as is the case in refraction or reflection. Optomechanics is defined as that part of optical engineering concerned with maintaining the shape and position of the surfaces of an optical system.

Deviation from a stress-free condition is defined as deflection. Deflection affects the shape and position of surfaces in an optical system. Very small deflections, sometimes as small as one part per million or less, are important in optomechanical engineering. Unlike ordinary mechanical engineering practice, emphasis in design of optomechanical systems is on deflection, or strain, rather than strength, or stress.

This chapter discusses engineering design methods to counteract the effects of deflection on the performance of an optomechanical system. The following topics are covered in this chapter:

1. Service environments
2. Structural design
3. Kinematic design
4. Athermalization
5. Vibration control

2.2 Service Environments

Often the exact service environment of the system is not well understood. To overcome lack of knowledge about the actual working environment for the system, standard environment specifications are used. Military standards are used in the U.S. engineering community as reference environments. One such military standard for environments is Military Standard 810, Environmental Test Methods and Engineering Guidelines.[1] Not all applications require the use of military environmental standards. Systems intended for use in laboratory environments need not accommodate the very severe specifications typical of military systems. Laboratory environments are normally assumed to be similar to those described in the standards on environment control[2] such as ASHRAE Standard 55-81 and on vibration control.[3] Table 2.1 provides examples of service environments, from mild to severe.

TABLE 2.1 Service Environments

Environment	Normal	Severe	Extreme	Example of Extreme
Low temperature	293 K	222 K	2.4 K	Cryogenic satellite telescope
High temperature	300 K	344 K	423 K	White cell for combustion studies
Low pressure	88 KPa	57 KPa	0	Satellite telescope
High pressure	108 KPa	1 MPa	138 MPa	Submersible window
Humidity	25–75% RH	100% RH	(Underwater)	Submersible window
Acceleration	2 g	12 g	11×10^3 g	Gun-launched projectile
Vibration	200×10^{-6} m/sec RMS, $f \geq 8$ Hz	0.04 g^2/Hz $20 \leq f \leq 100$ Hz	0.13 g^2/Hz $30 \leq f \leq 1500$ Hz	Satellite launch vehicle

2.3 Structural Design

The support structure of an optical system must maintain the position of the optical components relative to each other within design tolerances. The most common load on optical systems is self-weight, due to gravity. Structures with a self-weight deflection that is less than the alignment tolerance are considered stiff. If the self-weight deflection exceeds the alignment tolerances, the optical support structure is compliant.

Conventional definitions of structural efficiency in mechanical engineering are based on strength-to-weight ratios. In contrast, structural efficiency in optomechanical support structures is determined by stiffness-to-weight. Stress in an optomechanical support structure is at a low level in comparison with structures used in other mechanical engineering applications.

One index of structural efficiency for optomechanical support structures is fundamental frequency. Fundamental frequency of a support structure is determined by both stiffness and weight. The fundamental frequency of a structure is a measure of the stiffness-to-weight ratio of a structure. Self-weight deflection is related to fundamental frequency by:[4]

$$f_n = \frac{1}{2\pi} \sqrt{\frac{g}{\delta}}$$

where f_n = the fundamental frequency, in Hz
  g = the acceleration due to the Earth's gravity
  δ = the self-weight deflection of the support structure

Fundamental frequency of a self-weight loaded beam is given by:[5]

$$f_n = \frac{\lambda_i^2}{2\pi L^2} \left(\frac{EI}{m} \right)^{\frac{1}{2}}$$

where f_n = the fundamental frequency, in Hz
  λ_i = a dimensionless constant depending on the type of beam support
  L = the beam length
  E = the elastic modulus of the beam material
  I = the cross-section area moment of inertia of the beam

m = the mass per unit length of the beam

The above equation is more useful if it is expanded to show the relationships between material properties and the geometrical efficiency of the beam shape. In the above equation, the mass per unit length is equivalent to the beam cross section multiplied by the mass density of the beam material. Using this relationship, the above equation for fundamental frequency becomes

$$ f_n = \frac{\lambda_i^2}{2\pi L^2} \left(\frac{E}{\rho} \right)^{\frac{1}{2}} \left(\frac{I}{A} \right)^{\frac{1}{2}} $$

where E/ρ = the specific stiffness, the ratio of material elastic modulus to material elastic modulus
   I/A = the ratio of area moment of inertia of the beam to cross-section area

The highest fundamental frequency is produced by selecting a material with greatest values of E/ρ and I/A. The ratio E/ρ is a measure of the structural efficiency of the material, while the ratio I/A is a measure of the geometric efficiency of the beam cross section. Efficiency is independent of the strength of the material. There is no reason in optomechanical design to use high-strength materials to provide structural efficiency.

The specific stiffness, or ratio of elastic modulus to density, is of considerable importance in the design of optomechanical support structures. For most common materials, the ratio E/ρ is constant,[6] with a value of about 386×10^{-9} m^{-1}. There are materials such as beryllium and silicon carbide with ratios of E/ρ below this constant. Such materials are considered lightweight materials.

Selection of materials for optomechanical support structures involves a variety of properties other than specific stiffness. Damping capacity, which is a property associated with the dissipation of energy in the material when excited by vibration, is important for dynamic environments. Materials with high damping capacities are often selected for use in optomechanical systems subjected to vibration. Damping capacities of materials are given in Lazan[7] and other references.[8-10] Other material properties of interest are discussed in standard engineering materials selection handbooks. Ashby[11] presents material properties in the form of selection charts. These charts are useful in selecting materials with optimum properties for optomechanical applications.

In cases where the weight of the support structure is limited, efficiency of the support is improved through the use of optimization. An example of the usefulness of optimization is the placement of two simple supports for a beam. Careful selection of the supports substantially reduces beam deflection.

If the beam is of uniform cross section and is carrying a uniform load, the optimum location of the support points to minimize the slope from end to end is at a distance of 0.2222 times the overall length from each end. The maximum slope for a beam supported by two points at this optimum distance is given by:

$$ \theta_b = 0.001543 \frac{wL^3}{EI} $$

where θ_b = the maximum slope change of the beam
   w = the weight per unit length of the beam (self-weight and load)
   L = the beam length
   E = the elastic modulus of the beam material
   I = the area moment of inertia of the beam

If the beam is of uniform cross section and is carrying a uniform load, the optimum location of the support points to minimize the deflection from end to end is at a distance of 0.2232 times the overall length from each end. The maximum deflection for a beam supported by two points at this optimum distance is given by:

$$\delta_b = 0.0002698 \frac{wL^4}{EI}$$

where δ_b is the maximum deflection from end to end of the beam.

If the beam is of uniform cross section and is carrying a moving load, the location of the two simple supports to minimize the slope is at 0.1189 times the overall beam length from the end. The change in slope for a beam supported by two points at this optimum distance is given by:

$$\theta_m = 0.03727 \frac{WL^2}{EI}$$

where θ_m = the change in slope under the weight of the moving load
           W = the weight of the moving load

If the beam is of uniform cross section and is carrying a moving load, the optimum location for the two simple supports to minimize deflection from end to end is at 0.1556 times the overall length of the beam from the end. The maximum change in deflection under the weight of the moving load for a beam on simple supports at these optimum locations is given by:

$$\delta_m = 0.006811 \frac{WL^3}{EI}$$

where δ_m is the change in deflection under the weight of the moving load.

Many optical support structures consist of a beam of uniform cross section with an optical component at each end. A laser cavity is an example of such a structure. Alignment of the optical components at each end is maintained with accuracy through the use of a special category of optimum locations for the beam supports. By placing the two simple supports for the beam at the Airy points, the end slope of both ends of a uniformly loaded beam becomes zero. With both supports at the same distance from the beam ends, and the end slopes equal to zero, the beam ends are in alignment. This alignment is preserved regardless of the orientation of gravity with respect to the beam. The location of the Airy points is at 0.2113 times the overall length of the beam from the ends. This location for the simple supports assumes that the weight of the optical components at each end of the beam is small in comparison with weight of the beam.

If the weight of the optical components at each end of the beam is not small in comparison with the weight of the beam, the location of the two Airy points is found using the following equation:

$$L_s = \frac{L_b}{2} \left(1 - \left[1 - \frac{W_b}{6 \left(\dfrac{W_o}{2} + \dfrac{W_b}{4} \right)} \right]^{\frac{1}{2}} \right)$$

where L_s = the location of the beam supports from one end of the beam

L_b = the beam length
W_b = the weight of the beam
W_o = the weight of the optics at one end of the beam

In some applications multiple supports are required for a beam. This is a difficult optimization problem; however, in many cases deflection is reduced to a low level through the use of a distributed set of simple supports at the Airy points. The optimum spacing required between two adjacent support points for a uniformly loaded beam to maintain the Airy condition with n simple supports is given by:

$$L_s = \frac{L_b}{\sqrt{n^2 - 1}}$$

where
L_s = the spacing between adjacent support points
L_b = the beam length
n = the number of support points

Efficiency of structural connections is often overlooked in the design of optical support structures. A bolted connection designed using normal strength of materials practice typically transmits only about 25% of the moment applied to the connection. Increasing the number of bolts in the connection by a factor of 2 or 3 increases the amount of moment transmitted to about 50%. For maximum stiffness it is desirable to limit the number of connections in a structure. The stiffness of bolted and riveted joints, as well as pin connections, is determined using methods given in Levina[12] and Rivin.[13]

The following types of structural connections are listed in order of decreasing stiffness:

1. Cast or machined from solid ("hog out" construction)
2. Welding
3. Bolting, using maximum number of bolts possible due to geometry ("overbolting")
4. Riveting
5. Conventional bolting
6. Adhesives

In design of an optical support structure, it is not necessary to reduce deflection at all parts of the structure. As shown by the Airy point-supported beam discussed above, deflection of the support structure is acceptable if optical alignment is not affected. The principle of equal and parallel end deflections used in the Airy point-supported beam is extended to a truss through the use of a Serrurier truss.[14] The Serrurier truss consists of a truss supported at its center of gravity, with optical components at each end of the truss. The truss ends remain equal and parallel to each other regardless of the direction of the gravity vector. For equal and parallel deflections of the end rings of the Serrurier truss, the following condition must be satisfied:

$$\frac{w_1}{E_1 A_1}\left(\frac{4L_1^2}{b^2} + 1\right)^{\frac{3}{2}} = \frac{w_2}{E_2 A_2}\left(\frac{4L_2^2}{b^2} + 1\right)^{\frac{3}{2}}$$

where
w_1, w_2 = the end ring loads
E_1, E_2 = the elastic modulii of the truss member
A_1, A_2 = the cross-section areas of the truss members
L_1, L_2 = the distances from the center of gravity to the end ring loads
b = the end ring diameter

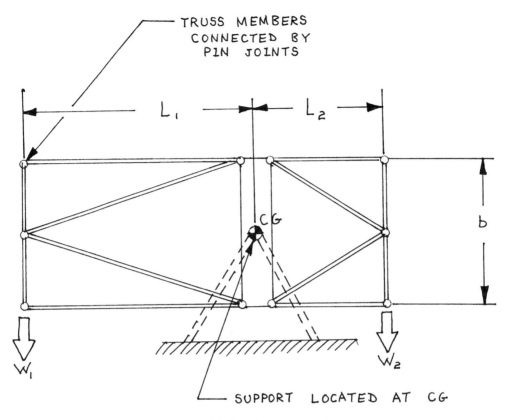

FIGURE 2.1 Serrurier truss geometry.

Figure 2.1 shows a side view of the Serrurier truss geometry. The maximum deflection of either end ring is given by:

$$\delta_{max} = \frac{Wb}{4EA}\left(\frac{4L^2}{b^2}+1\right)^{\frac{3}{2}}$$

In some cases it is necessary to locate optical components in portions of the structure which experience greater deflection than the allowable alignment tolerance. One possible solution to this problem is an insensitive optical design. Fold mirrors are replaced with penta-prisms which are insensitive to rotation in one plane. Similar in function to penta-prisms are penta-mirror combinations.

For an afocal system, single lens components are replaced by a pair of components, one positive and one negative.[15] The positive and negative components are placed so that the nodal point of the combination lies on the intersection of two lines. One line is a perpendicular to the focal plane of the system. The other line is drawn through the centers of the two components when the structure is undergoing deflection. Locating the nodal point at this intersection produces a rotation of the two-lens combination about its nodal point. Rotation about the nodal point does not cause motion in the final image plane.

For such an insensitive combination, the following equation must be satisfied:

$$f^2 = p(p+s)$$

where   f = the focal length of the afocal component
        p = the location of the rear nodal point, measured from the rear vertex of the second
            lens element
        s = the separation of the two optical elements

The focal lengths f_1 and f_2 of the two optical elements replacing the single lens are given by:

$$f_2 = f\left(\frac{s - f_1}{f - f_1}\right)$$

$$f = \frac{fs^2}{f_1^2} - \frac{s^2}{f_1}$$

Weight relief reduces the self-weight deflection to a very low level. Weight relief is obtained through the use of "off-loading" mechanisms. These mechanisms are soft in all but the direction of the gravity vector. Examples of this type of mechanism include springs, counterweighted levers, air bags, and mercury floats. Weight relief was common prior to the development of high capacity precision bearings. Although not common today for support structures, weight relief is still used for large optical mounts. Air bag support is provided for the primary mirrors of the multiple mirror telescope (MMT) at Mt. Hopkins Observatory.

A simple example of weight relief is the counterweight system used on older refractor telescopes.[16] A lever arm is parallel to the telescope tube and pivoted at the center of gravity of the tube. One end of the pivoted arm is attached to the objective end of the telescope tube, while the other floats free with a counterweight equivalent to the weight of the objective. The lever arm relieves the telescope tube of the weight of the objective, which reduces the deflection of the objective relative to the focal plane. The counterweight location at the other end of the telescope tube moves the center of gravity toward the focal plane, reducing the swing of the eyepiece.

More complex off-loading schemes were developed to reduce the load on precision bearings. Both whiffle tree mechanisms and mercury floats are used for this purpose. The 2.5-m Hooker telescope at Mt. Wilson observatory is provided with a mercury float to reduce the load on the polar axis bearings.[17]

2.4 Kinematic Design

It is desirable to hold optical components in a way that is repeatable and low in stress. The tolerances associated with optical systems require extremely accurate mechanical mounting surfaces to achieve these goals. Such extremely accurate mounting surfaces are very expensive to produce. Kinematic methods provide accurate and repeatable mounting of optical components, in a low stress condition, at much lower costs than conventional precision mechanical methods.[18]

Kinematic methods are derived from the principle of constraint of a rigid body. Every rigid body possesses six degrees of freedom in translation and rotation about each of three mutually perpendicular axes. Perfectly rigid bodies can touch only at infinitely small points. A perfectly rigid body has six N degrees of freedom, where N is the number of contact points.[19] Any rigid body with more than six contact points is overconstrained. An overconstrained rigid body is likely to have an uncertain position and be distorted from its stress-free condition.

The application of kinematic theory consists of selecting no more than six contact points to provide the type of support or motion required. Kinematic supports with less than six contact points permit motion of the supported body. The motion is a "degree of freedom" and is correlated with the number and geometry of the support points.

Real kinematic mounts violate the assumption of an infinite elastic modulus associated with a rigid body. Kinematic designs attempt to approximate the ideal point contact through the use of Hertz contact. In a Hertz contact, the bodies are assumed to be ellipsoidal, with differing radii of curvature. The area of contact developed is due to the radii of the contacting solids, the force pushing the solids into contact, and the elastic properties of the materials.[20] Approximate methods are used to calculate the deflection and stress developed in the Hertz contacts associated with kinematic design.[21]

The simplest type of kinematic mount is used to fix the location of a body. This requires six contact points. Often illustrated in texts on instrument design is the "Kelvin clamp".[22] The Kelvin clamp employs three spherical locating features attached to the body. One spherical feature is located in a trihedral socket, one in a v-groove, and one is placed on a flat. There is one contact point for each flat surface in contact with a sphere: three for the trihedral contact, two for the v-groove, and one for the flat. Six contact points uniquely locate the body without overconstraint.

Although theoretically attractive, the Kelvin clamp is not a physically practical design due to the difficulty of making trihedral sockets. One method of producing three-point contact between a sphere and socket replaces the trihedral socket with a compound socket. The compound socket consists of three spheres forced into a common cylindrical hole or socket.[23] The locating spherical feature rests against the three spheres. Another method is the replacement of the trihedral socket with a conical socket. A conical socket is readily produced, but is not truly kinematic since line contact develops between the conical surface and spherical locating feature.

An alternate to the Kelvin clamp is the three-groove kinematic mounting or coupling.[24] The three-groove kinematic coupling retains the spherical locating features attached to the body. In place of the socket, v-groove and plane of the Kelvin clamp are three v-grooves. Each v-groove provides two points of contact. In a planar assembly, the long axes of the three v-grooves intersect at a common point and are about 120° apart. Figure 2.2 shows a typical three v-groove kinematic coupling. Three-dimensional configurations are also possible.[25]

Guidance of motion is provided by use of kinematic principles without any play or backlash. Use of kinematic design provides unique location of the body at all times, with a resulting absence of lost motion in all but the desired direction. Both translation and rotation are possible in kinematic designs.

A kinematic linear translation guiding mechanism consists of two parallel cylinders in contact, forming a "v-groove" and an adjacent flat guide way. There are three hemispherical feet located below and attached to the moving carriage. Two of the hemispheres ride in the trough formed by the two cylinders in contact. Each hemisphere makes a contact with each cylinder, so there are four contacts between the two cylinders and hemispheres. The third hemisphere is in contact with the flat guide way, providing a fifth contact point. There are 6-N remaining degrees of freedom. The number of contact points, N, is equal to 5. Linear translation, in the direction of the axis of the trough formed between the two cylinders, is this degree of freedom.

Rotation and translation are provided in a single kinematic mechanism by placing a cylinder between two sets of adjacent cones. There are four contact points between the cylindrical surface and the four conical surfaces. Rotation of the cylinder about its axis is possible, as is linear translation of the cylinder along its long axis. This type of mount is used to support alignment telescopes and collimators.

Pure rotation is obtained in the above example by adding a fifth contact point. This contact point is a hemisphere bearing against the flat end of the cylinder. Alternate kinematic rotation guides use curved tracks or combinations of cylindrical and conical ended shafts. A variety of kinematic mechanisms are possible for guiding motion.[26]

A serious limitation on the use of kinematic design is stress in the elastic contacts. Hertz contact stress between curving surfaces is usually much higher than the usual stresses in a support structure. The high stress level associated with the elastic contacts in kinematic design often lead to the use of hard, high strength materials such as sapphire and tungsten carbide. Rapid wear in the area of the elastic contact is a potential problem associated with kinematic guides. Potential wear is assessed

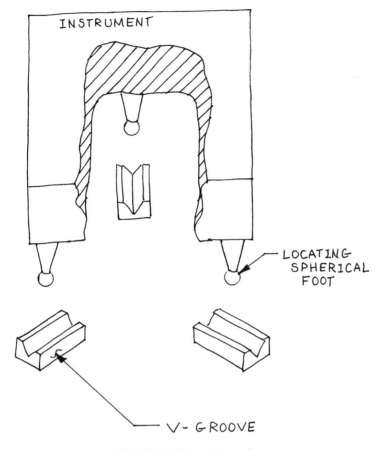

FIGURE 2.2 Kinematic coupling.

using the contact stress models developed by Bayer.[27] Wear is reduced to "zero" using the IBM zero wear theory.[28] This wear model states that zero wear is defined as wear equivalent to the surface finish of a polished metal part. Zero wear occurs when:

$$\frac{\tau_{max}}{\tau_{ys}} \leq 0.2$$

where τ_{max} = the maximum shear stress in the elastic contact
τ_{ys} = the yield stress in shear of the material

Semi-kinematic design reduces the elastic contact stress by replacing the point contacts of a true semi-kinematic design with small contact areas. These areas are sized to reduce the stress to acceptable levels. Semi-kinematic design also provides for increasing the number of supports without overconstraint. Increasing the number of supports is sometimes necessary to decrease self-weight deflection or to improve stiffness.

In semi-kinematic design, supports are located using kinematic theory. Each support point in the kinematic design is replaced with a small contact area. Errors in the fit between the two surfaces in contact at each support induces moments in the supported body. Successful semi-kinematic design requires careful attention to the quality of the support areas to minimize the support-induced moments.

There are two methods for providing high quality support areas in semi-kinematic design: tight tolerances on the surfaces, and surfaces with rotational compliance. Semi-kinematic mounts are used in supporting optics during diamond turning. In this application the tolerances of the mounting surfaces must be comparable to the tolerances of the optical surface. This condition is obtained in diamond turning of optics, but is difficult to provide for most other applications. A more reasonable tolerance is that the flatness and co-planarity of the mounting surfaces must be less than the elastic deflection of the mounting surface. For most applications, even this tolerance is well below what is normally obtained with conventional production methods.

An alternative to tight tolerances is to introduce rotational compliance into the mounting surface. This is common in the design of tooling fixtures. One commercial component that is used for semi-kinematic mounts is the tooling (spherical) washer set. A tooling washer set consists of a concave and convex set of mating washers, with an oversized central hole. During assembly the washers tilt and de-center to reduce induced moments in the part.

A sphere in cone geometry is used for semi-kinematic mounts. A line contact develops between the sphere and cone. For optimum performance, the profile error on the conical surface must be comparable to the profile error (roundness) of the sphere. Trylinski[29] gives equations for calculating the stress and residual moment for a sphere in a conical socket:

$$\sigma_a = \frac{1}{\pi r}\left[\frac{F_a}{(\sin 2\theta)\left(\frac{1-v_s^2}{E_s}\right)+\left(\frac{1-v_c^2}{E_c}\right)}\right]^{\frac{1}{2}}$$

$$T_a = F_a \mu r \cos\theta$$

$$\tau_t = \frac{1}{\pi r \cos\theta}\left[\frac{2.5F_t}{\left(\frac{1-v_s^2}{E_s}\right)+\left(\frac{1-v_c^2}{E_c}\right)}\right]^{\frac{1}{2}}$$

$$T_t = F_t \mu r$$

where F_a = the axial force acting on the socket
F_t = the shear force acting on the socket
σ_a = the stress due to the axial force
τ_t = the stress due to the shear force
r = the radius of the sphere
θ = the vertex angle of the conical socket
E_s = the elastic modulus of the sphere
E_c = the elastic modulus of the conical socket
v_s = the Poisson's ratio of the sphere
v_c = the Poisson's ratio of the conical socket
μ = the coefficient of breakaway friction between sphere and conical socket
T_a = the maximum moment due to the axial force
T_t = the maximum moment due to the shear force

Semi-kinematic design is used to provide for multiple point support of bodies without causing overconstraint. Multiple support points are tied together in groups; each group acts as a single kinematic support. The support points in a group are connected by pivots, with each pivot located at the center of gravity of the support points. Residual moments induced in the mounted body using a whiffle tree mechanism arise from friction in the pivots. It is desirable to reduce the pivot friction to a very low level. Since the pivots provide rotation to balance forces, static friction is important.

The simplest example of a whiffle tree is a support for a beam. Each of the two simple supports for the beam is provided with a rocker. At each end of the rockers is a contact point with the bottom of the beam, so there are four points in contact with the beam. This type of mechanism is self-adjusting for irregularity in the beam.

More complex whiffle trees are created by adding pivots to the ends of the balance beam carrying subrockers. The whiffle tree is cascaded using the approach to provide 8, 16, or more support points. Performance of the whiffle tree is determined by the number of supports, pivot friction, and rocker stiffness. The distance between adjacent support points in a whiffle tree for a uniformly loaded beam is determined using the Airy point equation given above. Figure 2.3 shows a typical whiffle tree design for a beam.

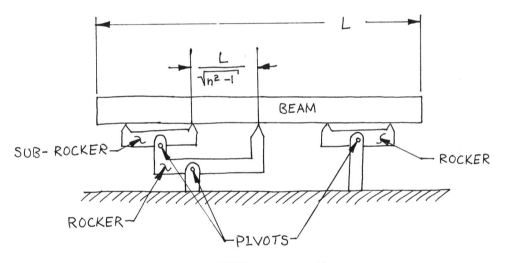

FIGURE 2.3 Whiffle tree-supported beam.

A similar approach is used in whiffle trees supporting plates. In this application, the pivots are attached to subplates and provide rotation about two different axes. The subplates are normally triangular, with contacts at the tips of the triangles. The simplest type of plate support uses three pivots below the plate, located using kinematic principles. Each pivot carries a triangular rocker plate with three contact points. The optimum location for the support points is determined using complex structural analysis methods. Alternately, the support points are located using the principle of equal areas. Each support point carries the same plate area and is located at the centroid of the area. In axisymmetric plates (optical mirrors) the support points in the whiffle tree are equally spaced on concentric rings.

2.5 Athermalization

There are three types of thermal effects on optomechanical systems:

1. A temperature change induces distortion in the optical element due to the material properties of the element. Important thermal properties determining the amount of distortion

are the thermal coefficient of expansion (α), spatial variation in thermal coefficient of expansion ($\Delta\alpha$), thermal distortion index (ratio of thermal coefficient of expansion to thermal conductivity, α/k), and thermal diffusivity (D).

2. A temperature change induces distortion in the optical element due to the way in which the element is mounted. This distortion is due primarily to the difference in thermal coefficients of expansion of the mount and optical element materials. This type of thermal distortion is discussed in the section on optical mounts.

3. A temperature change induces distortion in the optical support structure. Distortion of the optical support structure causes a loss of optical alignment. The most important effect of this loss of alignment is a change in system focus. Athermalization means that the optical system maintains focus when the temperature changes.

A window is a zero power optical component used in transmission. Windows provide an introduction to optical effects of temperature changes. Consider a circular, plane parallel window subjected to a linear temperature gradient through its thickness (linear axial gradient). The hot side of the window expands relative to the cold side. The change in area of the two sides of the window causes the window to spring out of shape and become curved. Since the surfaces are curved, the window acts as a meniscus lens with a power (power is defined as the reciprocal of the focal length) given by[30,31]

$$\frac{1}{f} = \frac{n-1}{n}\left(\frac{\alpha}{k}\right)^2 hq^2$$

where $1/f$ = the power of the distorted window

   n = the index of refraction of the window material

   α = the thermal coefficient of expansion of the window material

   k = the thermal conductivity of the window material

   h = the axial thickness of the window

   q = the heat flux per unit area absorbed by the window

Distortion of a window due to axial temperature gradients is a weak effect. More serious is a radial gradient. The radial temperature profile of a window is determined by heat transfer at the window surfaces. Heat transfer occurs due to conduction, convection, and radiation. Complex temperature profiles develop due to different kinds of heat transfer occurring simultaneously.

A parabolic radial temperature distribution is a good approximation of many types of window heat transfer.[32] The power of a window with a parabolic radial temperature profile is given by:[33,34]

$$\frac{1}{f} = 8\frac{h}{D^2}\Delta T\left[(n-1)(1+v)\alpha + \frac{dn}{dT}\right]$$

where $1/f$ = the power of the distorted window

   h = the window axial thickness

   D = the window diameter

   ΔT = the radial difference in temperature, from center to edge

   n = the index of refraction of the window material

   v = the Poisson's ratio of the window material

   α = the thermal coefficient of expansion of the window material

   dn/dT = the thermoptic coefficient of the window material (change in index of refraction with temperature)

In most applications the effects of radial gradients are much larger than those due to axial gradients. It is desirable to reduce radial temperature differences in optical elements to limit such effects due to gradient. The size of the gradient is reduced by insulating the edge of the window. Titanium is a good material for this application. The thermal conductivity of titanium is much lower than that of most common metals, limiting heat transfer into the mount. In addition, the thermal coefficient of expansion of titanium is a good match with that of many optical materials. An alternate method is heating or cooling the edge of the window to reduce heat transfer. In some cases the radial profile is represented by a polynomial expression of order greater than three. The temperature change associated with this type of gradient is most rapid near the edge of the window. Making the window oversize with respect to the optical clear aperture limits the effect of such higher-order gradients. As a rule of thumb, the window should be about 25% larger in diameter than the optical clear aperture.

The index of refraction of a lens changes with temperature. This change in temperature is due to the thermoptic property ("dn/dT") of the lens material. A change in index of refraction of the lens alters the focal length of the lens. The change in lens focal length with temperature is given by:[35]

$$\beta = \left(\frac{1}{f}\right)\left(\frac{df}{dT}\right) = \alpha - \left(\frac{1}{n - n_{air}}\right)\left(\frac{dn}{dT} - n\frac{dn_{air}}{dT}\right)$$

where β = the optothermal expansion coefficient
 f = the lens focal length
 df/dT = the change in lens focal length with temperature
 α = the lens material thermal coefficient of expansion
 n = the lens material refractive index
 n_{air} = the index of refraction of air
 dn/dT = the lens material thermoptic coefficient
 dn_{air}/dT = the thermoptic coefficient of air

For ordinary optical glass used in visible wavelengths, $n_{air} \approx 1$ and $dn_{air}/dT \approx 0$, so the above equation becomes

$$\beta \approx \alpha - \left(\frac{1}{n-1}\right)\left(\frac{dn}{dT}\right)$$

The above equation indicates that β, the change in lens power with temperature, is a material property and is independent of lens surface curvature. Below is a table (Table 2.2) of the optothermal expansion coefficients, β, of a variety of materials.

TABLE 2.2 Optothermal Expansion Coefficients

Glass Type	β (m/m-k $\times 10^{-6}$)
TiF6	20.94
BK1	3.28
LaKN9	0.32
BAK4	−0.23
KzFS1	−2.89
ZnSe	−28.24
Silicon	−64.10
Germanium	−85.19

In optical systems with multiple lens elements, the optothermal expansion coefficient of the system, β_S, is given by the following equation:

$$\beta_s = \sum_{i=1}^{n} \frac{K_i}{K_s} \beta_i$$

where β_s = the system optothermal expansion coefficient
         K_i = the individual element power (reciprocal of focal length)
         K_s = the system power
         β_i = the individual element optothermal expansion coefficient

For athermalization, the focus of the system must not change with temperature. This condition is obtained if the optothermal expansion coefficient, which represents the change in focus with temperature, is the same as the thermal expansion coefficient of the system. Or

$$\beta_s = \alpha_s$$

The range of thermal expansion coefficients is limited by the availability of adequate materials for optical support structures. It is rare for the optothermal expansion coefficient to be nearly the same as that of a common structural material. It is possible to adjust the optical design of the system to produce a β_s that is the same as a selected thermal coefficient of expansion.

Another structural technique for athermalization is a bi-metallic compensator. A bi-metallic compensator is made of materials with different thermal coefficients of expansion. By adjustment of the lengths of the two types of materials, a good match between the effective thermal coefficient of expansion of the structure along the optical axis and the optothermal expansion coefficient of the structure is obtained. For athermalization using a bi-metallic compensator:

$$\alpha_1 L_1 + \alpha_2 L_2 = \beta_s f$$

where α_1, α_2 = the thermal coefficients of expansion of the two materials in the bi-metallic struc-
              ture
      L_1, L_2 = the respective lengths of the two materials in the bi-metallic structure
        β_s = the system optothermal expansion coefficient
         f = the system focal length

Figure 2.4 shows two types of bi-metallic compensators. Below is a table (Table 2.3) of some bi-metallic compensator combinations for different types of lens materials:

If the optothermal expansion coefficient of the system is small, a metering structure is used to athermalize the system. A metering structure consists of a structure made of conventional materials that provides stiffness in location in all directions except along the optical axis, and an inner, low thermal coefficient of expansion structure that maintains spacing and alignment with temperature. Typically, the inner structure consists of rods made of a low thermal coefficient of expansion material such as invar. The rods are attached to the optical components and are connected to the optical support structure by linear translation bearings. Low thermal coefficient of expansion materials such as invar often are expensive and low in structural efficiency.[36,37] Metering structures avoid these disadvantages.

Methods used to athermalize reflective systems are similar in some respects to those used for refractive systems. Like lenses, mirrors are sensitive to temperature gradients. A linear axial temperature gradient in a mirror along the optical axis will cause the optical surface of the mirror to change its radius of curvature. The change in surface curvature is determined by the same material

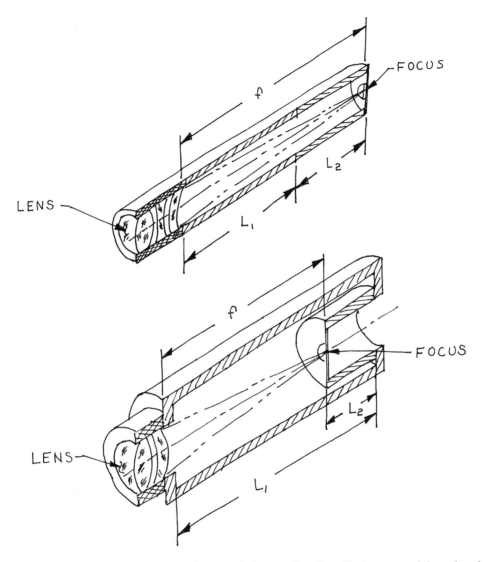

FIGURE 2.4 Bi-metallic compensators. The upper design provides a low effective structural thermal coefficient of expansion and the lower design provides a near-zero to negative effective structural thermal coefficient of expansion.

TABLE 2.3 Bi-Metallic Compensators

Glass Type	β_S (m/m-k $\times 10^{-6}$)	Material 1	α_1 (m/m-k $\times 10^{-6}$)	$L_1 \times f$	Material 2	α_2 (m/m-k $\times 10^{-6}$)	$L_2 \times f$
TiF6	20.94	Aluminum	23	0.678	Stainless steel	16.6	0.322
BK1	3.28	Invar	0.54	0.829	Stainless steel	16.6	0.171
LaKN9	0.32	Invar	0.54	1.01	Aluminum	23	−0.01
BAK4	−0.23	Invar	0.54	1.034	Aluminum	23	−0.034
KzFS1	−2.89	Invar	0.54	1.153	Aluminum	23	−0.153
ZnSe	−28.24	Stainless steel	16.6	1.233	Plastic (ABS) (polyurethane)	209	−0.233
Silicon	−64.10	Stainless steel	16.6	1.419	Plastic (ABS) (polyurethane)	209	−0.419
Germanium	−85.19	Stainless steel	16.6	1.358	Plastic (polyethylene)	301	−0.358

property, the thermal distortion index, that determines the change for a refractive component. The change in surface curvature of a mirror subjected to a linear axial temperature gradient is given by:

$$\frac{1}{R_o} - \frac{1}{R} = \frac{\alpha}{k} q$$

where R_o = the original surface radius of curvature
   R = the radius of curvature of the surface due to the gradient
   α = the thermal coefficient of expansion of the mirror
   k = the thermal conductivity of the mirror material
   q = the heat flux per unit area through the mirror

Analysis of the effects of more complex gradients requires the use of a method developed by Pearson and Stepp.[38] This set of equations provides a means of evaluating the effect of a global temperature change as well as linear temperature gradients. The linear temperature gradients are assumed to lie along each axis of a Cartesian coordinate system, with the origin of the coordinate system at the vertex of the mirror, with the z axis coincident with the optical axis. These equations are applicable for both concave and convex mirrors. For convex mirrors a sign change is necessary. The temperature distribution in the mirror is given by:

$$T(x, y, z) = c_o + c_1(x) + c_2(y) + c_3(z)$$

where c_o = the global change in temperature of the mirror
   c_1 = the linear temperature gradient along the x axis
   c_2 = the linear temperature gradient along the y axis
   c_3 = the linear temperature gradient along the z axis (the optical axis)

The surface deformations due to the above gradients are given by:

$$W(r, \theta) = \frac{\alpha c_3}{8R^2} r^4 \quad \text{(spherical)}$$

$$+ \frac{\alpha c_1}{2R} r^3 \cos\theta + \frac{\alpha c_2}{2R} r^3 \sin\theta \quad \text{(coma)}$$

$$+ \left(\frac{\alpha c_3 h}{2R} - \frac{\alpha c_3}{2} + \frac{\alpha c_o}{2R} \right) r^2 \quad \text{(focus)}$$

$$+ \alpha c_1 h \cos\theta + \alpha c_2 h \sin\theta \quad \text{(tilt)}$$

$$+ \frac{\alpha c_3 h^2}{2} + \alpha c_o h \quad \text{(piston)}$$

where r = the radius position on the mirror surface
   θ = the angular position on the mirror surface
   α = the mirror material thermal coefficient of expansion
   R = the optical radius of curvature of the mirror surface
   h = the mirror axial thickness

The above equations suggest that temperature gradients are the source of potential problems in mirrors. A mirror exposed to a sudden change in temperature is likely to develop temperature gradients from its surface to interior. The time required for the mirror to reach thermal equilibrium is estimated using a simple one-degree-of-freedom model. This model gives the time required for the mirror interior to approach the surface temperature after an instantaneous temperature change. The mirror interior temperature after some time is given by:

$$T' \approx T - \Delta T \exp\left(-\frac{\pi^2 D t}{h^2}\right)$$

where
- T' = the temperature of the mirror after some time t
- T = the initial mirror temperature
- ΔT = the temperature change at the mirror surface
- t = the time after the sudden temperature change
- h = the mirror axial thickness
- D = the thermal diffusivity of the mirror material, where:

$$D = \frac{k}{\rho c_\rho}$$

- k = the thermal conductivity of the mirror material
- ρ = the mirror material density
- c_ρ = the mirror material specific heat

If the thermal coefficient of expansion of the mirror material is not uniform, the mirror distorts when the temperature is changed. This effect occurs even if the temperature changes globally without any gradients, and is due to the spatial variation of properties in the mirror. As a rule of thumb, a spatial variation of 3 to 5% of the thermal coefficient of expansion should be expected in most materials.[39] Low thermal coefficient of expansion materials, with an α near zero, are less affected by this spatial variation. There may be difficulties due to spatial variation of α with high thermal coefficient of expansion materials (such as aluminum and beryllium) used for mirror substrates. If the mirror thermal coefficient of expansion varies linearly along the axis of the mirror, the surface deformation is given by:

$$\delta = \frac{r^2}{2h} \Delta T \ \Delta \alpha$$

If the mirror thermal coefficient of expansion varies linearly across the diameter of the mirror, the surface deformation is given by:

$$\delta = \frac{r^2}{4h} \Delta T \ \Delta \alpha$$

where in the above equations:
- δ = the surface deformation
- r = the mirror radius
- h = the mirror axial thickness
- ΔT = the change in temperature
- $\Delta \alpha$ = the spatial variation in thermal coefficient of expansion

Making all components, optical and structural, of a system out of the same material is an important method of athermalization of reflective optical systems. This method of athermalization is called "same material athermalization". Although all-glass optical systems have been built,[40] such systems are expensive and fragile. Same material athermalization is commonly used with metal optics. Cost and strength of metal optics and structure are not as much an issue as when glass is used. Cryogenic optical systems often employ same material athermalization.[41,42]

2.6 Vibration Control

Vibration is a source of performance degradation in optomechanical systems. Very low levels of vibration induce a blur in the focus. This vibration-induced blur is sometimes mistaken for blur due to a system misalignment, or an out-of-focus condition. Higher levels of vibration create a time-variant blur, which is at least easy to diagnose. Very high levels of vibration carry the potential for structural failure of the system. In general, operation is not expected at such levels, only survival.

There are two important types of vibration that affect optomechanical systems: periodic and random. Periodic vibration is characterized by a period and amplitude. The amplitude of complex periodic vibration is characterized, statistically, by quantities such as the root-mean-square of the amplitude. Random vibration is also characterized by statistical methods. Random vibration contains all frequencies. One statistical quantity that is often used to describe random vibration is the power spectral density (PSD). The PSD is a measure of the amplitude of vibration contained within some bandwidth, typically 1 Hz. Since PSD is a measure of the area under a curve, it is given in units of area per bandwidth. One such measure is "g^2/Hz", where "g" is a dimensionless acceleration unit (1 g = acceleration of Earth's surface gravity).

Response of systems to vibration is a complex topic. Considerable insight is derived from the use of a simple, single-degree-of-freedom (SDOF) model. This model is used to determine response of systems to both periodic and random vibration. The most important property of a system exposed to vibration is the natural frequency. The natural frequency of a system is that frequency at which the system will oscillate if perturbed from equilibrium. For a simple SDOF system, the fundamental frequency is given by:

$$f_n = \frac{1}{2\pi}\sqrt{\frac{k}{m}}$$

where f_n = the natural frequency, in Hz
          k = the system spring stiffness
          m = the system mass

Many optomechanical systems are mounted kinematically. Certain types of athermalized kinematic mounts employ flexural elements between the optical component and the structure. These flexures act as springs and reduce the fundamental frequency of the mounted component. For athermalization the flexures are compliant in the radial direction and stiff in all other directions. In this case the stiffness of the mounted optic in the radial direction is given by:

$$k_r = \frac{n}{2}\left(k_{rf} + k_{tf}\right)$$

And the stiffness of the mounted optic in the axial direction is given by:

$$k_a = nk_{af}$$

where k_r = the radial stiffness
       n = the number of mounting flexures
       k_{rf} = the radial stiffness of an individual flexure
       k_{tf} = the tangential stiffness of an individual flexure
       k_a = the axial stiffness
       k_{af} = the axial stiffness of an individual flexure

If a system is perturbed from equilibrium, the amplitude of each successive cycle of vibration is less than that of the preceding cycle. This decrease in amplitude of vibration with time is due to energy lost during each cycle. The process of energy loss in a vibrating system is called damping. A system is said to be critically damped if there is no vibration when the system is perturbed from equilibrium. For a SDOF system, the critical damping coefficient C_c is given by:

$$C_c = 2(km)^{1/2}$$

where k = the stiffness
       m = the mass

Real systems are usually not critically damped. It is common to give the system damping in terms of the critical damping ratio. The critical damping ratio C_R is the ratio of the system damping to the amount of damping necessary to make the system critically damped, or C/C_c. The critical damping ratio is a dimensionless number, but is usually given as a percentage. The critical damping ratio of optomechanical systems is often less than 5%.[43]

The response of a SDOF system to a sinusoidal excitation is given by:

$$\frac{X_0}{X_1} = \left[\frac{1 + 2\dfrac{f}{f_n} c_R}{\left(1 - \dfrac{f^2}{f_n^2}\right)^2 + \left(2\dfrac{f}{f_n} c_R\right)^2} \right]^{\frac{1}{2}}$$

where X_0 = the amplitude of oscillation of the SDOF system
       X_1 = the amplitude of oscillation of the exciting force
       f = the frequency of the exciting force
       f_n = the natural frequency of the SDOF system
       c_R = the critical damping ratio of the SDOF system

There are three special cases of the above equation: spring, damper, and mass-controlled cases. In the mass-controlled case, the fundamental frequency is much less than the frequency of excitation ($f \gg f_n$). The response is determined by the amount of mass of the system and the frequency ratio. In the damper-controlled case, the fundamental frequency is near the frequency of excitation ($f \cong f_n$), and the response is determined by the amount of damping in the system. In the spring-controlled case, the fundamental frequency is much greater than the frequency of excitation ($f \ll f_n$), and the response if determined by the spring stiffness of the system.

The ratio of X_0/X_1 is the transmissibility of the system and is a dimensionless number. A transmissibility of less than unity means that the amplitude of response of the SDOF system is less than the exciting force amplitude. Transmissibilities of greater than unity mean that the amplitude of response of the SDOF system is greater than the exciting force amplitude. A transmissibility of greater than unity is very undesirable in optomechanical systems. The Q of a system is the transmissibility at resonance.

A vibration isolation system operates in the mass-controlled domain, where the frequency of excitation is always at least $\sqrt{2}$ higher than that of the fundamental frequency of the isolation system.[44] In the mass-controlled case, the transmissibility T $(T = X_0/X_1)$ is given approximately by:

$$T = \frac{X_0}{X_1} \approx \left(\frac{f_n}{f}\right)^2$$

where   T = the transmissibility
$\quad\quad\quad\quad$ X_0 = the amplitude of oscillation of the SDOF system
$\quad\quad\quad\quad$ X_1 = the amplitude of oscillation of the exciting force
$\quad\quad\quad\quad$ f = the frequency of the exciting force
$\quad\quad\quad\quad$ f_n = the natural frequency of the SDOF system

At high frequency ratios, the above equation indicates that transmissibility is small. For example, a typical vertical fundamental frequency for an isolation system is 2 Hz. If this isolation system is subjected to a 60-Hz excitation, the transmissibility is about 0.001. At low frequency ratios, the transmissibility will approach unity, and isolation suffers. Damping increases the transmissibility of an isolation system at high frequency ratios and is, therefore, undesirable. Some damping is necessary to prevent damage to the isolation system if exposed to excitation at the natural frequency of the system. Nonlinear damping response is provided in isolation system through the use of surge tanks connected to the cylinders of air springs by metering orifices. Near resonance, the surge tank damped air spring is high in damping, and response is limited. At high frequencies, the surge tank is not effective in damping and the air spring isolator operates as though it were undamped.[45]

Isolation systems consist of a stiff platform supported on isolators. Platform stiffness is much higher than that of the isolators. Due to the high platform stiffness, the fundamental frequency of the isolation system is determined by combining the stiffness of the independent isolators with the inertial properties of the isolators. An isolation system has three natural frequencies in translation, one in each axis, and three natural frequencies in rotation, one about each axis. Natural frequencies of the isolation system in translation and rotation are given by:[46]

$$f_{nt} = \frac{1}{2\pi}\left(\frac{1}{m}\sum_{i=1}^{n}k_{it}\right)^{\frac{1}{2}}$$

where f_{nt} = the natural frequency in translation
$\quad\quad\quad\quad$ f_{nr} = the natural frequency in rotation
$\quad\quad\quad\quad$ m = the isolation platform mass
$\quad\quad\quad\quad$ n = the number of isolators

$$f_{nr} = \frac{1}{2\pi}\left(\frac{1}{I_r}\sum_{i=1}^{n}r_i^2 k_{ir}\right)^{\frac{1}{2}}$$

$\quad\quad\quad\quad$ k_{it} = the isolator stiffness in translation in the axis of the natural frequency
$\quad\quad\quad\quad$ I_r = the platform moment of inertia about the axis of the natural frequency, with respect
$\quad\quad\quad\quad\quad\quad$ to the center of gravity
$\quad\quad\quad\quad$ r_i = the distance of the ith isolator from the center of gravity
$\quad\quad\quad\quad$ k_{ir} = the isolator stiffness in rotation in a direction perpendicular to r_i

Isolation systems must provide protection against excitation from vibration produced within the isolation system. Normally random vibration is produced within the isolation system. For example, in a vibration isolation table used in a laboratory, a blow to the surface of the table produces both an impulse and random vibration. The effect of the random vibration produced within the system is to produce relative motion of the components. This relative motion is given approximately by:

$$X_{RM} = g \left(\frac{1}{32\pi^3} \right)^{\frac{1}{2}} \left(\frac{Q}{f_n^3} \right)^{\frac{1}{2}} PSD^{\frac{1}{2}}$$

where X_{RM} = the maximum relative motion to the excitation within the vibration isolation system
g = the acceleration due to Earth's gravity field (in metric units, 9.81 m/sec^2)
Q = the transmissibility at resonance of the vibration isolation platform
f_n = the fundamental frequency of the vibration isolation platform
PSD = the random vibration excitation (power spectral density) within the vibration isolation system, in units of dimensionless "g^2" per hertz

In applying the above equation it is important to note that the fundamental frequency and Q are for the platform, not the entire vibration isolation system. It is also important to use the proper units: g in the equation has the units of length over time^2, while the PSD is given in units of dimensionless "g^2" per hertz. The relative motion should be in units of length. This equation indicates that the platform used in a vibration isolation system should be very stiff and well damped (low "Q" value). This is exactly the opposite of the optimum characteristics of the entire system, which are low frequency and high "Q".

A system exposed to random vibration vibrates at its fundamental frequency, since random vibration contains all frequencies. The response amplitude of the system is determined by a statistical process. The average or "root-mean-square" amplitude of response of a simple SDOF system exposed to random vibration is given by:[47]

$$g_{rms} = \left(\frac{\pi}{2} f_n Q PSD \right)^{\frac{1}{2}}$$

where g_{rms} = the "root-mean-square" acceleration response, in dimensionless "g"
Q = the transmissibility at resonance of the system
f_n = the fundamental frequency of the system
PSD = the random vibration excitation (power spectral density) of the system, in units of dimensionless "g^2" per hertz

It is common in vibration engineering to assume that most structural damage is done by the "3-sigma" peak acceleration. The "3-sigma" acceleration is found by multiplying g_{rms} by a factor of 3. The displacement response of the system is given by:

$$\delta = \frac{g_{rms}}{\left(2\pi f_n \right)^2}$$

where δ = the displacement response
f_n = the fundamental frequency of the system

Table 2.4 gives the power spectral density of some representative environments.

TABLE 2.4   Power Spectral Densities (PSDs)

Environment	Frequency f (Hz)	Power Spectral Density (PSD)
Navy warships	1–50	0.001 g²/Hz
Minimum integrity test	20–1000	0.04 g²/Hz
(MIL-STD-810E)	1000–2000	−6 dB/octave
Typical aircraft	15–100	0.03 g²/Hz
	100–300	+4 dB/octave
	300–1000	0.17 g²/Hz
	≥1000	−3 dB/octave
Thor-Delta launch vehicle	20–200	0.07 g²/Hz
Titan launch vehicle	10–30	+6 dB/octave
	30–1500	0.13 g²/Hz
	1500–2000	−6 dB/octave
Ariane launch vehicle	5–150	+6 dB/octave
	150–700	0.04 g²/Hz
	700–2000	−3 dB/octave
Space shuttle	15–100	+6 dB/octave
(orbiter keel location)	100–400	0.10 g²/Hz
	400–2000	−6 dB/octave

REFERENCES

1. July 14, 1989. *Environmental Test Methods and Engineeringlines.* MIL-STD-810E.
2. Parker, J.D. and McQuiston, F.C. 1982. *Heating, Ventilating and Air Conditioning,* 2nd. Ed., John Wiley & Sons, New York, NY.
3. Ungar, E.E., Sturz, D.H., and Amick, C.H. July 1990. Vibration control design of high technology facilities, *Sound and Vibration,* Vol. 24, No. 7, 20.
4. Blevins, R.D. 1979. *Formulas for Natural Frequency and Mode Shape,* Van Nostrand Reinhold. Co., New York, NY.
5. Blevins, R.D. 1979. *Formulas for Natural Frequency and Mode Shape,* Van Nostrand Reinhold, New York, NY.
6. Wrigley, W., Hollister, W.M., and Denhard, W.C. 1969. *Gyroscopic Theory, Design, and Instrumentation,* The M.I.T. Press, Cambridge, MA.
7. Lazan, B.J. 1968. *Damping of Materials and Members in Structural Mechanics,* Pergamon Press, New York, NY.
8. Schetky, L.M. and Perkins, J. April 6, 1978. The 'quiet' alloys, *Machine Design,* Vol. 50, No. 8, 202.
9. James, D.W. 1969. High damping materials for engineering applications, *Mater. Sci. Eng.,* Vol. 4. 1.
10. Adams, R.D. 1972. The damping characteristics of certain steels, cast irons and other metals, *J. Sound and Vibration,* Vol. 23, No. 2, 199.
11. Ashby, M.F. 1992. *Materials Selection in Mechanical Design,* Pergamon Press, New York, NY.
12. Levina, Z.M. 1968. Research on the Static Stiffness of Joints in Machine Tools, in *Advances in Machine Tool Design and Research,* S.A. Tobia and F. Koenigs Berger, Eds., Pergamon Press, London, UK.
13. Rivin, E.I. 1988. *Mechanical Design of Robots,* McGraw Hill, New York, NY.
14. Serrurier, M. August 1938. Structural features of the 200-inch telescope for Mt. Palomar observatory, *Civil Eng.,* Vol. 8. No. 8. 524.
15. Moffltt, G.W. 1947. Compensation of Flexure in Range Finders and Sighting Instruments, *J. Optical Soc. Am.,* Vol. 37, 582.
16. King, H.C. 1955. *The History of the Telescope,* Dover Publications, New York, NY.
17. Manly, P.L. 1991. *Unusual Telescopes,* Cambridge University Press, Cambridge, England.

18. Kamm, L.J. 1990. *DesigningCost-Efficient Mechanisms,* McGraw-Hill, New York, NY.

19. Blanding, D.L. 1992. *Principles of Exact Constraint Mechanical Design*, Eastman Kodak Co., Rochester. NY 14650.

20. Hills, D.A., Nowell, D., and Sackfield, A. 1993. *Mechanics of Elastic Contacts,* Butterworth Heinemann .

21. Brewe, D.E. and Hamrock, B.J. October 1977. Simplified Solution for Elliptical-Contact Deformation Between Two Elastic Solids, *J. Lubrication Technol.*, Vol. 99, 485.

22. Braddick, H.J.J. 1963. *The Physics of Experimental Method,* Chapman & Hall, Ltd.

23. Moore, J.H., Davis, C.C., and Coplan, M.A. 1983. *Building Scientific Apparatus,* Addison-Wesley Publishing Co., Reading, MA.

24. Slocum, A.H. April 1992. The Design of Three Groove Kinematic Couplings, *Precision Engineering,* Vol. 14, No. 2, 67.

25. Slocum, A.H. April 1988. Kinematic Coupling for Precision Fixturing - Part I - Formulation of Design Parameters, *Precision Engineering*, Vol. 10, No. 2, 85.

26. Pollard, A.F.C. 1951. *The Kinematical Design of Couplings in Instrument Mechanisms,* Hilger and Watts. Ltd.

27. Bayer, R.G. 1994. *Mechanical Wear Prediction,* Marcel Dekker, Inc., New York.

28. Bayer, R.G., Shalkey, A.T., and Wayson, January 9, 1969. Designing for zero wear, *Machine Design* 142.

29. Trylinski, W. 1971. *Fine Mechanisms and Precision Instruments,* Pergamon Press, Warsaw.

30. Barnes, Jr., W.P. 1966. Some effects of aerospace thermal environments on high-acuity optical systems, *Appl. Optics,* Vol. 5, 701.

31. Ramsay, J.V. 1961. The optical performance of windows with axial temperature gradients, *Optica Acta,* Vol. 8, 169.

32. Kohler, H. and Strahle, F., 1974. Design of Athermal Lens Systems, in *Space Optics.* Thompson, B.J. and Shannon, R.R., Eds., National Academy of Sciences, Washington, D.C.

33. Sparks, M. November 1971. Optical Distortion of Heated Windows in High-Power Laser Systems, *J. Applied Physics,* Vol. 42, No. 12, 5029.

34. Sliusarev, G.G. February 1959. The Influence of Temperature Gradient of Glass of Optical Systems on the Image Produced by the Latter, *Optics and Spectroscopy,* Vol. VI, No. 2, 134.

35. Jamieson, T. H. April 1981. Thermal effects in optical systems, *Optical Eng.,* Vol. 20, No. 2, 156.

36. Wan Lai, et al. 1986. Design Characteristics ofthe 1.56 m Astrometric Telescope and its Usage in Astrometry, in *Astrometric Techniques*, Eichhorm, H. K. and Leacock, R. J., Eds., IAU.

37. Zurmely, G.E. and Hookman, R.A. 1989. Thermal/Optical Test Setup for the Geostationary Operational Environmental Satellite Telescope, *Proc. SPIE 1 167,* 360.

38. Pearson, E. and Stepp, L. 1987. Response of large optical mirrors to thermal distributions, *Proc. SPIE 748,* 164.

39. Pellerin, C.J. et al. 1985. New opportunities for materials selection trade-offs for high precision space mirrors, *Proc. SPIE 542,* 5.

40. Everitt, C.W.F., Davidson, D.E. and Van Patten, R.A. 1986. Cryogenic star-tracking telescope for Gravity Probe B. *Proc. SPE 619,* 89.

41. Erickson, E.F. et al. 1984. All-aluminum optical system for a large cryogenically cooled infrared echel1e spectrometer, *Proc. SPIE 509,* 129.

42. McGlynn, J.B. 1989. Design and development of a rugged airborne scanning optical assembly for a calibrated IR imaging radiometer, *Proc. SPE 1 167,* 183.

43. Lazan, B.J. 1968. *Damping of Materials and Members in Structural Mechanics,* Pergamon Press, Elmsford, NY.

44. Ungar, E.E. 1992. Vibration isolation, in *Noise and Vibration Control Engineering: Principles and Applications,* Beranek, L.L. and Ver, I.L., Eds., John Wiley & Sons, Inc., New York, NY.

45. Vukobratovich, D. 1987. Principles of vibration isolation, Eds., *Proc. SPE 732,* 27.

46. Macinante, J.A. 1984. *Seismic Mounting for Vibration Isolation,* John Wiley & Sons, New York, NY.

47. Harris, C.M. and Crede, C.E. 1976. *Shock and Vibration Handbook*, 2nd. ed., McGraw-Hill Book Co., New York, NY.

3

Materials for Optical Systems

Roger A. Paquin

3.1 Introduction

Any optical system, of necessity, consists of many components, often fabricated from and joined by a variety of materials. The choice of materials depends on the system performance requirements and many other factors including size, weight, mechanical loading, environment, number of systems required, and, of course, cost. This chapter provides both a discussion of the importance of various properties and figures of merit, and a listing of many of them. Typical application requirements for the various classes of materials are given along with brief discussions of selection methods and cost comparisons. A discussion of dimensional stability is deferred to Chapters 4 and 10.

3.2 Applications

In general, materials for optical systems include almost all available materials, the choice depending on the requirements of the particular application. For simplicity, these materials can be divided into four applications categories: refractors, reflectors, structural components, and adhesives. Typical applications are discussed below.

Refractors

The refractors are generally defined as those optical elements which are transmissive to light. These may include image-forming lenses, which generally have one or both surfaces curved to a spherical or aspherical shape. Another class of transmissive optical elements includes optical windows, which are commonly used at the front end of an optical system to protect and seal the critical components of an optical assembly from adverse environmental effects such as dirt, dust, and humidity. Usually, optical windows are plane-parallel plates of optical quality glass, but sometimes a small wedge may be introduced between the two surfaces to correct the errors introduced by the window itself. In some applications involving large field of view, optical windows are shaped like a shell or dome with a significant curvature. Although optical windows are not a part of the image-forming optics, these can have a significant effect on the wavefront and image quality of the system if they have a significant thickness and are located in a converging or diverging beam. Therefore, the selection of an appropriate material and thickness of the windows is critical to optimize the performance of the system. Another important class of refractive optics includes filters, which are extensively employed in photography, spectrometers, and other chemical analysis equipment. Such absorption filters may be made of glass or optical grade plastics. The glass absorption filters with multilayer coatings can be designed to isolate specific transmission bands in environmental monitoring instruments to detect the presence of specific gases and chemicals.

The choice of material used for making refractors depends on the wavelength and application. Although hundreds of optical glasses are available from major manufacturers in the U.S. and Europe, only 50 or so are most commonly used for making refracting components. Most of the other glasses tend to stain, have poor machinability or thermal properties, or are too expensive. The commonly used glasses are available in various formulations of SiO_2 plus small amounts of the oxides and fluorides of barium, boron, calcium, lanthanum, sodium, and potassium. A number of lightweight glasses have also been developed for head-mounted displays, binoculars, and other airborne and space applications, where the overall weight of the system may be critical. Most of these lightweight glasses also have good hardness and better resistance to acids and alkalis.

The transmission of different glasses varies greatly over the spectral region from UV to IR. The crown glasses have good transmission at shorter wavelengths, while flints have good transmission in the near-IR region. Fused silica, Schott Ultran 30, and a few crystals transmit well in the near-UV region between 200 and 350 nm. A large variety of synthetic crystalline materials are available for UV and IR applications. These materials include alkali halides (KCl, NaCl, LiF, etc.) and alkaline earth halides (BaF_2, MgF_2, etc.), oxides (quartz, fused silica, etc.), semiconductors (Si, GaAs, Ge, diamond, etc.), and calcogenides (CdTe, ZnS, ZnSe, etc). New optical quality plastics are becoming more readily available. Plastics are lightweight, have low fabrication cost, and are resistant to mechanical shock and vibrations. The plastics do have low scratch resistance and softening temperature, and may be difficult to coat. They often exhibit birefringence due to stresses from the molding process. The plastics, in general, have low refractive index and not as many optical grade plastics are available as compared to glasses. Some of the commonly used polymers are polycarbonates, acrylics, and polystyrenes.

Reflectors

Reflective components are all mirrors, but include scanners, reflecting prisms, diffraction gratings, and other specialized components. The reflecting surface of a mirror can be bare, as for certain infrared telescopes of beryllium, or have an optical coating for specific wavelengths. All glass mirrors are coated. A mirror consists then of the reflective surface and the substrate that supports it. That substrate can be anything from a simple plane-parallel flat disk to a lightweighted, off-axis asphere of nonsymmetrical geometric form. They range in size from millimeters to meters and can be made from glasses, ceramics, metals, composites, or plastics. The classical reflective optical system, such as an astronomical telescope, usually consists of glass mirrors and a metal support structure. For light weight, whether for space applications or thermal considerations, the glass can be lightweighted, or other materials such as aluminum (Al) or silicon carbide (SiC) can be used. The

Hubble Space Telescope has a lightweight ULE™ fused silica (ULE™*) primary mirror, a Zerodur®** secondary, and a graphite/epoxy (Gr/Ep) structure. For ultralightweight systems, typically for space applications, beryllium (Be) or metal matrix composites can be used for both the mirrors and structure.

For high heat load applications, such as synchrotron or laser optics, actively cooled mirrors of copper (Cu), molybdenum (Mo), silicon (Si), or SiC are usually specified. These mirrors are fabricated with internal cooling channels, the complexity of which depends on the incident heat flux. Cooled mirrors have also been successfully fabricated with internal heat pipes. For lower heat loads, the low expansion materials such as ULE™, Zerodur®, or invar can be used. Heat absorption is minimized with high efficiency optical coatings and/or by using the optical surface at grazing incidence.

At the opposite end of the temperature scale, cryogenic mirrors are typically fabricated from fused quartz/fused silica. Beryllium and SiC are also an option, because of their unique properties. The Infrared Astronomical Satellite (IRAS) was a cryogenic all-Be system that successfully provided a far-infrared survey of the galaxies.

For production systems where cost is critical, replicated optics are popular. These mirrors are manufactured with Al or glass substrates on which a thin polymer layer forms the mirror surface and is pulled, with the optical coating already in place, from a polished master surface. This technique is used extensively for small aspheric mirrors and for diffraction gratings. In the latter application, a master grating is ruled into a metal surface, often plated or otherwise consisting of deposited gold (Au) or Al.

Structural Components

While optical components, both reflective and refractive, may have to be designed as structures, the components referred to here are those that mechanically support and connect the optical components. Typical examples are optical benches, metering structures, mounting hardware, lens housings, fasteners, etc. These components must be relatively stiff, dimensionally stable (but not necessarily to the same tolerance as optical components), and should be thermally matched to the optics in both expansion and conductivity.

In many production systems that are used primarily at room temperature, Al is the preferred material because of low cost and fabricability. Wrought products such as rod, bar, tube, plate, or extrusions are used as well as castings. For systems where weight is critical, such as space systems or inertia-critical systems, Be and Gr/Ep are the preferred materials. Metal matrix composites (MMCs) can provide intermediate properties and can be cost effective in production applications. While Gr/Ep has become a common material in optical structures, each application requires a custom design and fabrication process. As with the MMCs, production quantities can be cost effective for demanding applications. For extremely stable and/or controlled expansion applications such as optical benches and metering structures, the low expansion materials such as invar and graphite epoxy composites are most often used, but Zerodur is also used in critical metering applications.

Fastening of structural components can be accomplished in many ways. All metals, including Be and the MMCs, can be attached with conventional fasteners such as screws. Some, such as Al, Cu, and steels, can be brazed or welded as well. Others like Be and Mo can be brazed, but just about all materials can be adhesively bonded.

---

* ULE™ is a registered trademark of the Corning Glass Works, Corning, NY.
** Zerodur® is a registered trademark of Schott Glaswerke, Mainz, Germany.

Adhesives and Cements

Adhesives can either be structural or optical. Structural adhesives have no optical requirements and are used strictly to mechanically attach components to each other such as a baffle to a support structure or a lens to its housing. When used in optical applications, even in the supporting structures, adhesives should have low thermal expansion and relatively low residual stress due to shrinkage during curing. Optical cements are part of the optical train since they are in the light path where they join refractive components such as cemented doublets, and as such must have high transmission and index homogeneity.

A structure in which optical and mechanical components are secured together by adhesives as opposed to tradition fasteners is lighter in weight and usually less expensive to fabricate because the machining of threaded and through holes for screws is not required. Moreover, the required machining tolerances (flatness, parallelism, etc.) for the bonded components are, in general, much looser compared to parts that must be rigidly bolted together. Also, bonded joints are flexible to a degree, thereby providing a better stress distribution under high loads, damping in vibration and shock environments, and allowing differential expansion between parts made from dissimilar materials. The silicone elastomers can also be used for sealing and damping. All adhesives have good shear strength, but have lower strength in tension and peel modes.[1]

Adhesives and cements are formulated from many different polymers. The most common structural adhesives are epoxies, polyurethanes, modified acrylics, cyanoacrylates, and anaerobics. Silicones are used in structural applications where resiliency is required, such as for joining of materials with disparate thermal expansions.

Optical cements can be epoxies, silicones, or other polymers. They can be thermosets, i.e., two-part systems, thermoplastics that are heated to a liquid state and applied, photosetting, e.g., UV curing, or solvent loss cements. The latter are seldom used in modern optical systems because of the stress induced in the components by shrinkage during and after curing.

3.3 Properties

Important Properties and Figures of Merit

Important properties vary with the type of material: refractor, reflector, structure, or adhesive. For all materials, the properties fall into four categories: physical, mechanical, thermal, and optical. The most significant of these properties are discussed here, and those properties more appropriate for a specific type of material are discusssed under that section. All material properties vary with temperature, some in a linear fashion, but most nonlinear. For systems that operate at temperatures other than room temperature, great care is required in selecting and matching materials in order to ensure that the system will meet specifications over the required temperature range.

General references for properties include *Handbook of Optics*, 2nd ed., Vol. 2;[2] *The Infrared Handbook*, 2nd ed.;[3] *Handbook of Infrared Optical Materials*;[4] *CRC Handbook of Laser Science and Technology*, Vol. 4, Part 2;[5] *ASM Handbook*, Vol. 1 and 2 (early printings are titled Metals Handbook, 10th ed.);[6,7] and *Engineered Materials Handbook*, Vol. 1 to 4.[8-11]

Physical

For all materials under consideration here, the physical properties of interest are mass density ρ, electrical conductivity, and/or electrical resistivity. Electrical conductivity is inversely proportional to electrical resistivity and for most materials, one or the other is normally reported. These properties vary with temperature, but density varies slowly.

Mechanical

The design of optical components often involves some structural aspects where mechanical properties can be used as a basis for comparison. Deflection in any application is a function of five parameters: support conditions, materials, structural efficiency of the design, size (i.e., diameter), and loading. For static conditions, deflection is proportional to the fourth power of diameter of a circular plate, while for dynamic conditions it is proportional to the fifth power. This means that for many large components, keeping edge roll-off allowables to a minimum is essential for good performance.

The easiest of the five parameters to control is the material, the subject of this chapter. The important mechanical properties include elastic and/or plastic, strength, and fracture. Figures of merit for structural efficiency are used to rapidly compare materials for a given structural application, particularly in the design of lightweight reflective systems.

The elastic properties of crystalline materials can be described by a 6 × 6 matrix of constants called elastic stiffness constants.[12] From these constants, the elastic properties of the material: Young's modulus E (the elastic modulus in tension), bulk modulus K, modulus of rigidity G (also called shear modulus), and Poisson's ratio ν, can be calculated. The constants, and consequently the properties, vary as functions of temperature. Young's modulus of elasticity is the measure of stiffness or rigidity of a material; the ratio of stress, in the completely elastic region, to the corresponding strain. Bulk modulus is the measure of resistance to change in volume; the ratio of hydrostatic stress to the corresponding change in volume. Shear modulus, or modulus of rigidity, is the ratio of shear stress to the corresponding shear strain under completely elastic conditions. Poisson's ratio is the ratio of the absolute value of the rate of transverse (lateral) strain to the corresponding axial strain resulting from uniformly distributed axial stress in the elastic deformation region. For isotropic materials, the properties are interrelated by the following equations:

$$G = \frac{E}{2(1+\nu)} \tag{1}$$

$$K = \frac{E}{3(1-2\nu)} \tag{2}$$

A group of structural figures of merit, all utilizing combinations of density and Young's modulus, have been used to compare the structural efficiency of materials. The most commonly used term is specific stiffness, E/ρ. For many applications, a simple comparison of specific stiffness shows that the highest structural efficiency material will have the lowest mass or self-weight deflection for identical geometry. For more realistic comparisons, other related proportionality factors are more appropriate. For example, ρ^3/E should be compared if mass is a specified parameter and minimum self-weight deflection is desired. The optimum geometries for each material are appropriate with this factor as they are with $(\rho^3/E)^{1/2}$, the factor to be compared when self-weight deflection is specified in a mass-critical application. For most applications, the resonant frequency of a mirror is an important design consideration and since natural frequency is proportional to the square root of specific stiffness, a simple comparison is also possible. More detailed discussions of design issues are given in Chapters 2 and 5.

Mechanical strength and fracture properties are important for structural aspects of the optical system. The components in the system must be able to support loads with no permanent deformation within the limits set by the error budget and certainly with no fracture. For ductile materials, the yield and/or microyield strength may be most important, but for brittle or near-brittle materials fracture toughness and/or modulus of rupture are more significant. A listing of definitions for each of these and other related terms follows:

Creep strength — the stress that will cause a given time-dependent plastic strain in a creep test for a given time

Ductility — the ability of a material to deform plastically before fracture

Fatigue strength — the maximum stress that can be sustained for a specific number of cycles without failure

Fracture toughness — a generic term for measures of resistance to extension of a crack

Hardness — a measure of the resistance of a material to surface indentation

Microcreep strength — the stress that will cause 1 ppm of permanent strain in a given time, usually less than the microyield strength

Microstrain — a deformation of 10^{-6} m/m (1 ppm)

Microyield strength — the stress that will cause 1 ppm of permanent strain in a short time; also called precision elastic limit, PEL

Ultimate strength — the maximum stress a material can withstand without fracture

Yield strength — the stress at which a material exhibits a specified deviation from elastic behavior (proportionality of stress and strain), usually 2×10^{-3} m/m (0.2%)

Hysteresis is a term that has more than one meaning. In terms of mechanical behavior, it is the time-dependent strain from an applied mechanical or thermal load, also referred to as anelasticity. In this case, removal of the load causes the strain to eventually return to zero. It also refers to the accumulated strain when a component is subjected to cyclic loading and unloading, a factor in fatigue failures. Another use of the word refers to the residual plastic strain in a component that has been thermally or mechanically cycled. This type of hysteresis is due to combined applied and residual stresses that exceed the microyield strength of the material. Further discussion of residual stress and hysteresis is given in Chapter 4.

Thermal

The significant thermal properties are coefficient of linear thermal expansion α (or CTE), thermal conductivity k, and specific heat C_p. Diffusivity D, a derived property equal to $k/\rho C_p$, is also important. All of these properties vary with temperature, α and C_p tending to vary directly with temperature and k and D varying inversely.

Thermal expansion is a generic term for a change in length for a specific temperature change, but there are more precise terms that describe specific aspects of this material property.[13] CTE is the most generally applicable version and is defined as:

$$\alpha \equiv \frac{1}{L} \frac{\Delta L}{\Delta T} \tag{3}$$

Many materials are also anisotropic in thermal expansion. This is particularly true in polycrystalline materials and fiber-reinforced composites. Lower CTE is better for optical system performance as it minimizes the effect of thermal gradients on dimensional changes of components. It is important to match CTE of adjacent components to minimize thermally induced strain in the system.

Thermal conductivity is the quantity of heat transmitted per unit of time through a unit of area per unit of temperature gradient. Higher thermal conductivity is desirable to minimize temperature gradients when there is a heat source in or close to the optical system. Specific heat, also called heat capacity per unit mass, is the quantity of heat required to change the temperature of a unit mass of material 1° under conditions of constant pressure. A material with high specific heat requires more heat to cause a temperature change that might cause a distortion. High specific heat also means that more energy is required to force a temperature change (e.g., in cooling an infrared telescope assembly to cryogenic temperatures). Thermal diffusivity determines the rate at which a nonuniform temperature distribution reaches equilibrium.

The two common thermal figures of merit are the steady-state and transient distortion coefficients, α/k and α/D. The former is a measure of the total thermal displacement for a given steady-state thermal input. The latter indicates the time for a thermal distortion to dissipate per unit of temperature gradient. Note that for actively cooled mirror applications such as laser mirrors, there is no appropriate figure of merit, since the single most important factor is the coefficient of linear thermal expansion, α.

Optical

Optical properties of solids are complex tensors, and as such will not be described in depth here. For a more complete treatise, see Wooten[14] or Born and Wolf.[15]

The most important optical property used in geometric optics is the index of refraction, n. The index of refraction is the ratio of the velocity of light in a vacuum to that in the material. In its general form it is a complex quantity expressed as:

$$\bar{n} = n - ik \tag{4}$$

where n is the real index and k is the imaginary part, called the index of absorption or extinction coefficient. For normal incidence, only the real part is important in optical design. For isotropic and cubic materials, there is one index, but for more complex crystals, the index varies with crystallographic direction. For an in-depth treatment of the optical properties of crystals and glasses see Tropf et al.[16]

The index of refraction varies with wavelength, temperature, and applied stress. The variation of refractive index with wavelength is called dispersion. The index of all transmitting materials increases with decreasing wavelength as shown in Figure 3.1. One way to characterize the dispersion, as devised for optical glasses, is with the Abbé number, v_d, where:

$$v_d = \frac{n_d - 1}{n_F - n_C} \tag{5}$$

The subscripts d, F, and C refer to the wavelengths for the emission lines of hydrogen and helium at 587.56, 486.13, and 656.27, respectively. There are other dispersion parameters, such as partial dispersions, that are discussed in more detail in Yoder[1] and in optical glass catalogues.

The variation of refractive index with temperature, dn/dT, is positive for most glasses, but negative for a few. When combined with CTE, the change in optical path length with temperature can be obtained from the thermo-optical constant G, equal to $\alpha(n - 1) + dn/dT$. The change in path length is then $t \cdot G \cdot \Delta T$, where t is the mechanical thickness of the element. Note that a material can be athermal if α and dn/dT have opposite signs.

Properties of Refractive Materials

The refractive materials commonly used for making lenses, prisms, optical windows, and filters can be broadly classified into three distinct categories, namely: glasses, optical crystals, and plastics and semiconductor materials. The physical, mechanical, and thermal properties of selected materials, which are most commonly used for optical and mechanical components, are covered in the subsequent sections. To keep the material property tables concise, only the nominal values at room temperature are listed, and therefore must only be used for preliminary evaluation and comparison purposes. Since the mechanical and thermal properties of materials can vary from one manufacturer to another and even from lot to lot for the same material from the same manufacturer, it is advisable to contact the manufacturer for obtaining more exact values of these properties for critical applications.

The optical properties of materials such as refractive index, Abbe value, reflectivity and transmittance, and variations of these properties as funtion of wavelength and temperature have deliberately been left out of these tables to avoid duplication of property tables from other sources. Some excellent and comprehensive references for optical properties of materials are *Handbook of Optics*,[2] *The Infrared Handbook*, 2nd ed.,[3] and Yoder.[1]

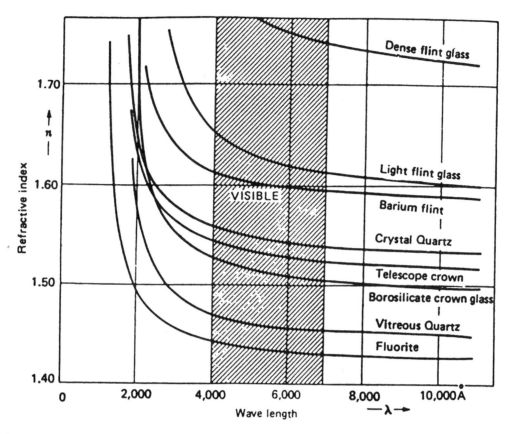

FIGURE 3.1 Dispersion curves for several materials commonly used for refracting optical components. *Source:* Jenkins, F.A. and White, H.E., *Fundamentals of Optics*, McGraw-Hill, New York.)

Glasses

Glasses are the most commonly used class of refracting material in optical systems. The most important property of a particular glass is its ability to transmit light over a desired range of wavelengths. An optical designer would primarily select a particular type of glass, which would have maximum transmission over the entire spectral region for which the instrument is being designed. Although the mechanical properties of the glass being used in a system may be of secondary importance, they do play a critical role in ensuring dependable performance during operation. Mechanical and thermal properties of the selected refractive materials such as density, elastic modulus, microyield strength, coefficient of thermal expansion, and thermal conductivity are of special significance if the designed optical system must be lightweight, rugged, and capable of retaining its performance over a large temperature range. Therefore, rather than selecting a particular glass merely on the basis of its optical properties, due consideration must also be given to its mechanical, and thermal properties before finalizing the choice. Table 3.1 lists physical, mechanical and thermal properties of selected optical and specialty glasses which are most commonly used in a majority of the optical systems.[1,17] For a clear explanation of the various terms used in describing the properties of optical glasses, see Marker[18] and the catalog Schott Optical Glass.[19]

There is a difference between fused silica and fused quartz, or quartzglass.[20] Fused silica is manufactured by the pyrolytic decomposition of reactive gases and usually has high water content and no metallic impurities. Fused quartz is made by fusing crystalline quartz to form a glass. Fused quartz has some level of metallic impurities that can cause UV fluorescence, and the water content depends on the firing method. Fused quartz can have some granularity, a residual of the original

TABLE 3.1   Properties of Selected Optical Glasses

Glass Code	Schott Type	Density ρ (g/cm³)	Young's Modulus E (Gpa)	Poisson's Ratio υ	Knoop Hardness HK	Thermal Expansion α (ppm/K)	Specific Heat C_p (J/kg K)	Thermal Conductivity k or λ (W/m K)
487 704	FK 5	2.45	62	0.232	520	9.2	808	0.925
517 642	BK 7	2.51	82	0.206	610	7.1	858	1.114
518 651	PK 2	2.51	84	0.209	640	6.9	736	0.755
522 595	K 5	2.59	71	0.224	530	8.2	783	0.950
523 515	KF 9	2.71	67	0.202	490	6.8	720	1.160
532 488	LLF 6	2.81	63	0.203	470	7.5	700	1.000
540 597	BaK 2	2.86	71	0.233	530	8.0	708	0.715
548 458	LLF 1	2.94	60	0.208	450	8.1	650	0.738
573 576	BaK 1	3.19	73	0.252	530	7.6	687	0.795
575 415	LF 7	3.20	58	0.213	440	7.9	660	0.879
589 613	SK 5	3.30	84	0.256	590	5.5	560	0.990
617 366	F4	3.58	56	0.222	420	8.3	553	0.768
620 364	F 2	3.61	57	0.220	420	8.2	557	0.780
620 603	SK 16	3.58	89	0.264	600	6.3	578	0.818
626 390	BaSF 1	3.66	62	0.242	460	8.5	553	0.741
636 353	F 6	3.76	57	0.224	410	8.5	510	0.755
648 339	SF 2	3.86	55	0.227	410	8.4	498	0.735
651 559	LaK N22	3.73	90	0.266	600	6.6	550	0.797
652 449	BaF 51	3.42	89	0.266	590	8.4	620	0.870
658 509	SSK N5	3.71	88	0.278	590	6.8	574	0.773
670 471	BaF N10	3.76	89	0.281	590	6.8	595	0.798
673 322	SF 5	4.07	56	0.233	410	8.2	488	0.738
699 301	SF 15	4.06	60	0.235	420	7.9	464	0.741
702 410	BaSF 52	3.96	86	0.283	550	5.2	540	0.737
717 295	SF 1	4.46	56	0.232	390	8.1	415	0.702
717 480	LaF 3	4.14	95	0.286	580	7.6	465	0.690
720 504	LaK 10	3.81	111	0.288	720	5.7	580	0.840
740 282	SF 3	4.64	56	0.236	380	8.4	423	0.706
744 448	LaF 2	4.34	93	0.289	560	8.1	480	0.648
750 350	LaF N7	4.38	80	0.280	520	5.3	440	0.770
755 276	SF 4	4.79	56	0.241	390	8.0	410	0.650
785 258	SF 11	4.74	66	0.235	450	6.1	431	0.737

Source: Schott Glass Technologies, Inc., Duryea, PA.

quartz crystal structure. Properties of these silica materials depend to some extent on their thermal history, and therefore nominally identical materials will have slightly different *n* and α from different manufacturers.

Crystals and Semiconductors

Optical crystals and ceramics are widely used in broadband optical systems and include both synthetic and naturally occurring materials. Optical crystal materials are available as single and polycrystalline forms. The polycrystalline form of a material consists of small, randomly oriented individual crystals, and are manufactured by various methods such as by hot pressing of powders, sintering, and chemical vapor deposition (CVD). On the other hand, single crystals are typically grown from dissolved and molten materials. Polycrystalline materials, in general, have higher strength and hardness as compared to single crystal materials.[16] Polycrystalline materials generally have isotropic properties while single crystals have directionally dependent anisotropic properties.

Optical crystals are widely used in infrared applications but there are a number of crystals which have good tranmission over a wide band from ultraviolet (UV) to far-infrared (IR) wavelengths; e.g., CaF_2 and LiF are extensively used for achromitized lenses in far-UV to mid-IR (0.11 to 10 μm) applications. CaF_2 has the best strength and moisture resistance of all fluoride crystals and has a very low thermo-optic coefficient.

Another very useful material for high temperature application is sapphire, which is the single crystal form of aluminum oxide. It has high strength and hardness, and excellent thermal shock resistance. It is widely used in specialized optical systems subjected to severe environments. Another useful naturally occurring optical crystal is quartz, which is commonly used for UV prisms and windows and in IR applications to 4 μm. Quartz and sapphire are also grown artificially to improve transmission by controlling the amount of impurities. The thermal properties, such as CTE, of these materials are direction dependent and quite sensitive to thermal shock.[21]

Silicon and germanium (Ge) are extensively used in IR systems for lenses, windows, and domes. Silicon is very suitable for missile domes because of its good mechanical and thermal properties. Germanium is quite hard but is susceptible to brittle fracture. Both materials have high index of refraction and are therefore very suitable for making multiple lens assemblies to keep the thicknesses and weight within reasonable limits. Due to the high index of these materials, efficient antireflection coatings are required to minimize internal reflection losses. The optical properties of Ge, such as index of refraction and absorption, are quite sensitive to temperature. Table 3.2 lists optical, physical, mechanical, and thermal properties of selected crystalline materials, while Tables 3.3 and 3.4 list the same properties for IR-transmitting materials. A more extensive list of materials and properties can be found in Tropf.[16]

TABLE 3.2   Properties of Selected Alkali Halide Crystalline Materials

Material Name and Symbol	Useful Spectral Range (μm)	Refractive Index n at λ (μm)	Density ρ (g/cm³)	Young's Modulus E (GPa)	Poisson's Ratio υ	Thermal Expansion coeff. α (ppm/K)	Specific Heat C_p (J/kg K)	Thermal Conductivity k (W/mM)
Barium fluoride, BaF_2	0.13–15	1.463 (0.6) 1.458 (3.8) 1.449 (5.3) 1.396 (10.6)	4.89	53	0.343	6.7 at 75 K 19.9 at 300 K 24.7 at 500 K	402	11.7
Calcium fluoride, CaF_2	0.13–10	1.431 (0.7) 1.420 (2.7) 1.411 (3.8) 1.395 (5.3)	3.18	96	0.28	18.9	854	10
Calcium fluoride, CaF (Kodak Irtran 3)	0.15–11.8	1.434 (0.6) 1.407 (4.3)	3.18	99		18.9	853	9.7
Potassium chloride, KCl	0.21–25	1.474 (2.7) 1.472 (3.8) 1.469 (5.3) 1.454 (10.6)	1.98	30	0.216	37.1	678	6.7
Lithium fluoride, LiF	0.12–8.5	1.394 (0.5) 1.367 (3.0) 1.327 (5.0)	2.63	65	0.33	5.5	1548	11.3
Magnesium fluoride, MgF_2	0.15–9.6	1.384 (0.4)o 1.356 (3.8)o 1.333 (5.3)o	3.18	169	0.308	14.0 (P) 8.9 (N)	1004	21
Magnesium fluoride, (Kodak Irtran 1)	0.6–8	1.378 (1.0) 1.364 (3.0) 1.337 (5.0)	3.18	73	0.25–0.36	10.7	502	14.6
Sodium chloride, NaCl	0.17–18	1.525 (2.7) 1.522 (3.8) 1.517 (5.3) 1.488 (10.6)	2.16	40	0.28	39.6	837	6.5

Source: Adapted from Yoder, P.R., Jr. 1993. *Optomechanical Systems Design*, 2nd ed., pp. 108–109. Marcel Dekker, New York.

TABLE 3.3 Properties of Selected IR Materials

Material Name and Symbol	Useful Spectral Range (μm)	Refractive Index n at λ (μm)	Density ρ (g/cm³)	Young's Modulus E (GPa)	Poisson's Ratio υ	Thermal Expansion coeff. α (ppm/K)	Specific Heat C_p (J/kg K)	Thermal Conductivity k (W/mK)
Semiconductors								
Diamond C	0.25–200	2.382 (2.5) 2.381 (5.0) 2.381 (10.6)	3.51	1050	0.16	–0.1 at 25 K 0.8 at 293 K 5.8 at 1600 K	108	2600
Gallium arsenide, GaAs	1–16	3.1 (10.6)	5.32	83	0.31	5.7	326	48–55
	1.8–23	4.055 (2.7) 4.026 (3.8) 4.015 (5.3) 4.00 (10.6)	5.33	104	0.278	5.7–6.0	310	59
Silicon, SI	1.2–15	3.436 (2.7) 3.427 (3.8) 3.422 (5.3) 3.148 (10.6)	2.329	131	0.22	2.7–3.1	53	150–163
Calcogenides								
Cadmium telluride, CdTe	0.9–15	2.682 (6) 2.650 (10) 2.637 (14)	5.85	37	0.41	5.9 at 300 K	234	6.3
Zinc sulfide, ZnS (Raytheon Standard grade)	0.5–12	2.36 (0.6) 2.257 (3.0) 2.246 (5.0) 2.192 (10.6)	4.08	75	0.29	4.6 at 173 K 6.6 at 273 K 7.7 at 473 K	469	17 at 296 K
Zinc sulfide, ZnS (Kodak Irtran 2)	0.5–14	2.37 (0.6) 2.249 (4.3)	4.09	96	0.25–0.36	6.6	962	15.4
Zinc selenide, ZnSe (Raytheon CVD Raytran)	0.6–20	2.61 (0.6) 2.438 (3.0) 2.429 (5.0) 2.403 (10.6) 2.376 (14)	5.27	70	0.28	5.6 at 163 K 7.1 at 273 K 8.3 at 473K	339 at 0296 K	18 at 300 K
Zinc selenide, ZnSe (Kodak Irtran 4)	0.5–20	2.485 (1.0) 2.440 (3.0) 2.432 (5.0) 2.418 (8.0) 2.407 (10.0) 2.394 (12.0)	5.27	71		7.7	335	

Source: Adapted from Yoder, P.R., Jr. 1993. *Optomechanical Systems Design*, 2nd ed., pp. 112–115. Marcel Dekker, New York.

Plastics

Optical plastics are used in a small fraction of optical systems as compared to optical glasses and crystal materials. The largest market for plastic optics is the high volume consumer items such as ophthalmic lenses (eyeglasses) and low end camera, microscope, and binocular lenses. The number of optical plastics available is quite limited as compared to the number of optical glasses. The optical plastics can be classified into two broad categories: thermoplastics and thermosets. The term thermoplastic means a material which flows when heated, but there are some thermoplastics that do not flow when heated. Thermoset plastics can be set by heating these materials. Thermoplastics, or linear plastics as they are sometimes called, do not undergo any chemical change during the molding process and therefore can be remolded several times without affecting their properties. On the other hand, thermosets, also known as cross-linked plastics, start with a linear polmer chain, which gets cross-linked permanently in the presence of heat during molding.[22]

TABLE 3.4   Properties of Selected IR-Transmitting Glasses and Other Oxide Materials

Material Name and Symbol	Useful Spectral Range (μm)	Refractive Index n at λ (μm)	Density ρ (g/cm³)	Young's Modulus E (GPa)	Poisson's Ratio υ	Thermal Expansion coeff. α (ppm/K)	Specific Heat C_p (J/kg K)	Thermal Conductivity k (W/mK)
Calcium alumino-silicate (Schott IRG 11)	0.5–5	1.684 (0.55) 1.635 (3.3) 1.608 (4.6)	3.12	108	0.284	8.2 at 293–573 K	749 at 293–373 K	1.13
Calcium alumino-silicate (Corning 9753)	0.6–4.2	1.61 (0.5) 1.57 (2.5)	2.798	99	0.28	5.95 at 293–573 K	837 at 373 K	2.5 at 373 K
Calcium alumino-silicate (Schott IRGN6)	0.4–4.5	1.592 (0.55) 1.562 (2.3) 1.521 (4.3)	2.81	103	0.276	6.3 at 293–573 K	808 at 293–373K	1.36
Fluoro phosphate (Schott IRG9)	0.4–4	1.488 (0.55) 1.469 (2.3) 1.458 (3.3)	3.63	77	0.288	6.1 at 293–573 K	694 at 293–373 K	0.88
Germanate (Corning 9754)	0.4–5	1.67 (0.5) 1.63 (2.5) 1.61 (4.0)	3.581	84	0.290	6.2 at 293–573 K	544	1.0
Germanate (Schott IRG 2)	0.4–5	1.899 (0.55) 1.841 (2.3) 1.825 (3.3)	5.00	96	0.282	8.8 at 293–573 K	452 at 293–373 K	0.91
Lanthanum-dense flint (Schott IRG3)	0.4–4	1.851 (0.55) 1.796 (2.3) 1.776 (3.3)	4.47	100	0.287	8.1 at 293–573 K		0.87
Lead silicate (Schott IRG7)	0.4–4	1.573 (0.55) 1.534 (2.3) 1.516 (3.3)	3.06	60	0.216	9.6 at 293–573 K	632 at 293–373 K	0.73
Magnesium oxide, MgO_2 (Kodak Irtran 5)	0.2–6	1.723 (1.0) 1.692 (3.0) 1.637 (5.0)	3.58	332	0.2	11.5	879	50
Sapphire, Al_2O_3	0.15–7.5	1684 (3.8) 1.586 (5.8)	3.97	400	0.27	5.6 (P) 5.0 (N)	753	25–33
Silica, fused, SiO_2 (Corning 7940)	0.15–3.6	1.566 (0.19) 1.460 (0.55) 1.433 (2.3) 1.412 (3.3)	2.202	73	0.17	−0.6 at 73 K 0.52 at 278–308 K 0.57 at 273–473 K	108.8	13.8

Source: Adapted from Yoder, P.R., Jr. 1993. *Optomechanical Systems Design*, 2nd ed., pp. 110–111. Marcel Dekker, New York.

The most widely used optical plastic is acrylic, specifically known as polymethyl methacrylate (PMMA). It is a low-cost plastic that can be easily molded, machined, and polished, and has the best combination of optical properties. It has a low thermal conductivity and a high linear coefficient of thermal expansion (70 ppm/K), which is eight to ten times greater than that of typical optical glass. It has a shrinkage of 0.2 to 0.6% and a good optical memory, which is the ability to return to its original shape after exposure to heat. Acrylic has very good transmission (92%) and low internal scattering, and its refractive index varies from 1.483 for λ = 1 μm to 1.510 for λ = 380 nm. The index varies from about 1.492 to 1.480 over a temperature range of 20 to 90°C.[22]

Polystyrene is a second common optical plastic that can be combined with acrylic to obtain highly corrected achromatic lens designs. Its index of refraction is 1.590 and transmission is about 90%. It can be easily injection molded and it is the lowest cost optical plastic. It has a lower moisture absorption than acrylic, but it is more difficult to machine and polish as compared to acrylic.

Polycarbonate is another widely used optical plastic for ophthalmic lenses, street lights, and automotive tail light lenses due to its high durability and impact resistance. It is more expensive than acrylic and styrene because it is more difficult to mold, machine, and polish, and it scratches easily. Its refractive index ranges from 1.560 to 1.654, and the transmission is about 85%. It retains its performance over a broad range of temperature (−137 to 121°C).

The only thermoset optical plastic used is allyl diglycol carbonate (ADC), commonly known as CR 39. It is extensively used in making cast ophthalmic lenses, which are subsequently machined and polished, which makes it more expensive. It cannot be injection molded. It has excellent optical and mechanical properties such as clarity, impact, and abrasion resistance. It can withstand continuous temperature up to 100°C, and up to 150°C for short periods, but it also has a high shrinkage rate of up to 14%.[22]

Other less commonly used optical copolymers of styrene and acrylic include methyl methacrylate styrene (NAS), styrene acrylonitrile (SAN), methyl pentene (TPX), and clear acrylonitrile butadiene styrene (ABS). Table 3.5 lists physical, mechanical, and thermal properties of some of the optical plastics discussed here. A more detailed discussion of optical properties and design and fabrication methods for plastic lenses can be found in Lytle[23] and Welham.[24]

TABLE 3.5   Selected Properties of Optical Plastics

Material	Density (gm/cc) ρ	CTE (ppm/K) α	Max. Service Temp (°C) T_v	Thermal Conductivity k	Water Absorption (%) A	Luminous Transmittance (%)
P-methylmethacrylate (acrylic)	1.18	6.0	85	4–6	0.3	92
P-styrene	1.05	6.4–6.7	80	2.4–3.3	0.03	88
Methyl methacrylate styrene (NAS)	1.13	5.6	85	4.5	0.15	90
Styrene acrylonitrile (SAN)	1.07	6.4	75	2.8	0.28	88
P-carbonate	1.25	6.7	120	4.7	0.2–0.3	89
P-methyl pentene (TPX)	0.835	11.7	115	4.0	0.01	90
P-amide (Nylon)	1.185	8.2	80	5.1–5.8	1.5–3.0	
P-arylate	1.21	6.3		7.1	0.26	
P-sulfone	1.24	2.5	160	2.8	0.1–0.6	
P-styrene co-butadiene	1.01	7.8–12			0.08	
P-cyclohexyl methacrylate	1.11					
P-allyl diglycol carbonate (ADC)	1.32		100	4.9		93
Cellulose acetate butyrate	1.20			4.0–8.0		
P-ethersulfone	1.37	5.5	200	3.2–4.4		
P-chloro-trifluoroethelyne	2.2	4.7	200	6.2	0.003	
P-vinylidene fluoride	1.78	7.4–13	150		0.05	
P-etherimide	1.27	5.6	170		0.25	

Source: Adapted from Lytle, J.D. 1995. *Handbook of Optics*, Vol. 2, 2nd ed., p. 34.1. McGraw-Hill, New York; and Wolpert, H.D. 1988. *Engineered Materials Handbook*, Vol. 2, pp. 481–486. ASM International, Metals Park, OH.

Properties of Mirror and Structural Materials

A number of metals, ceramics, and composites can be used for making the substrates of mirrors and the structural components of an optical system including the housings, optical benches, and metering structures. Table 3.6 lists physical, mechanical, and thermal properties of a number of commonly used mirror and structural materials at room temperature. Some temperature dependencies for these materials are given in Chapter 2, but for a more comprehensive treatment, see Paquin.[25] An examination of the table shows that there is no one material that is best in all categories. Choice of a material for a particular application is always a trade-off.

Table 3.7 lists some of the more important figures of merit for these same materials. The thermal properties and consequently the distortion coefficients are strongly temperature dependent and care should be taken to ensure that appropriate values are used if the mirror application is other than at room temperature. These figures of merit can be used for selecting one or more materials best suited for a particular application during the preliminary design phase. For example, if the weight of an optical system has to be minimized for a given deflection, then a quick look at the

TABLE 3.6   Properties of Selected Mirror and Structural Materials

Preferred	ρ Density (g/cm³)	E Young's Modulus (GPa) Large	ν Poisson's Ratio Small	K_{Ic} Fracture Toughness (MPa[m]$^{1/2}$) Large	MYS Microyield Strength (MPa) Large	α Thermal Expansion (10^{-6}/K) Small	k Thermal Conductivity (W/m K) Large	Cp Specific Heat (W sec/kg K)	D Thermal Diffusivity (10^{-6} m²/sec) Large
Pyrex 7740	2.23	63	0.2			3.3	1.13	1050	0.65
Fused silica	2.19	72	0.17	<1.0	—	0.5	1.4	750	0.85
ULE fused silica	2.21	67	0.17			0.03	1.31	766	0.78
Zerodur	2.53	91	0.24			0.05	1.64	821	0.77
Aluminum: 6061	2.70	68	0.33	—	140.	22.5	167.	896	69.
MMC: 30% SiC/Al	2.91	117	0.29	>10.	>200.	12.4	123.	870	57.
Beryllium: I-70-H	1.85	287	0.043	12.	35.	11.3	216.	1925	57.2
Beryllium: I-220-H	1.85	287	0.043	12.	50.	11.3	216.	1925	57.2
Copper: OFC	8.94	117	0.343	—	12.	16.5	391.	385	115.5
Invar 36	8.05	141	0.259	—	41.	1.0	10.4	515	2.6
Super Invar	8.13	148	0.26	—	41.	0.3	10.5	515	2.5
Molybdenum	10.21	324	0.293	—	280.	5.0	140.	247	55.5
Silicon	2.33	131	0.42	1.0	—	2.6	156.	710	94.3
SiC: HP alpha	3.2	455	0.14	5.2	—	2.4	155.	650	74.5
SiC: CVD beta	3.21	465	0.21	2.7	—	2.4	198.	733	84.2
SiC: RB-30% Si	2.89	330	0.24	2.5	—	2.5	155.	670	80.0
Stainless steel: 304	8.00	193	0.27	—	65.	14.7	16.2	500	4.1
Stainless steel: 416	7.80	215	0.283	—	—	8.5	24.9	460	6.9
Titanium: 6Al4V	4.43	114	0.31	—	50.	8.8	7.3	560	2.9
Gr/Ep (GY-70/×30)	1.78	93	—			0.02	35.0		

numbers in the ρ/E column reveals that SiC and beryllium may be the best materials for such an application, while the mirror substrate or structure made out of aluminum or titanium would have the maximum weight. It must be emphasized that the final selection of a material for a particular application must also be based on a number of other factors such as fabrication cost, microyield strength, and thermal properties.

Properties of Adhesives and Cements

When using optical cements and adhesives in optical systems, a number of their characteristics and properties must be considered carefully, including shrinkage, outgassing, shear and peel strengths, and curing time and temperature. Tables in the following sections list physical, mechanical, and thermal properties of a number of commonly used structural adhesives and optical cements at room temperature. These properties' values are nominal values and therefore must be used for comparison and preliminary design purposes. For critical applications, it is advisable to obtain the latest data and specification sheets from the manufacturers.

Structural Adhesives

Structural adhesives and elastomers can be used to bond structural components to each other or to bond optical components such as mirrors and lenses to their cells or mounts. The three main classes of structural adhesives are epoxies, urethanes, and cyanoacrylate adhesives. The thermosetting epoxy adhesives have high bonding strengths and good thermal properties. The epoxies are available in one- or two-part types, and some are room temperature curable. The urethanes or polyurethanes have fairly high strength and can be used to bond together a variety of materials.

TABLE 3.7 Figures of Merit for Selected Mirror and Structural Materials

	Weight and Self-Weight Deflection Proportionality Factors				Thermal Distortion Coefficients	
	$(E/\rho)^{1/2}$ Resonant Frequency for Same Geometry (arb. units)	ρ/E Mass or Deflection for Same Geometry (arb. units)	ρ^3/E Deflection for Same Mass (arb. units)	$(\rho^3/E)^{1/2}$ Mass for Same Deflection (arb. units)	α/k Steady State (μm/W)	α/D Transient (sec/m² K)
ferred	Large	Small	Small	Small	Small	Small
ex	5.3	3.53	1.76	0.420	2.92	5.08
ed silica	5.7	3.04	1.46	0.382	0.36	0.59
E fused silica	5.5	3.30	1.61	0.401	0.02	0.04
odur	6.0	2.78	1.78	0.422	0.03	0.07
minum: 6061	5.0	3.97	2.90	0.538	0.13	0.33
1C: 30% SiC/Al	6.3	2.49	2.11	0.459	0.10	0.22
yllium: I-70-H	12.5	0.64	0.22	0.149	0.05	0.20
yllium: I-220-H	12.5	0.64	0.22	0.149	0.05	0.20
per: OFC	3.6	7.64	61.1	2.471	0.53	0.14
ar 36	4.2	5.71	37.0	1.924	0.10	0.38
er Invar	4.3	5.49	36.3	1.906	0.03	0.12
ybdenum	5.6	3.15	32.8	1.812	0.04	0.09
con	7.5	1.78	0.97	0.311	0.02	0.03
HP alpha	11.9	0.70	0.72	0.268	0.02	0.03
CVD	12.0	0.69	0.71	0.267	0.02	0.03
RB-30% Si	10.7	0.88	0.73	0.270	0.01	0.03
nless steel: 304	4.9	4.15	26.5	1.629	0.91	3.68
nless steel: 416	5.2	3.63	22.1	1.486	0.34	1.23
nium: 6Al4V	5.1	3.89	7.63	0.873	1.21	3.03

They are flexible and therefore susceptable to creep and not suitable for high temperature (>100°C) applications, but are well suited for cryogenic applications.

The one-part cyanoacrylate adhesives have low viscosity and are suitable for bonding smooth surfaces with very thin bond joints. They have cure times of less than 30 sec, so proper fixturing is a requirement and care must be taken to protect the skin from accidental bonding. These materials outgas more than other adhesives and are suitable for applications where the humidity is low and the temperatue stays belows 70°C. The key physical, mechanical, and thermal properties of some commonly used structural adhesives in optical applications are listed in Table 3.8.[1]

The two-part room temperature vulcanizing rubbers (RTVs) available from GE and Dow Corning are extensively used to bond mirrors, lenses, filters, and optical windows to their mounts. These silicone rubber elastomers are inert chemically and can tolerate a temperature variation of −80 to 200°C or more. The two main reasons for their popularity are the low cost and ability to accommodate differential thermal expansion between high-expansion metal mounts and low-expansion optical elements. Since a fairly thick bond layer of RTV is needed, the edges or diameters of the optical elements and their mounts do not need to be machined to close tolerances, thereby reducing their fabrication cost. Moreover, retainers, clips, and screws for securing the optics in their mounts are also eliminated resulting in a much simpler design. RTV is resilient and allows for the differential expansion between the optic and its metal mount when the ambient temperatrure changes, without introducing any adverse stresses in the optic. The key physical, mechanical, and thermal properties of some commonly used silicone rubber-type elastomers in optical applications are listed in Table 3.9.[1]

Optical Cements

Optical cements are the adhesives used for bonding the refracting optical elements to each other. Therefore, these adhesives must have good transmission and homogeneity over the desired spectral wavelengths in addition to the desirable mechanical properties such as low shrinkage and outgas-

TABLE 3.8   Properties of Representative Structural Adhesives

Material	Mfr. Code[a]	Recommended Cure (time at °C)	Cured Joint Strength MPa[psi] at °C	Temperature Range of Use (°C)	Thermal Expansion Coeff. (ppm/°C at °C)	Recommended Joint Thickness (mm)	Density at 25°C (g/cm³)
One-part epoxies 2214Hi-Temp[b]	3M	40 min at 121	13.8 (2000) at 24 20.7 (3000) at 82	−55 to 177			
Two-part epoxies Milbond	SL	7 days at 25	14.5 (2099) at 25 6.8 (992) at 70		0.381 ± 0.025		
EC-2216B/A Gray	3M	Fix: 2 hr at 65 Full: 7 days at 75	17.2 (2500) at 24 2.8 (400) at 82	−55 to 150	102 at 0–40 134 at 40–80	0.102 ± 0.025	
Translucent		Fix: 6 hr at 65 Full: 30 days at 75	8.3 (1200) at 24 1.4 (200) at 82		81 at 50–0 207 at 60–150	0.102 ± 0.025	
27A/B	EC	24 hr at 25	13.8 (2000)	−65 to 105			
45	EC	24 hr at 25	21.4 (3100)	−40 to 90			1.34–1.18
324	L	Fix: 3 min at 20 Full: 24 hr at 20	10.3 (1500)	−54 to 135	12.6 (7.0)	>1.0 (0.040)	1.1
3532B/A	3M	3 days at 24	13.8 (2000) at 24 2.1 (300) at 82				
UV — curable UV-900	EC	UV cure: 15 sec at 200 W/in.; heat cure: 10 min at 120		−25 to 125		<3.2 (0.125)	1.04
349		Fix: 20 sec at 0.25 Full: 36 sec at 0.25 gap	5.2 (750)	−54 to 130			
460	L	Fix: 1 min at 22 Full: 7 days at 22	12.1 (1750)	−54 to 71	100 (56)	0.100 max	1.06

[a] Mfr. code: 3M, SL = Summers Laboratories; EC = Emerson & Cummings; L = Loctite.

[b] Also available in varieties with higher density, greater flexibility, or increased electrical resistivity.

Source: Adapted from Yoder, P.R., Jr. 1993. *Optomechanical Systems Design*, 2nd ed., pp. 144–146. Marcel Dekker, New York.

TABLE 3.9   Properties of Representative Elastomeric Sealants

Material	Mfr. Code[a]	Suggested Cure Time at °C	Elastomer Tensile Strength, MPa (psi)	Temperature Range of Use (°C [°F])	Thermal Expansion Coeff. (ppm/°C)	Density at 25°C (g/cm³)	Shrinkage after 3 Days at 25°C (%)
732	DC	24 hr at 25	2.2 (325)	−55 to 200	310	1.04	
RTV112	GE	3 days at 2	2.2 (325)	<204 (400)	270 293	1.05	1.0
NUVA-SIL	L	UV cure 1 min or 7 days at 25	3.8 (550)	−70 to 260	167	1.35	0.4
3112	DC		4.5 (650)	−55 to 250	300	1.02	0.25
93–500	DC	7 days at 25 4 hr at 65	69 (100)	−55 to 155			
RTV88	GE	<24 hr at 25	5.9 (850)	−54 to 260	210	1.48	0.6
RTV8111	GE	<24 hr at 25	2.4 (350)	−54 to 204	250	1.18	0.6
RTV8262	GE	<24 hr at 25	5.2 (750)	−54 to 260	210	1.47	0.6

[a] Mfr. code: 3M = 3M Company; DC = Dow Corning; GE = General Electric; L = Loctite.

Source: Adapted from Yoder, P.R., Jr. 1993. *Optomechanical Systems Design*, 2nd ed., pp. 148–149. Marcel Dekker, New York.

sing, good strength and stability, and resistance to adverse environmental effects such as humidity, temperature variations, and UV exposure. The optical cements come in four basic types: solvent loss, thermoplastic, thermosetting, and photosetting cements.

The solvent-loss cements, such as Canada balsam, have a high viscosity and are heat cured by elimination of solvent to a refractive index ~1.53. They have a poor bond strength and can introduce distortion in the bonded optical surfaces due to high shrinkage on curing and are, therefore, seldom used in precision optical systems. The thermoplastic cements, such as cellulose carpate with n ~ 1.48, are colorless or lightly colored solids that liquify when heated to about 120°C. Their principle advantage is that the bonded elements can be separated by applying heat, which is easy and risk free. The thermosetting cements are two-part adhesives, which can be cured at room temperature by addition of an appropriate catalyst. The room temperature curing time for this type of adhesive varies from 3 to 7 days. The cure time can be reduced to a few hours with a low elevated temperature cure, typically 70°C. Summer's C-59, M-69, F-65, RD3 to 74, Lens Bond's, and Kodak's Eastman HE-80 are some of the commercially available thermosetting cements with n of ~1.55.[1]

Photosetting optical cements are generally one-part clear adhesives that are cured by exposure to UV light of 250 to 380 nm wavelength. These cements are suitable for bonding small low-mass optics that have transmission in this spectral region. Bondline thickness must be kept small to prevent excessive stress due to shrinkage. Norland's NOA-61 (n = 1.56) and Summer's UV-69 (n = 1.55) and UV-74 are some of the UV-curing optical cements available. A two-step curing process, a short exposure for 20 sec, followed by a long exposure of up to 60 min, is used for some of these cements. The bonded parts can be gently handled and cleaned, or debonded if needed, after the short exposure. Once the adhesive is fully cured after the long exposure, it becomes quite difficult to separate the parts. The entire area of the bond joint must be completely exposed to a uniform intensity UV illumination to obtain complete curing of the joint to prevent surface distortions. If feasible, the strength of the bond joint can be improved by heating the bonded parts to 40°C.

Typical properties of optical cements are shown in Table 3.10. For specific properties, manufacturer's data sheets should be obtained.

TABLE 3.10 Typical Properties of Optical Cements

Refractive index n after cure	1.48–1.56
Thermal expansion coefficient	
27 to 100°C	63 ppm/°C (35 ppm/°F)
100 to –200°C	56 ppm/°C (31 ppm/°F)
Young's modulus E	430 GPa (62 × 10⁶ lb/in.²)
Shear strength	360 GPa (5200 lb/in.²)
Specific heat K	837 J/kg K
	(0.2–0.4 BTU/lb °F)
Water absorption (bulk material)	0.3% after 24 hr at 25°C
Shrinkage during cure	Approximately 6%
Viscosity	200–320 cps
Density	1.22 g/cm³ (0.044 lb/in.³)
Hardness (shore D)	Approximately 90
Total mass loss in vacuum	3–6%

Source: Adapted from Yoder, P.R., Jr. 1993. *Optomechanical Systems Design*, 2nd ed., p. 137. Marcel Dekker, New York.

3.4 Material Selection Criteria

Material selection for any particular optical system is a trade-off process that starts at the initial design stage and involves the system performance and environmental specifications, structural and optical designs, material properties of candidate materials, and potential substrate and optical fabrication methods. A discussion of selection of material candidates is included in Section 10.4 of Chapter 10.

A schematic diagram of the trade-off process for lightweight metal mirrors is given in Figure 3.2 as an example of the interactive process that includes material selection. The process is basically one of addressing a number of options that may meet the requirements and performing rough analyses to narrow the choices. More detailed analyses can then determine the best choice of material, structural design, and fabrication methods. The process is iterative and multidisciplinary, with continuous feedback to continually refine the choices.

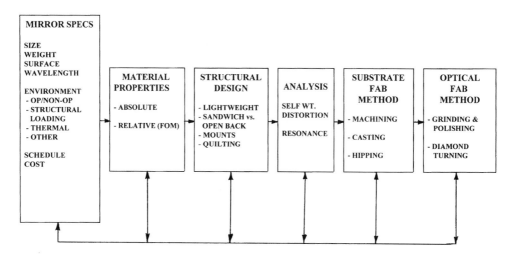

FIGURE 3.2 Schematic diagram showing trade-off process for material selection and structural design of a lightweight mirror.

3.5 Summary

This chapter provides guidelines for use of materials in optical systems by providing a description of the types of materials used — refractors, reflectors, structural, and adhesives — typical applications of each type, and detailed descriptions and properties of the most representative of each type.

After a brief introduction, typical applications are discussed with the materials commonly used for each type of application. Under refractors, the various types of glasses, crystals, and other materials are discussed for typical applications in various wavelength ranges. Reflectors include all types of mirrors from 8-m astronomical telescopes to small scanners and cooled mirrors, with the range of materials typically used for each. Structural components can be metals, ceramics, or composites and some typical examples are discussed. Athermalization of the optical system is accomplished through use of matched expansion materials for optics and structures. Structural adhesives and optical cements with their typical requirements for most applications are discussed.

Definitions of the important properties and figures of merit and their relevance to optical systems are given. Tables of properties of the most used reflector, refractor, and structural materials are given with sources for more extensive property data including the temperature dependence of properties.

A brief discussion of material selection methods completes the chapter.

References

1. Yoder, P. R., Jr. 1993. Optomechanical characteristics of materials. In *Optomechanical Systems Design*, 2nd ed., pp. 78–154. Marcel Dekker, New York.
2. Bass, M., ed.-in-chief. 1995. *Handbook of Optics*, Vol. 2, Devices, Measurements, and Properties, 2nd ed., McGraw-Hill, New York.
3. *The Infrared Handbook,* 2nd ed. 1994. SPIE Optical Engineering Press, Bellingham, WA.
4. Klocek, P., ed. 1991. *Handbook of Infrared Optical Materials*. Marcel Dekker, New York.
5. Weber, M. J., ed. 1986. *CRC Handbook of Laser Science and Technology*, Vol. 4, Optical Materials, Part 2: Properties. CRC Press, Boca Raton, FL.
6. *ASM Handbook,* Vol. 1. 1990. Properties and Selection: Irons, Steels and High Performance Alloys. ASM International, Materials Park, OH.
7. *ASM Handbook,* Vol. 2. 1990 Properties and Selection: Nonferrous Alloys and Special-Purpose Materials. ASM International, Materials Park, OH.
8. *Engineered Materials Handbook:* Vol. 1. 1987. Composites. ASM International, Metals Park, OH.
9. *Engineered Materials Handbook:* Vol. 2. 1988. Engineering Plastics. ASM International, Metals Park, OH.
10. *Engineered Materials Handbook:* Vol. 3. 1990. Adhesives and Sealants. ASM International, Metals Park, OH.
11. *Engineered Materials Handbook:* Vol. 4. 1991. Ceramics and Glasses. ASM International, Metals Park, OH.
12. Reisman, H. and Pawlik, P. S. 1980. *Elasticity*, pp. 111–129. Wiley, New York.
13. Jacobs, S. F. 1992. Variable invariables: dimensional instability with time and temperature. In *Optomechanical Design*, Vol. CR43, Yoder, P. R., Jr., ed., p. 181. SPIE Optical Engineering Press, Bellingham, WA.
14. Wooten, F. 1972. *Optical Properties of Solids*. Academic Press, New York.
15. Born, M. and Wolf, E. 1975. *Principles of Optics*, 5th ed. Pergamon Press, London.
16. Tropf, W. J., Thomas, M. E., and Harris, T. J. 1995. Properties of crystals and glasses. In *Handbook of Optics,* Vol. 2. Devices, Measurements, and Properties, 2nd ed. p. 33.3. McGraw-Hill, New York.

17. Zhang, S. and Shannon, R. S. 1995. Lens design using a minimum number of glasses, *Opt. Eng.*, Vol. 34, 3536–3544.
18. Marker, A. M., III. 1991. Optical properties: a trip through the glass map. In *Passive Materials for Optical Elements*, Wilkerson, G. W., ed., Proc. SPIE Vol. 1535, pp. 60–65.
19. Schott Optical Glass, 1992. Schott Glass Technologies, Duryea, PA.
20. Hahn, T. A. and Kirby, R. K. 1972. Thermal expansion of fused silica from 80 to 1000 K — standard reference material 739, Thermal Expansion — 1971, AIP Conf. Proc. No. 3, Graham, M. G. and Hogy, H. E., eds., American Institute of Physics, New York.
21. Parker, C. J. 1979. Optical Materials — Refractive. In *Applied Optics and Optical Engineering*, Vol. 7, R. R. Shannon and J. C. Wyant, eds., Academic Press, New York.
22. Wolpert, H. D. 1988. Optical properties. In *Engineered Materials Handbook*. Vol, 2, Engineering Plastics, pp. 481–486. ASM International, Metals Park, OH.
23. Lytle, J. D. 1995. Polymeric optics. In *Handbook of Optics*, Vol. 2, Devices, Measurements, and Properties, 2nd ed., p. 34.1. McGraw-Hill, New York.
24. Welham, B. 1979. Plastic optical components. In *Applied Optics and Optical Engineering*, Vol. 7, pp. 79–96. Academic Press, New York.
25. Paquin, R. A. 1995. Properties of metals. In *Handbook of Optics*, Vol. 2, Devices, Measurements, and Properties, 2nd ed., p. 35.1. McGraw-Hill, New York.

4

Metal Mirrors

Roger A. Paquin

4.1 Introduction

Mirrors come in all shapes and sizes, but from an engineering point of view, just what is a mirror? There are large mirrors such as the highly structured lightweight Hubble Space Telescope's primary mirror, the segmented 10-m mirror of the Keck telescope, and the active thin shell primary mirror of the 8-m Gemini telescope. There are small scanner mirrors, flip mirrors in cameras, and cooled mirrors in high energy systems. Is the mirror the reflecting surface or the structure that contains the reflecting surface? A good definition is that the mirror consists of the reflecting surface and whatever it takes to support that surface. A mirror is then an optomechanical system in itself.

Mirrors can come in many forms, but generally fall into one of three categories:

1. Passive mirrors — self-supporting for minimum deflection in a 1-g field on a simple, nonadjustable mount; can be solid or lightweighted
2. Semiactive mirrors — semirigid, usually lightweighted mirrors with more complex mounts, usually with multipoint mounts that are in dynamic systems that require occasional figure correction
3. Active mirrors — thin facesheets with fully adjustable, computer-controlled, multipoint actuator/mounts for wavefront correction in real time; the number of actuators and actuator density varies with application

Cooled mirrors can take any of these forms and can be cooled in any number of ways — from passive heat sinks to heat pipes to high flow, pumped fluid heat exchangers. The majority of cooled mirrors are small to moderate size and passive, but there are a growing number of relatively small, high actuator density, high speed mirrors for wavefront correction in high energy systems.

The mirror assembly includes the mirror surface (reflective coating and/or polished faceplate), substrate, mounts/actuators, and, for active mirrors, the reaction structure. For passive mirrors, the mounts may incorporate actuators to provide tip/tilt or chopping functions where the mirror

0-8493-0133-5/97/$0.00+$.50

surface is not distorted by the motion. In semiactive mirrors, the actuators may either be in addition to or replace the mounts. They are used for final and/or periodic figure adjustment to a relatively flimsy mirror.

A successful mirror design meets the performance criteria for the specified environment and lifetime for a reasonable cost. This requires trade-offs among materials, fabrication methods, structural considerations such as mirror design and mounting scheme, and, of course, cost. The problem in achieving this is the multitude of constraints that impact the design. Typical among these constraints are

- Environmental, such as operating temperature range and incident thermal/energy fluxes
- Mechanical, such as applied forces and dynamic conditions
- Weight and deflection (figure) limits
- Schedule and cost

The following sections discuss dimensional stability issues for these materials and mirror material characteristics, primarily metal and silicon carbide (SiC), blank fabrication methods, and the kinds of lightweighting achievable with each of them.

4.2 Dimensional Stability

Dimensional stability of a component is actually the degree to which instabilities are controlled. Therefore, any discussion of dimensional stability is really a discussion of instabilities, and dimensional instability is simply the dimensional change that occurs in response to internal or external influences. All materials are dimensionally unstable to some degree. In preparing to design and fabricate dimensionally stable mirrors, it is important to realize that this implies controlling the sources of dimensional instabilities to a level such that any dimensional changes that occur are kept within specified tolerances. To be able to accomplish this requires an understanding of the sources of these instabilities.

The key to stability is knowing the performance requirements. Stable materials can then be chosen from which mirrors can be fabricated utilizing the methods that minimize introduction of dimensional instabilities. The challenge is to control the dimensional change, a distortion or strain in the mirror, to levels that will not compromise the performance requirements. If required stability is on the order of machining tolerances, strain of approximately 10^{-3}, there is no serious problem. If dimensional tolerances must be maintained to parts per million, e.g., microns/meter, then care and consideration must be given to materials selection and processing steps. This is the realm of precision instruments and optics. If nano-tolerances are required, e.g., nanometers/meter, there is little help available, although this has recently become the subject of increasing study. For nanostability, the principles of controlling the sources of dimensional instability must be diligently applied — and then hope for the best.

Once a tolerance has been specified to bound the amount of allowable instability, the next step is determining the potential sources of instability and controlling them to meet the specification. The balance of this section contains examples of common types of instabilities and their sources, and gives some suggestions as to how they can be controlled. While this section is an overview, the following references are recommend for further study:

- Marschall and Maringer,[1] an excellent book on the subject, although it is unfortunately now out of print
- Paquin[2] and Paquin and Vukobratovich,[3] the two volumes of SPIE proceedings dealing specifically with dimensional stability
- Paquin,[4] the paper on which this section is based

Types of Instability

Instabilities can be categorized as:

- Temporal instability
- Thermal/mechanical hysteresis
- Thermal instability
- Other instabilities

Each of these factors can have magnitudes ranging from nanostrain to very large numbers and is described in the following section.

Temporal

Temporal instability is the change that takes place in a component as a function of time in a fixed environment. It is a permanent change. For example, two sets of nominally similar 1-in.-gage blocks were tested at NBS over a period of roughly 30 years.[5] One set exhibited a positive and relatively constant rate of change of dimensions of as much as 10^{-6} m/m/year. That is a very small amount and yet it is totally unacceptable for the application. The other set typically changed only 25 nm in 22 years. This kind of dimensional instability is generally associated with relaxation of residual stress.

Hysteresis

Thermal/mechanical hysteresis is the change measured in a fixed environment after exposure to a variable environment, i.e., measured in a laboratory environment before and after exposure to changes in temperature and/or mechanical loading. It, too, is a permanent dimensional change. A common example is the dimensional change that takes place in fiber-reinforced composites when subjected to thermal cycling over a wide temperature range. The behavior typically shows a substantial change in length of up to 1% on the first cycle, but the amount of change decreases with each succeeding cycle, approaching an asymptote. This kind of behavior is discussed later in this section for other materials. For composites, the cause for the dimensional changes is usually internal microcracking of the fibers, while in single-phase materials it is usually some other form of internal stress relief. Similar behavior has been observed with mechanical cycling and vibration.

Thermal

Thermal instability is the dimensional change measured in one fixed environment after a change from another fixed environment, independent of the environmental path. This dimensional change is reversible upon returning to the original conditions. Figure 4.1 shows evidence of just such a change. This beryllium (Be) mirror was made from an experimental billet produced in the late 1960s which had a substantial amount of thermal expansion inhomogeneity.[6] It was interferometrically tested many times over a period of almost 10 years and exhibited the same distortion shown in the figure when heated, and always returned to the same optical figure at room temperature, within the 0.02 wave accuracy of the instrument. This behavior has been virtually eliminated in modern Be materials.

Other Instabilities

The principle "other" type of instability is the change measured in a fixed environment after being exposed to a variable environment where the change is dependent on the environmental path between the fixed environment measurements. This type of distortion can be permanent or reversible. For example, in Figure 4.2, the length of Zerodur® on cooling from 300 to 20°C depends on the cooling rate.[7] This is typical behavior for glasses containing MgO. But note that the curves are parallel below 150°C, indicating that the temperature range of sensitivity is between 150 and 300°C. This behavior has been eliminated in a new version of this material called Zerodur® M. This type

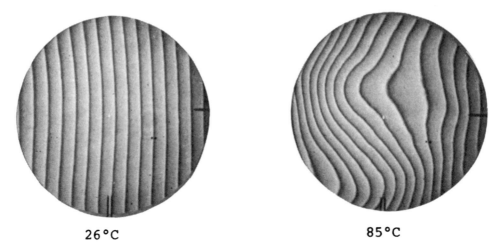

26°C   85°C

FIGURE 4.1   Optical interferograms of an electroless nickel-coated experimental beryllium alloy mirror (circa 1968) showing a reversible thermal instability of approximately two waves. *Source:* Paquin, R.A. 1981. Workshop on Optical Fab. and Test, Technical Digest. Optic Society of America, TB-1.

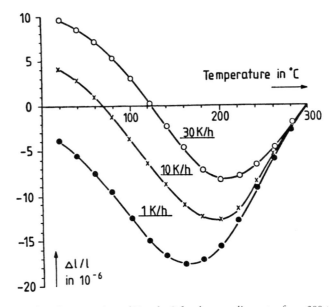

FIGURE 4.2   Thermal length contraction of Zerodur® for three cooling rates from 300 to 20°C illustrating a hysteresis type of dimensional instability. *Source:* Lindig, O. and Pannhorst, W. 1985. *Appl. Optics*, 24, 3330.

of behavior is rarely observed in metals. These are the major types of dimensional instability that can be encountered in optics and precision instruments. Many of the other commonly observed instabilities can be placed into one or more of these four categories.

Sources of Dimensional Change

The sources of dimensional changes such as those described above can be attributed to one or more of the following factors:

- Externally applied stress
- Changes in internal stresses

- Microstructural changes
- Inhomogeneity/anisotropy of properties

External Stress

When an external stress is applied to a component, if it behaves according to Hooke's law, it should deform elastically no matter how long the stress is applied, and return exactly to its original shape when the stress is removed. But this being an imperfect world, and most materials not being perfect, there are other responses to externally applied stress. If a load is applied suddenly, held for a length of time and then released, the elastic response has exactly the same square wave shape as the applied load. An anelastic strain shows a time-dependent elastic response with respect to the applied load. For this type of behavior there is no strain when the load is first applied, but it increases toward an asymptote with time; when the load is removed, the strain asymptotically returns to zero. Anelastic behavior is rarely observed in metals and ceramics, has been observed in some glass ceramics at low temperature, but is more commonly observed in polymers. Plastic strain is permanent and does not decrease as the load is removed. The most common behavior for metals is a combination of elastic and plastic response to stress. Time-dependent plastic strain is called creep. Most of the time, many materials exhibit a combination of these elastic responses to externally applied loads.

There are a number of material properties that are important to dimensional stability, most of which are covered in Chapter 3. Among these are thermal properties such as the coefficient of thermal expansion (CTE) and thermal conductivity, and mechanical properties: elastic modulus (Young's modulus), a measure of stiffness, and the slope of a stress vs. strain curve; Poisson's ratio, the relationship between tensile (or compressive) and shear strain; yield strength (at 0.2% offset), the stress to cause 2×10^{-3} permanent or plastic strain; ultimate or fracture strength; microyield strength (MYS), the stress to cause 1×10^{-6} plastic strain (one microstrain); and microcreep strength, which has no acceptable definition other than that it is less than the MYS and is a constant stress that produces microstrain after some period of time.

Microyield behavior cannot be directly inferred from the macromechanical properties of either yield strength or modulus. For example, when the behaviors of I-400 Be and 2024-T4 aluminum, metals with approximately the same yield strength, are compared, Be exhibits a MYS of approximately 50 MPa, but with increased stress yields little more. However, the Al alloy resists yielding for a high MYS of 250 MPa, but then continues yielding readily. Recent analyses have shown, however, that for any given family of alloys of the same base material, MYS is proportional to yield strength.

Microyield strength is strongly dependent on the prior history of the material. If it has been annealed, the microyield will be lower than in almost any other condition. Conversely, if there has been prior straining, either through intentional or inadvertent applications, the MYS will be raised. While prestraining produces a stronger material, it also leaves a level of residual stress that may be detrimental. Residual stress is discussed further in the next section.

Since high MYS is a desirable property, and since many materials have relatively low MYS, it is important to know that there are methods for increasing it. Prestrain, as mentioned above, is one method, but it has its disadvantages. Many aluminum alloys, after rolling to plate form, are stretched a few percent to both straighten and level the stress through the thickness of the plate, which also increases MYS; but this process also seems to lower microcreep strength. Since the process of microyielding occurs, at least in the early stages, by movement of dislocations, anything that pins or prevents dislocation movement will increase MYS. By reducing the grain size of a material, dislocations are more readily pinned, as they are when particle or fiber reinforcement is added to a single phase material. Multiphase materials almost always have higher MYS than similar single phase alloys. Thermal treatments that precipitate a second phase or produce a metastable phase tend to increase strength, and alloying a pure material usually produces dislocations and lattice strains that likewise increase MYS.

Changes in Internal Stress

While external stress is applied and removed from a component and is readily observed and measured, internal stress is not obvious. A component can be free of external attachments, even floating in a zero-g environment, and have internal stresses. They are in equilibrium and consist of balanced tensile and compressive stresses.

There are two types of internal stress called short range and long range. The spatial extent of these is, as the name implies, microscopic and macroscopic, respectively. The long-range internal stress is better known as residual stress. To illustrate both types, consider a component machined from an inherently anisotropic metal like Be. If the component was annealed prior to machining, the bulk of the material will consist of Be grains that vary in crystallographic orientation, and therefore in CTE at the grain boundaries. This produces short-range internal stress at the grain boundaries as further described below. At the surface of the part, the machining will have plastically deformed a surface layer within which the residual or long-range stress will be approximately equal to the yield strength, approximately 275 MPa for Be. Below the deformed layer there would be a partially deformed transition layer where the residual stress level would decrease rapidly from 275 MPa (probably tensile), through zero, to a low level compressive stress.

Short-range internal stress can result from unequal amounts of distortion between neighboring crystals in plastically deformed material. It can also arise from inhomogeneous CTE: in a two-phase material, between adjacent crystals with anisotropic CTE as described above for Be or in a matrix with a dispersed phase or reinforcing particles, whiskers, or fibers. Table 4.1 lists the theoretical maximum values of thermally induced microstrain due to CTE mismatch between adjacent grains of a few noncubic materials.[8] In practice, the average values are approximately one third of these calculated maxima. It can be seen that for Be, one third of the 437-KPa/°C value over a 100°C temperature change results in a short-range stress of over 14 MPa, a value that exceeds the microyield strength of some Be alloys.

TABLE 4.1   Theoretical Maximum Values of Short-Range Internal Stresses Due to Thermal Expansion Anisotropy

Material	Lattice	Kpa/°C
Zinc	Hexagonal	1212.
Calcite	Rhombohedral	1130.
Cadmium	Hexagonal	626.
Beryllium	Hexagonal	437.
Quartz	Rhombohedral	295.
Indium	Tetragonal	223.
Magnesium	Hexagonal	19.
Graphite	Hexagonal	6.5

Source: Adapted from Likhachev, V. A. 1961. *Sov. Phys. Solid State*, 3, 1330.

Long-range internal stress, residual stress, is usually the result of processing operations such as forming, heat treating, welding, machining, or plating. As you would expect, the level of the stress is dependent on the severity of the operation, as shown in the following examples. Figure 4.3 shows how the temperature of quench water affects the residual stress in Al alloys. The yield strength also drops when the water quench is less severe. Polymer quenchants are available that can provide the low residual stress of a boiling water quench with the strength of the cold water quench.

The introduction of residual stress, or any change in the balance of the stress, will cause changes in dimensions of the component. This means that removal of a stressed layer will cause dimensional changes as demonstrated in Figure 4.4. When two specimens with surface residual stress are acid etched, they both shrink, the one with the higher stress (deeper cut) shrinking more. This principle is used in the manufacture of dimensionally stable components where after heavy material removal

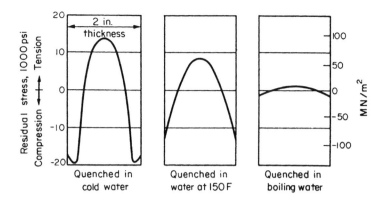

FIGURE 4.3 Residual stresses in specimens of 7075-T6 aluminum plate quenched in water at different temperatures. *Source:* Barker, R.S. and Sutton, J.C. 1967. *Aluminum, Vol. III: Fabrication and Finishing,* Van Horn, K.R., ed., chap. 10. American Society for Metals, Metals Park, OH.

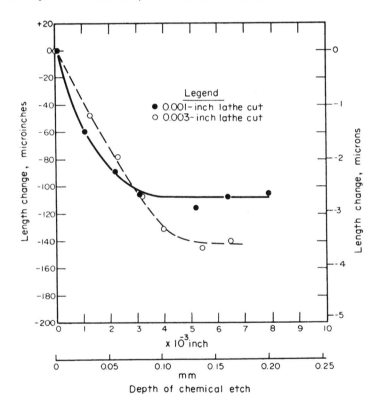

FIGURE 4.4 Dimensional changes in 3.0-in.-long specimens of Ni-Span-C on chemical removal of residual stresses due to machining. *Source:* Marschall, C.W. and Maringer, R.E. 1971. *J. Mater.,* 6, 374.

operations the surfaces are acid etched to restore unstressed dimensions to the part. Residual stress can also decrease spontaneously with time with a related change in dimensions. This effect is called stress relaxation and the decrease in stress is proportional to the stress level as shown in Equation (1), where s is stress, t is time, and τ is the relaxation time.

$$\left(-ds/dt\right)\tau = s \tag{1}$$

But stress also decreases exponentially with time as shown by Equation 2.

$$s = s_o \, e^{\upsilon\tau} \tag{2}$$

Note that when time, t, is equal to the relaxation time, τ, the ratio of stress to original stress, s/s_o, is equal to $1/e = 0.37$. This behavior is shown in Figure 4.5, where a Be mirror was fabricated with no treatment for stress relief after annealing the rough blank. Note the exponential shape to the curve for optical figure change.

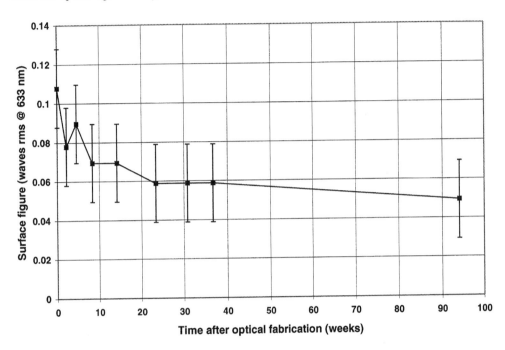

FIGURE 4.5 Temporal stability of a commercial 13-in.-diameter beryllium mirror fabricated from vacuum hot pressed block. *Source:* Paquin, R.A. 1992. *Optomechanical Design*, Yoder, P.R., ed., p. 160. CR43, SPIE Optical Engineering Press, Bellingham, WA.

Stress relaxation is also a thermally sensitive process, behaving according to the Arrhenius relationship of Equation 3, where E is the activation energy, k is Boltzmann's constant, and T is absolute temperature.

$$1/\tau \propto e^{-E/kT} \tag{3}$$

This phenomenon can be used to reduce the stress level with thermal treatment, i.e., stress relief. However, the question is often raised whether an isothermal treatment for some reasonable time or thermal cycling between elevated and reduced temperatures is a more effective stress relief treatment. Much has been written on this subject as summarized in Chapter 6 of Marschall and Maringer,[1] but the best answer is, "it depends." It depends on the crystal structure and purity of the material; it depends on the prior thermomechanical history of the component; it depends on the temperature, time, and rate of change of temperature; and, of course, it depends on the level, type, and distribution of the internal stresses.

One example of what can happen is given in Figure 4.6, where both isothermal and thermal cycling treatments were given to Be specimens previously stressed to 77 MPa.[9] In this case, the low annealing temperatures of 100 and 190°C do very little but relieve peak stresses. The best treatment is a 600°C stress relief treatment, but this is higher than most designers would want to subject a semifinished optic to. A temperature of 400°C still only removes 40% of the stress, but note that

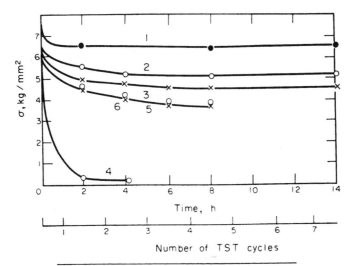

FIGURE 4.6 Comparison of the effects of thermal cycling and isothermal exposure on stress relaxation of pure beryllium. *Source:* Lokshin, I.Kh. 1970. *Metal Sci. Heat Treat. (U.S.S.R.)*, 426.

thermal cycling from 400 to either –70 or –196°C provides a 55% reduction. The cycling is more effective than the isothermal treatment to the same upper temperature. This cycling effect may only hold for noncubic materials with reasonably high expansion anisotropy as listed above in Table 4.1, or for materials with more than one phase such as composite materials. There is conflicting evidence in the literature on the effect of thermal cycling on stress relief of homogeneous cubic or amorphous materials.

For those materials where thermal cycling is more effective than isothermal exposure at the same elevated temperature for reduction of residual stress, the proposed mechanism is as follows. These materials develop short-range internal stress when the temperature is changed due to the expansion mismatch between adjacent grains and/or phases. When this stress is added to the long-range, or residual, stress, the yield, or microyield strength, is exceeded locally and plastic strain results. When the temperature is returned to room temperature, the plastic strain remains, but the level of the residual stress is reduced. Holding at the elevated temperature provides no further benefit unless it is close to either the creep or annealing temperature. When the temperature is then reduced below room temperature, the sign of the short-range stress is reversed, exercising the material further and providing more stress relief. In a similar manner, vibration, or mechanical cycling, can provide stress relief, although, as for thermal cycling, there are resulting dimensional changes.

For reducing stress levels in critical components we then have a number of options:

1. Thermal treatments such as isothermal exposure or thermal cycling
2. Mechanical treatments such as vibration or mechanical working
3. Removal of surface material by chemical etching, controlled grinding and/or polishing or other stress-free methods
4. Time

Microstructural Changes

Microstructural changes in materials can result in both induced dimensional change and internal stress. The type of response depends on the material type and the kind of microstructural change. Changes can take place in mirror materials: phase transformations, recrystallization, and grain growth in metals and ceramics; and devitrification, phase transformations, recrystallization, and grain growth in coatings. This cause of dimensional instability is quite common, but cannot be covered adequately here. An example illustrates the principle.

The dimensional change that takes place in heat-treatable aluminum alloys during precipitation heat treatment, also called precipitation hardening, is illustrated in Figure 4.7. This shows that a component aged from the solution-treated and quenched condition to obtain maximum mechanical properties, normally from 4 to 8 hr, will undergo a dimensional change due to the precipitation of the second phase. The change is small for 6061, a significant shrinkage for 7075, and a significant expansion for 2014. But notice that additional hours of aging, often performed for stress relief, induce additional significant dimensional change, particularly for the 2014 alloy. The 6061 alloy, most often used for precision optical structures and mirrors, changes the least, verifying its applicability for these applications.

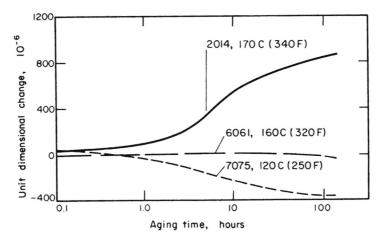

FIGURE 4.7 Dimensional change as a function of time at the precipitation heat treating temperature employed to produce the T6 temper for three aluminum alloys. *Source:* Hunsicker, H.Y. 1967. *Aluminum, Vol. I: Properties, Physical Metallurgy and Phase Diagrams*, Van Horn, K.R., ed., chap. 5. American Society for Metals, Metals Park, OH.

Electroless nickel coatings are used extensively, both for polishable coatings and for corrosion protection of Al, Be, and iron alloy components. The coatings are nickel–phosphorous alloys that may be amorphous and are thermodynamically unstable as deposited. Coatings can have significant residual stress[10] depending on phosphorous content and substrate CTE, and annealing changes the stress level as shown in Figure 4.8. During thermal treatment (annealing) immediately after plating, hydrogen is driven off, adhesion improves, hardness increases, and low phosphorous coatings can devitrify (change from amorphous to polycrystalline). There is shrinkage and a decrease in CTE that takes place during thermal treatment, the magnitude of which depends on the annealing conditions.[11] For higher temperatures and longer times, nickel phosphide (NiP), which has a smaller specific volume than pure nickel, forms as a precipitate in the coating. The annealing temperature used in the referenced studies is 190°C, with 4 hr at temperature for Be[11] and 1 hr for the other materials. Stress goes from tensile to compressive with increasing phosphorous content and increasing substrate expansion coefficient. Annealing changes the stress toward compression for high expansion materials and toward tension for low expansion materials. What this all means is that for a particular substrate, to obtain zero stress at room temperature after

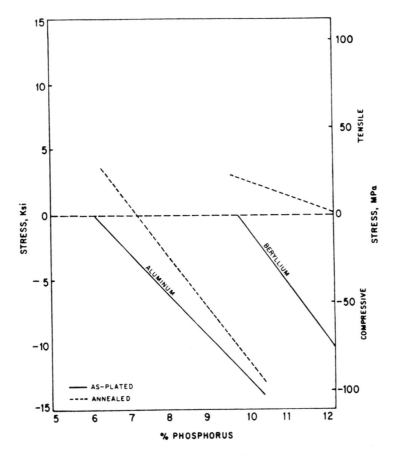

FIGURE 4.8 Stress of electroless nickel on aluminum and beryllium as a function of phosphorous content. *Source:* Parker, K. 1987. *Testing of Metallic and Inorganic Coatings*, ASTM STP 947, Harding, W.B. and DiBari, G.A., eds., p. 111. American Society for Testing and Materials, Philadelphia.

annealing, the right phosphorous content must be chosen, e.g., 5% on Al and >11.5% for Be. For use at lower or higher temperatures, different phosphorous contents should be chosen to minimize the stress.

Inhomogeneity/Anisotropy of Properties

Most materials, as fabricated, are neither completely isotropic nor homogeneous; they are to some level anisotropic, i.e., having some preferred directionality of properties, and/or inhomogeneous, i.e., having a spatial variation in properties. Anisotropy of properties exists in pure single crystals of materials. Inhomogeneity of properties occurs in bulk and is a function of raw material fabrication processes.

Cubic materials generally have anisotropy in their elastic properties. For example, the Young's modulus of elasticity of pure iron varies with crystallographic direction from 132 to 282 GPa. Similar variations in shear modulus and Poisson's ratio are also present. Comparable variations in these properties occur in other cubic materials such as Cu, Ni, Si, beta SiC, etc. When these materials are used in polycrystalline form, the variations average out and are not noticed. In components that are fabricated from single crystals, or applications that have crystallographic texture such as plated or chemically vapor-deposited (CVD) materials, there can be substantial elastic property anisotropy and this should be included in any detailed deformation modeling of such components. Thermal properties such as CTE are isotropic for cubic materials. Thermal conductivity, which is also isotropic, is affected by grain size and grain boundaries so that for plated or CVD materials the anisotropy is present. For example, CVD SiC has a deposition texture and both elastic modulus and thermal conductivity have approximately 15% anisotropy.

In a similar manner to the cubic materials, there are variations in elastic properties in noncubic materials, i.e., hexagonal, rhombohedral, tetragonal, orthorhombic, etc. However, the thermal properties of the noncubic materials are anisotropic. For example, the CTE of Be is 38% higher in the basal plane than it is in the axial direction of a Be crystal or grain. This anisotropy leads to the microstructural strains listed in Table 4.1. Some materials such as graphite and quartz have negative CTE in some directions and positive in others. The effects of thermal properties anisotropy, for the various crystalline materials that exhibit it, can be minimized with a fine-grain, randomly oriented microstructure.

Inhomogeneity can be attributed to spatial variations in chemistry, grain size, and/or grain orientation, and many other factors. In general, CTE inhomogeneity in metals and ceramics is due to compositional or microstructural variations where the latter can be due to crystal orientation differences and/or the presence of other phases. For composites, CTE inhomogeneity is a given due to the presence of multiple phases. The inhomogeneity can be caused by variations in orientation of the reinforcement, or from variations in the concentration of the reinforcement. Care should be exercised in the selection of multiphase materials that may be used over a wide temperature range to ensure dimensional stability.

Components having CTE inhomogeneity can behave in the manner shown in Figure 4.1. In this particular case, the cause is a combination of Be powder with basal plane cleavage, a poor distribution of grain sizes, and inappropriate consolidation parameters. Current Be fabrication processes have virtually eliminated these types of inhomogeneities through the use of impact ground or spherical powder with well-controlled chemistries combined with hot isostatic pressing of the powder.[12-14]

Low Expansion Materials

Low expansion materials have been used for many years for precision instrumentation to minimize thermal dimensional changes in service. For optics, fused silica and fused quartz have replaced borosilicate glasses in many applications, and ULE and the low expansion glass ceramics have been used extensively for more critical applications. Invar alloys have been used in precision metering applications for many years, especially in combination with fused silica optics because of the excellent CTE match at room temperature. More recently composite materials, particularly graphite-epoxy composites, have been used for precision structures since they can be tailored for near-zero expansion over a fairly wide temperature range. However, each of these material types has distinct advantages and disadvantages.

Composite materials, by their very nature, have the disadvantage of having built-in residual stress at the phase boundaries which usually leads to temporal dimensional instability. Graphite-epoxy composites can be designed and fabricated with near-zero CTE, low density, and high modulus. The properties can be tailored to the application, but are usually anisotropic due to the fabrication methods that require a preferred fiber direction or, at best, a pseudoisotropic lay-up. The biggest drawback of graphite-epoxy is its moisture sensitivity which affects dimensions, CTE, and mechanical properties.

Further information on low expansion material dimensional stability is available in Jacobs,[15,16] while further information on composites can be found in the papers of Session 2 of Paquin.[2]

Promoting Dimensional Stability

There are many potential pitfalls in the design and fabrication of dimensionally stable components. In order to avoid these pitfalls, there is a sequence of actions that can be taken that should lead to stable components.

The first step is to establish a budget for the allowable dimensional change for each component in the system and allocate a tolerance to each element. Structural components probably will not have as tight a tolerance as optical components and all components of each type will not necessarily have the same requirements. Then consider the sources of dimensional change as they relate to the components to be fabricated. For example, if the system is to operate at reduced or elevated temperatures, then both thermal instability and thermal cycling instability are potential types of instability, and the sources for such behavior are changes in internal stress, inhomogeneity and/or anisotropy, and microstructural changes. Consider the fabrication options and how they relate to the dimensional instability sources and the component performance requirements. With all these factors in hand, select the candidate materials and reevaluate the sources and fabrication methods for each material with respect to meeting the budgeted dimensional tolerances in the specified use environment.

In order to make the final materials and fabrication method selection, you need to demonstrate that external stresses will not cause excessive strain in the component; that internal stresses in the component due to fabrication methods or inhomogeneities and/or anisotropies will not change excessively; and that any microstructural changes will not cause excessive strains. While this method sounds complicated, once you understand the basic sources of potential instability and the magnitude of the possible changes for each of the common candidate materials and their respective fabrication methods, the selection process becomes almost intuitive. The difficult part is when you must produce a component that operates in an environment for which the material properties information is severely limited, or when designing components to nano tolerances.

4.3 Metal Mirror Materials

Mirrors can be fabricated from any number of materials. Traditionally, glasses such as fused silica and borosilicate glass have been used, with an emphasis for high performance applications on near-zero expansion materials such as ULE™ and Zerodur®. With modern fabrication techniques such as those discussed here and in Chapter 10, metals and ceramics are being used more frequently. When designing a mirror, there are many metals from which to choose. Of course the first step is to consider the properties of individual materials with regard to the requirements of the actual application. But beyond that there are a number of other considerations including fabricability, corrosion resistance, availability, cost, and probably many others. Paquin[17] has addressed some of these trade-off issues.

Aluminum and Aluminum Matrix Composites

Aluminum (Al) is a low cost, easily fabricated mirror material. The main drawbacks are high thermal expansion and low elastic modulus. This metal is available in many alloys and can be fabricated in many ways, including casting and the wrought processes that produce plate, bar, and other forms, and can be machined with many techniques as described in Chapter 10.

The common wrought and cast alloys are listed in Table 4.2 with their compositions and normal heat treatments. In general, there are two types of Al alloys: those that can be heat treated to improve strength and those that cannot. The common heat-treatable alloys have two or more phases and are usually rough machined, heat treated at a temperature near the melting point (solution treated), partially aged (a lower temperature heat treatment), finish machined, and further aged for near-maximum strength and dimensional stability. The nonheat-treatable alloys can be annealed and still retain some strength. These alloys usually do not have a second phase and are therefore better for diamond turning. They are also preferred for large components with disparate section thicknesses that do not respond in a uniform manner or have a substantial risk of distortion in heat treatment. Wrought alloys are usually machined to shape when the number of identical components is small. Very large components are sometimes forged from billets for

TABLE 4.2   Characteristics of Common Aluminum Alloys for Mirrors

Alloy No.	Form	Hardenable	Remarks
1100	Wrought	No	Relatively pure; low strength; can be diamond turned
2014/2024	Wrought	Yes	High strength and ductility; multiphase; must be plated
5086/5456	Wrought	No	Moderate strength when annealed; weldable; available in large plate
6061	Wrought	Yes	Low alloy, all purpose; reasonably high strength; weldable; can be diamond turned and/or plated; all forms readily available
7075	Wrought	Yes	Highest strength; usually plated; strength more temperature sensitive than others
B201	Cast	Yes	Sand or permanent mold cast; high strength; can be diamond turned
A356/357	Cast	Yes	Sand or permanent mold cast; moderate strength; most common; extensive processing for dimensional stability
713/Tenzalloy	Cast	Yes	Sand or permanent mold cast; moderate strength
771/Precedent 71A	Cast	Yes	Sand cast; moderate strength; very stable; expensive casting procedures required; easiest to machine

TABLE 4.3   Characteristics of Aluminum Matrix Composites

Property	Instrument Grade	Optical Grade	Structural Grade
Matrix alloy	6061-T6	2124-T6	2124-T6
Volume % SiC	40%	30%	20%
SiC form	Particulate	Particulate	Whisker
CTE (10^{-6}/K)	10.7	12.4	14.8
Thermal conductivity (W/m K)	127	123	n/a
Young's modulus (Mpa)	145	117	127
Density (g/cm³)	2.91	2.91	2.86

Source: Mohn, W.R. and Vukobratovich, D. 1988. *Opt. Eng.*, 27, 90.

better grain structure and stability. Castings are most often used when there are many identical parts. Aluminum foam has been effectively used as core material in lightweight mirrors.[18,19]

Aluminum mirrors are not normally finished bare by polishing, but are diamond turned and/or nickel plated and polished or diamond turned.[20] More complete information on Al alloys, their properties, and fabrication techniques can be found in the *ASM Handbook 2*[21] and optics-specific fabrication information can be found in Chapter 10.

Al matrix composites, usually SiC loaded, can be effective where a somewhat higher elastic modulus and/or lower CTE than can be obtained with the Al alloys are required. Typically 15 to 40% SiC, either whisker or particulate, is employed in a 2024 or 6061 matrix. Typical formulations and some properties are shown in Table 4.3. For optics, 30 vol% particulate in a 2124-T6 matrix has been most successful.[22]

Beryllium

Of the metals, beryllium (Be) has the most attractive properties for lightweight mirrors, but has high thermal expansion and relatively low microyield strength. The very high reflectance of bare Be[2] allows use without optical coatings in infrared systems. However, anomalous scatter observed on Be and some other bare polished metals may require a Be or other coating to meet system requirements.[23]

Beryllium is particularly suited for cryogenic systems since the thermal expansion drops rapidly below room temperature and approaches zero below approximately 80 K, while at the same time thermal conductivity increases to a maximum below 150 K. A unique property is its low absorption of X-rays, making it useful in nuclear systems and for X-ray windows. This property also means that Be mirrors can withstand high levels of nuclear energy without appreciable damage.

Beryllium has anisotropic properties, as do most crystalline materials. The thermal expansion of Be is approximately 30% higher in the basal plane of the hexagonal crystal than in the axial direction, normal to the basal plane. For this reason, to end up with isotropic bulk material, it is essential to have fine grains of average size ≤15 μm. Section 4.4 of this chapter contains additional information on this subject.

Beryllium is available in several grades suitable for mirrors as shown in Table 4.4. All Be mirrors should be made from hot isostatically pressed (HIPed) material for virtual elimination of both inhomogeneity and anisotropy. Of the Be grades, O-50 and I-70-B are most suitable for polishing bare because of their low impurity content. If nickel plating is to be used, then any of the structural grades are appropriate, with I-250 having the highest microyield strength. I-250, because of very fine powder particle size, is not available HIPed in very large sizes. I-400, due to the high BeO content and very fine powder particle size, is not normally HIPed and size is limited to approximately 40 cm (16 in.). Where ductility is critical, the lower oxide grades provide a reasonable amount, while the highest oxide material, I-400, has virtually none.

TABLE 4.4 Beryllium Grades and Their Properties

Property	O-50	I-70-H	I-220-H	I-250	S-200-FH
Max. beryllium oxide content (%)	0.5	0.7	2.2	2.5	1.5
Grain size (μm)	15	10	8	2.5	10
2% offset yield strength (Mpa)	172	207	345	544	296
Microyield strength (Mpa)	10	21	41	97	34
Elongation (%)	3.0	3.0	2.0	3.0	3.0

Source: Brush Wellman, Inc. Elmore, Ohio.

Beryllium has earned an unjustified reputation as being an extremely hazardous material. Studies that conclude that Be is a potential human carcinogen have been shown to be flawed, but the EPA continues to claim that Be is a "probable" human carcinogen. Recent medical data[24] have shown that only about 3% of workers exposed to Be appear to be susceptible to chronic Be disease. The disease results from inhalation of Be particles, is immunologically initiated, and has a genetic basis similar to an allergy. Simple exhaust systems at the source of the particle generation with absolute filters are very effective, and machine shops using these devices have had no incidences of Be disease. OSHA permissable exposure limits are

- 2 μg/m³ average for an 8-hr day (an amount not visible to the eye)
- 5 μg/m³ ceiling
- 25 μg/m³ peak for 30 min maximum duration in an 8-hr day

In loose abrasive grinding/polishing, there is no hazard as long as the slurry is captured and not allowed to dry. The contaminated waste is not considered hazardous waste under Federal Law. Disposal of the contaminated waste by a licensed disposal firm is usually not a problem. By contrast, inhalation of similar amounts of almost any fine particulate, but particularly SiO_2 and SiC, can also lead to a chronic respiratory disease commonly called silicosis.

Invars

Invars are alloys of primarily iron and nickel that exibit a minimum CTE at approximately 36% Ni. The invar alloys can have near-zero CTE with special processing, but only over a limited temperature range. While a few lightweight mirrors have been made from brazed invar sheet, the material is usually used for components and attachments for ULE™, Zerodur®, fused silica, silicon, and SiC mirrors.

Invar 36, an iron/36% Ni alloy with small amounts of manganese and silicon, is the most common of the invars. Invars with 39 and 42% nickel are less common, but have CTEs closely matched to that of silicon and silicon carbide. Super invar, an alloy with cobalt substituted for some of the nickel, has the lowest CTE but can have an irreversible phase transition at low temperatures that substantially changes properties if the composition departs much from the nominal.

Components can be readily fabricated using conventional metalworking processes. The materials have high fracture toughness but relatively low strength compared to other iron alloys, and almost all invar specimens measured have temporal dimensional instabilities. In addition, invar alloys are ferromagnetic, have high density and a relatively low elastic modulus, and one of the lowest specific stiffnesses of any material used for precision applications. All of the invars should be given a three-step heat treatment for the optimum combination of CTE, strength, and dimensional stability:

- Anneal at 830°C for 30 min/in. of thickness in inert atmosphere and quench in polyalkaline glycol/water solution.
- Reheat to 570°C for 1 hr and air cool.
- Heat at least 48 hr at 95°C and air cool.

Other Metals

Silicon (Si), copper (Cu), and molybdenum (Mo), as well as invar, are used for cooled mirrors because of the high thermal conductivity and relatively low CTE. Components are usually brazed together to form heat exchanger mirrors for high energy systems. While most of these mirrors are passive, and massive for the size of the aperture, fully active cooled mirrors of invar, Si, and Mo are used for wavefront correction in high energy optical systems. Titanium (Ti) is not a common mirror material, but has been used in applications where strength/weight is the important parameter. While Ti is polishable, it is usually electroless Ni plated. Electroforming is used for production of reflectors for lighting systems where Ni is the material of choice.

The most common form of Cu used is wrought oxygen-free copper, but where higher strength is needed without sacrificing thermal conductivity, chromium copper is preferred. The preferred form of Mo is low carbon, vacuum arc remelted material. A Mo alloy with small additions of Ti and zirconium called TZM has been used because of its higher recrystallization temperature, but it usually contains carbide inclusions that make optical fabrication difficult. Stainless steels are usually used for small mirrors in harsh environments where the exceptional corrosion resistance of the 300 series austenitic steels are preferred. In some cases the lower carbon martensitic stainless steels, such as 416, are used for their lower CTE and slightly higher thermal conductivity.

Silicon Carbide

Silicon carbide is a ceramic material that is available in many types and grades. It has high stiffness, strength, hardness, and thermal conductivity at low to high temperatures. It has moderate density and CTE, but low fracture toughness. Compared to other mirror materials it is stiffer and less fragile than glass, and has lower thermal expansion than Be. Silicon carbide in any of its forms is second only to Be in specific stiffness, but has much lower thermal distortion than Be. Pure forms have high reflectivity in the ultraviolet, but reflective coatings are required for most other wavelengths. The high Young's modulus and refractory properties of all forms mean that SiC can withstand very high heat loads with little or no damage and with very little distortion compared to most other mirror materials.[25]

There are two common types of crystal structure for pure SiC: hexagonal alpha (α) and cubic beta (β). Both structures have elastic anisotropies and α has CTE anisotropy.[26] Depending on crystal orientation, Young's modulus varies from 280 to 510 GPa for α and 340 to 510 GPa for β. The CTE for all directions of α and the basal plane of β are the same, while for all other directions in β CTE is up to 8% smaller. This means that if there is anisotropy in a formed, single-phase polycrystalline component of each material, when temperature is changed the thermal distortion will be up to 8% greater in β, but the internal stress at the grain boundaries will be as much as 40% higher in α.

There are four major types of SiC based on the fabrication method, listed in order of current volume of mirrors produced:

- Reaction bonded/sintered
- Chemically vapor deposited (CVD)
- Hot pressed
- Hot isostatically pressed (HIP)

The major characteristics of these types of SiC are given in Table 4.5 and fabrication methods are described in the next section. All except the reaction bonded are essentially pure SiC. Hot pressed and HIPed can be either all α, or β with some α that forms during processing. CVD is all β except that when deposited at lower temperatures some α is also formed. CVD SiC has preferred orientation that imparts higher thermal conductivity and lower Young's modulus perpendicular to the surface by approximately 15% compared to in-plane values. Hot pressed and HIPed SiC are normally very fine grained and essentially isotropic in all properties.

BLE 4.5 Characteristics of Major Silicon Carbide Types

Type	Structure/ Composition	Density	Fabrication Process	Properties	Remarks
t pressed	>98% alpha plus others	>98%	Powder pressed in heated dies	High E, ρ, K_{Ic}, MOR; lower k	Simple shapes only; size limited
t isostatic ressed	>98% alpha/beta plus others	>99%	Hot gas pressure on encapsulated preform	High E, ρ, K_{Ic}, MOR, lower k	Complex shapes possible; size limited
emically vapor eposited	100% beta	100%	Deposition on hot mandrel	High E, ρ, k, lower, K_{Ic}, MOR	Thin shell or plate forms; built-up shapes
action bonded	50-92% alpha plus silicon	100%	Cast, prefired, porous preform fired with silicon infiltration	Lower E, ρ, MOR, k; lowest K_{Ic}	Complex shapes readily formed; large sizes; properties are silicon content dependent

Reaction-bonded SiC can be formed by a number of related processes that determine the amount of SiC in the final component. The SiC content can be as low as 50% or as high as 92%, and consists of mostly α with some β formed during the final processing. The remaining material is Si, infiltrated into the SiC in final processing. Siliconized or reaction-sintered material is the simplest form of reaction-bonded SiC and has the highest Si content and, therefore, the lowest mechanical properties.

The highest quality surfaces can be obtained on CVD SiC because it is fully dense, fine grained, and single phase. Surface roughness less than 1 Å rms has been achieved on small mirrors.[27] HIP and hot-pressed SiC can be readily polished to low roughness, the level achieved depending on the grain size as well as porosity size and distribution. Reaction-bonded material can be finished to low roughness, the level depending on Si content and grain size.[28] Since the removal rates for Si and SiC are vastly different, the smaller the distance across Si between SiC grains that a grinding/polishing lap has to bridge, the better the finish. To reduce the difficulty in meeting figure and surface requirements, cladding of Si or SiC has been effectively used on reaction-bonded SiC.[28]

4.4 Fabrication Methods and Lightweighting

There are two basic methods for fabricating metal mirrors: machining from a solid or near-net-shape (NNS) forming with machining to finish. Machining causes surface damage and residual stress that must be removed for dimensional stability. This damage/stress condition also reduces strength and fatigue strength. Removal is normally through acid etching, annealing, and/or loose abrasive grinding/polishing. Some materials are often clad or plated for improved polishability.

Specific fabrication methods for each material or material type are presented in the following subsections.

Beryllium

Beryllium mirrors are typically fabricated from HIPed impact ground powders to produce homogeneous and isotropic blanks.[29] The vacuum hot-pressed Be used through the early 1970s was both anisotropic and inhomogeneous and often contained an unacceptable level of porosity. Electroless Ni was used in many applications to simplify the optical fabrication and obtain a better surface finish than could be obtained on the bare Be. Electroless Ni is still used in many applications where use temperature range is narrow enough so that the bimetallic distortion is small. For cryogenic (primarily infrared) applications, bare-polished Be is preferred. When all components of an on-axis telescope are HIPed from the same powder lot in the same HIP run, the telescope will be truly isothermal with only a magnification change as a function of temperature. Off-axis systems may require active alignment or passive compensation to remain in focus.

The majority of Be mirrors are now machined from solid HIPed billets. The HIP process consists of the following steps:

- Fabricate HIP container (can) of appropriate shape from low carbon steel sheet.
- Load Be powder into can with vibration assist for powder packing and weld lid onto can.
- Outgas the can at >670°C and seal off outgas tubes.
- HIP in autoclave at 103 MPa (15 Ksi) and 825 to 1000°C.
- Remove container and anneal at 790°C.

Near-net-shape mirrors, both open back[30] and closed back,[14] have been fabricated using a modification of this method where the container can be shaped and can contain precisely shaped and positioned formers of monel or copper.[31] Small mirrors have also been fabricated with replicated aspheric optical surfaces directly from the HIP process.[32]

For either machined or NNS mirrors, it is essential to provide adequate annealing and thermal cycling during the processing to ensure dimensional stability. A typical sequence for cryogenic mirrors, after HIP can (and formers) removal, consists of the following steps:

- Rough machine.
- Acid etch.
- Anneal at 790°C.
- Finish machine.
- Acid etch.
- Thermal cycle three to five times (limits determined by application, but at least −40°C to +100°C).
- Grind, etch, and thermal cycle.
- Figure and thermal cycle.
- Final polish and thermal cycle.

When electroless nickel or Be coatings are used, the coating should be deposited after the grind/etch/thermal cycle step. For thin Be coatings, deposition should be after the figure/thermal cycle step.

Other Metals

The metals such as Al, Cu, Mo, Ti, and stainless steels are fabricated by conventional methods as described in Chapter 10. Various plated layers can be used to enable diamond turning of the mirror face, sometimes after diamond turning of the bare substrate. Fluid-cooled mirrors are often brazed to assemble various elements of the heat exchanger to the mirror face. Anthony[33] contains an excellent overview of cooled mirror fabrication technology and Kittell and La Fiandra[34] describe a unique cooled deformable invar mirror. It is important to remember that appropriate annealing or other heat treatment, acid etching, and thermal cycling must also be applied to these materials to obtain dimensionally stabile mirrors. For precision components that must be quenched after heat treating, polyalkylene glycol solutions[35] are recommended to substantially reduce the distortion and residual stress while still maintaining optimum mechanical properties. This technique applies to all metals, but particularly Al alloys, invars, and steels.

Silicon

Silicon, being a semiconductor and a so-called semimetal, is fabricated by methods somewhat different than either metals or ceramics. It has primarily been used for the fabrication of fluid-cooled and/or deformable mirrors.[33,36] Mirrors have been fabricated from both single crystal and polycrystalline Si with great success. The components can be machined using small "conventional" tools or by ultrasonic machining. Brazing is the standard joining method, and both metal and glass have been used for this purpose. Acid etching of machined surfaces is essential since the fracture toughness is as low as glass in this brittle material. Optical finishing is accomplished using conventional metal polishing techniques, with alkaline slurries preferred.

Silicon Carbide

Section 4.3, "Silicon Carbide", of this chapter described four types of SiC: reaction bonded, hot pressed, HIPed, and CVDed. Each of these has distinctly different blank fabrication methods. All are optically finished with diamond tools and slurries unless Si cladding is used on the mirror surface.

Reaction Bonded

High tonnage quantities of reaction-bonded SiC are produced in many shapes, often complex, for industrial purposes. Of the four types, reaction bonded is by far the best choice for complex shapes. However, tooling costs can be substantial. The process for siliconized SiC, the simplest form of reaction-bonded SiC, consists of the following steps:

- Prepare a slurry of SiC powder.
- Form a porous shape (preform) by one of the methods described below.
- Fire to burn off non-SiC material and introduce Si at the high temperature that wicks into the resulting pores.
- Clean off the excess Si.
- Machine to final shape with diamond tools.

Forming can be accomplished by injection molding, dry pressing with a binder, or, the most appropriate method for mirrors, casting into precise molds. The casting is then dried and/or prefired to drive off volatiles to produce the porous preform. The component can then be siliconized or other options are available. If the casting is made oversize and prefired at a temperature high enough to provide some strength, then the preform can be machined to finish dimensions prior to final firing. Other options change the process from simple siliconizing to reaction bonding. For example, the original slurry can contain carbon that will react with Si in final firing to produce more SiC, a process that increases SiC content in the final part to 75 to 90%. Another method to obtain the high SiC is infiltration of the porous preform with a carbonaceous material that will later react with the Si, providing as much as 92% SiC.

Both open- and closed-back mirrors can be fabricated using this technology. The closed-back structure is formed with soluble tooling that is subsequently dissolved, melted, or burned out, forming the interior cavities. Structural components with complex geometries can also be formed in this manner.

Hot Pressed and Hot Isostatically Pressed

In these methods fine grain SiC usually is pressed at high temperature and pressure. Current hot-pressed blank fabrication methods employ relatively low pressure (compared to HIP) and several hours to consolidate in closed dies that are size limited to about 0.5 m. Sintering aids such as Al are usually added in small quantites to aid in densification. Simple shapes can be obtained by varying the shape of the dies, but unless production quantities are required, cost of the dies is probably not justified. Final shaping can be accomplished with diamond machining tools. In contrast, the HIP process uses pressures as high as 210 MPa to fully consolidate preformed, encapsulated components. Solid billets can be fabricated directly from powder in shaped containers, but NNS components can be fabricated by cold pressing the powder to shape in a flexible bag, encapsulating the preform, and then HIPing.

Chemical Vapor Deposited

The CVD process relies on the reactive decomposition of gases on a hot mandrel in a low pressure chamber to form the SiC.[27] The mandrel is normally a graphite chosen to match the contraction of SiC from the deposition temperature. The mandrel can be polished and treated with a release agent for replication of the mandrel in the SiC surface. Since the SiC is formed directly by decomposition of the reactive gases, the deposited material is very pure and dense. This method produces thin shells in the form of dishes, tubes, plates, and related shapes. To fabricate a light-weight mirror, the faceplate is formed, taken out of the chamber, and the rear surface is ground smooth. A graphite eggcrate-type rib structure with vent hole between cells is made from slotted sheet and is placed on the back of the facesheet in the chamber where deposition is continued. An additional layer of SiC forms on the graphite and back of the facesheet tying it all together. When cooled, the trapped graphite rib formers shrink away from the SiC, but are held in place at the rib vent holes. Mirrors have been fabricated up to 0.5 m, with a capacity somewhat larger than 1.0 m.

4.5 Summary

This chapter addresses the material selection, dimensional stability, and fabrication issues for various types of mirrors in common use including the passive, fully active, and actively cooled type of mirrors.

The dimensional stability section describes the four types of dimensional instability with examples, namely: temporal, thermal/mechanical hysteresis, thermal, and one other type that is related to hysteresis. The sources of these instabilities are presented in some detail, with the major source shown to be the residual stress. Methods for obtaining dimensionally stable mirrors are also described.

The characteristics of specific metal alloy types, including aluminum, beryllium, the invars, copper, molybdenum, aluminum metal matrix composites, silicon, and silicon carbide, are presented in some detail with the tables of comparative properties. The discussion on silicon carbide includes a description of various kinds of this common ceramic, and the advantages and disadvantages of each type for mirror applications.

A brief discussion of mirror blank fabrication methods is also given, with an emphasis on obtaining light-weighted blanks. Beryllium and various types of silicon carbide have been covered in more detail. The hot isostatic pressing (HIP) of beryllium is discussed, with the near-net-shape HIP method compared to the machined HIP billet method. An extended discussion of fabrication of the lightweight structures from the various types of silicon carbide concludes the chapter.

References

1. Marschall, C.W. and Maringer, R.E. 1977. *Dimensional Instability, An Introduction.* Pergamon Press, New York.
2. Paquin, R.A., ed. 1990. *Dimensional Stability,* SPIE Proc. 1335.
3. Paquin, R.A. and Vukobratovich, D.A., eds. 1991. *Optomechanics and Dimensional Stability,* SPIE Proc. 1533.
4. Paquin, R.A. 1992. Dimensional instability of materials: how critical is it in the design of optical instruments? In *Optomechanical Design,* Yoder, P.R. Jr., ed., p. 160. CR43, SPIE Optical Engineering Press, Bellingham, WA.
5. Emerson, W.B. 1957. Secular length changes of gage blocks during twenty-five years, p. 71. In Metrology of Gage Blocks, Circular 581. U.S. National Bureau of Standards.
6. Paquin, R.A. 1981. Processing metal mirrors for dimensional stability. In Workshop on Optical Fab. and Test, Technical Digest. Optic Society of America, TB-1.
7. Lindig, O. and Pannhorst, W. 1985. Thermal expansion and length stability of Zerodur® in dependence on temperature and time, *Appl. Optics,* 24, 3330.
8. Likhachev. V. A. 1961. Microstructural strains due to thermal anisotropy, *Sov. Phys. Solid State,* 3, 1330.
9. Lokshin, I.Kh. 1970. Heat treatment to reduce internal stresses in beryllium, *Metal Sci. Heat Treat.* (U.S.S.R.), 426.
10. Parker, K. and Shah, H. 1971. Residual stresses in electroless nickel plating, *Plating,* 58, 230.
11. Parker, K. and Shah, H. 1970. The stress of electroless nickel deposits on beryllium, *J. Electrochem. Soc.,* 117, 1091.
12. Paquin, R.A. 1986. Hot isostatic pressed beryllium for large optics, *Opt. Eng.,* 25, 1003.
13. Paquin, R.A. 1989. New technology for beryllium mirror production. In *Current Developments in Optical Engineering and Commercial Optics,* Fisher, R.E., Pollicove, H.M., and Smith, W.J., eds., SPIE Proc. 1168.
14. Paquin, R.A. and Gardopée, G.J. 1992. Fabrication and testing of a lightweight beryllium one meter f/0.58 ellipsoidal mirror. In *Large Optics II,* Parks, R.E., ed., SPIE Proc. 1618.
15. Jacobs, S.F. 1990. Unstable optics. In *Dimensional Stability,* Paquin, R.A. ed., p. 20. SPIE Proc. 1335.
16. Jacobs, S.F. 1992. Variable invariables: dimensional instability with time and temperature. In *Optomechanical Design,* Yoder, P.R., Jr., ed., p. 181. CR43, SPIE Optical Engineering Press, Bellingham, WA.
17. Paquin, R.A. 1975. Selection of Materials and Processes for Metal Optics. In *Design, Manufacture and Application of Metal Optics,* Barnes, W.P., Jr., ed., p. 12. Proc. SPIE 65; republished with corrections. 1985. *Selected Papers on Infrared Design,* p. 347. Proc. SPIE, Milestone Series 513; and 1987. *Selected Papers on Optomechanical Design,* p. 27. Proc. SPIE, Milestone Series 770.
18. Stone, R., Vukobratovich, D., and Richard, R. 1989. Shear modulii for cellular foam materials and its influence on the design of light-weight mirrors. In *Precision Engineering and Optomechanics,* Vukobratovich, D., ed., p. 37. SPIE Proc. 1167.
19. Vukobratovich, D. 1989. Lightweight Laser Communications Mirrors Made with Metal Foam Cores. SPIE Proc. 1044.
20. Rozelot, J.P. and Leblanc, J.-M. 1991. Metallic alternative to glass mirrors (active mirrors in aluminum). A review. In *Space Astronomical Telescopes and Instruments,* Bely, P.Y. and Breckinridge, J.B., eds., p. 481. Proc. SPIE 1494.
21. *ASM Handbook, 2.* 1990. Properties and Selection: Nonferrous Alloys and Special-Purpose Materials. ASM International, Materials Park, OH.
22. Mohn, W.R. and Vukobratovich, D. 1988. Recent applications of metal matrix composites in precision instruments and optical systems, *Opt. Eng.,* 27, 90.

23. Stover, J.C., ed. 1991. *Optical Scatter: Applications, Measurement, and Theory,* pp. 130–230. Proc. SPIE 1530, Session 3: "Scatter from Be mirrors".

24. Hoover, M.D., Seiler, F.A., Finch, G.L., Haley, P.J., Eidson, A.F., Mewhinney, J.A., Bice, D.E., Brooks A.L., and Jones, R.K. 1992. *Space Nuclear Power Systems 1989,* El-Genk, M.S. and Hoover, M.D., eds., p. 285. Orbit Book Co., Malabar, FL.

25. Sato, S., Maezawa, H., Yanagihara, M., Ishiguro, E., and Matsuo, S. 1995. High heat load vacuum ultraviolet mirror development in Japan, *Opt. Eng.,* 34, 377.

26. Li, Z. and Bradt, R.C. 1987. Thermal expansion and elastic anisotropies of SiC as related to polytype structure. In *Proc. Silicon Carbide Symp. 1987.* American Ceramic Society; see also Thermal expansion and thermal expansion anisotropy of SiC polytypes, *J. Ceram. Soc.,* 70, 445; Likhachev, V.A. 1961. Microstructural strains due to thermal anisotropy, *Sov. Phys. Solid State,* 3, 1330.

27. Goela, J.S., Pickering, M.A., Taylor, R.L., Murray, B.W., and Lompado, A. 1991. Properties of chemical-vapor-deposited silicon carbide for optics applications in severe environments, *Appl. Optics,* 30, 3166.

28. Paquin, R.A., Magida, M.B., and Vernold, C.L. 1991. Large optics from silicon carbide. In *Large Optics II,* Parks, R.E., ed., p. 53. Proc. SPIE 1618.

29. Paquin, R.A., Levenstein, H., Altadonna, L., and Gould, G. 1984. Advanced lightweight beryllium optics, *Opt. Eng.,* 23, 157; republished in 1985. *Selected Papers on Infrared Design,* p. 355. Proc. SPIE, Milestone Series 513.

30. Parsonage, T.B. 1990. Selecting mirror materials for high-performance optical systems. In *Dimensional Stability,* Paquin, R.A., ed., p. 119. SPIE Proc. 1335.

31. Gould, G. 1985. Method and Means for Making a Beryllium Mirror, U.S. Patent No. 4,492,669.

32. Gildner, D. and Marder, J. 1991. Creation of aspheric beryllium optical surfaces directly in the hot isostatic pressing consolidation process. In *Reflective and Refractive Optical Materials for Earth and Space Applications,* Riedl, M.J., Hale, R.R., and Parsonage, T.B., eds., p. 46. SPIE Proc. 1485.

33. Anthony, F.M. 1995. High heat load optics: an historical overview, *Opt. Eng.,* 34, 313.

34. Kittell, D. and La Fiandra, C. 1991. Cooled deformable mirror. In *Active and Adaptive Optical Components,* Ealey, M.A., ed., p. 101. Proc. SPIE, 1543.

35. Union Carbide Corp. 1988. UCON Quenchants for Ferrous and Non-Ferrous Metals, Brochure SC-955.

36. Arthur, J. 1995. Experience with microchannel and pin-post water cooling of silicon monochrometer crystals, *Opt. Eng.,* 34, 441.

37. Arakawa, E.T., Callcott, T.A., and Chang, Y.-C. 1991. Beryllium. In *Handbook of Optical Constants of Solids II,* Palik, E.D. ed., p. 421. Academic Press, Orlando.

38. Barker, R.S. and Sutton, J.C. 1967. *Aluminum, Vol. III: Fabrication and Finishing,* Van Horn, K.R. ed., chap. 10. American Society for Metals, Metals Park, OH.

39. Brush Wellman, Inc. (no date). Beryllium Optical Materials, brochure; also see the brochure Designing with Beryllium.

40. Hibbard, D.L. 1990. Dimensional stability of electroless Ni coatings. In *Dimensional Stability,* Paquin, R.A. ed., p. 180. SPIE Proc. 1335.

41. Hunsicker, H.Y. 1967. *Aluminum, Vol. I: Properties, Physical Metallurgy and Phase Diagrams,* Van Horn, K.R. ed., chap. 5. American Society for Metals, Metals Park, OH.

42. Marder, J.M. 1990. A comparison of microdeformation in I-70, O-50 and a new instrument grade of beryllium (I-250). In *Dimensional Stability,* Paquin, R.A. ed., p. 108. SPIE Proc. 1335.

43. Marschall, C.W. and Maringer, R.E. 1971. Stress relaxation as a source of dimensional instability, *J. Mater.,* 6, 374.

44. Parker, K. 1987. Internal stress measurements of electroless nickel coatings by the rigid strip method. In *Testing of Metallic and Inorganic Coatings,* ASTM STP 947, Harding, W.B. and DiBari, G.A., eds., p. 111. American Society for Testing and Materials, Philadelphia.

5

Lightweight Mirror Design

Daniel Vukobratovich

5.1 Introduction

Design of lightweight mirrors is a complex problem, involving optimization of both mirror and mount. The expense and complexity of lightweight mirrors require careful consideration of design requirements. Scaling laws provide rapid estimates of mirror weight during preliminary design. A significant issue is the self-weight deflection of lightweight mirrors. The simplest type of lightweight mirror is the contoured back mirror, of which there are three types: double concave, single arch, and double arch. The sandwich mirror offers the best stiffness-to-weight of any lightweight mirror, but is complex to design and fabricate. Open back mirrors are low in stiffness, but are relatively easy to fabricate. Mounting must be considered as part of the mirror design problem.

Lightweight mirrors are used in optical systems for a variety of reasons. Some advantages of lightweight mirrors include shorter thermal equilibrium times, reduced weight, and lower system cost. Reduced self-weight deflection, and higher fundamental frequency are additional reasons for the use of lightweight mirrors. Lightweight mirrors are often defined as mirrors that are lighter in weight than comparable-size conventional mirrors. This is often a difficult definition to apply, since there is considerable variation in the weight of "conventional" mirrors. One traditional rule of thumb first suggested by Ritchey is that "conventional" mirrors are right circular cylinders, with a diameter-to-thickness ratio of 6:1. In addition, this rule of thumb assumes that the mirror material is solid optical glass. This rule of thumb is easy to calculate and therefore is quite popular.

A more controversial definition based on structural efficiency is suggested by Schwesinger. Schwesinger suggests that a mirror is a "lightweight" if it has greater stiffness than a solid right circular cylinder mirror of the same weight.[1] If there is no improvement in stiffness, then the "lightweight" mirror does not have any advantage over the same weight solid mirror. This definition of a lightweight mirror requires considerable insight into the elastic behavior of the mirror, and is not as popular as the first rule of thumb suggested above.

-8493-0133-5/97/$0.00+$.50
© 1997 by CRC Press, Inc.

5.2 Estimating Mirror Weight

It is often desirable to estimate the weight of lightweight mirrors well in advance of detailed design. Scaling laws are used to estimate mirror weight based on mirror diameter. Caution is indicated in the use of scaling laws. Scaling laws are based on statistical analysis of existing mirrors. Attempts to extend the scaling laws beyond the range of statistical data are hazardous. There are often design constraints such as dynamic loads in the mirror that may cause the final design to depart significantly from the weight predicted by the scaling laws. Within these limitations, scaling laws are a useful tool, especially for performing parametric analysis.

Surveys of the open literature on mirrors that are in existence indicate that mirror weight is dependent on mirror diameter raised to a power. There is some controversy concerning the exponent in this scaling law. Ordinary engineering analysis suggests that mirror weight should scale as the cube of the mirror diameter. Weight per unit area is sometimes used as an index of lightweight mirror efficiency. Use of weight per unit area implies a scaling law based on the square of mirror diameter.

A survey of lightweight mirrors by Valente indicates that mirror weight varies approximately with the cube of the mirror diameter.[2] This survey included 61 mirrors from 0.24 to 7.5 m in diameter using a variety of materials. Valente's table of lightweight mirrors is shown in Table 5.1.

For conventional solid mirrors, Valente gives the following relationship:

$$W = 246 D^{2.92}$$

where W = mirror weight (kg)
           D = mirror diameter (m)

This relationship is shown in Figure 5.1.

For all lightweight mirrors, Valente gives the following relationship:

$$W = 82 D^{2.95}$$

This relationship is shown in Figure 5.2.

For specific mirror types, other scaling relationships are used. Contoured mirrors are mirrors with a back contoured to improve stiffness and reduce weight. The weight of contoured mirrors is shown in Figure 5.3 and is given by:

$$W = 106 D^{2.71}$$

Structured mirrors are mirrors with a sandwich or open back geometry. The weight of structured mirrors is given by:

$$W = 68 D^{2.90}$$

This relationship is shown in Figure 5.4.

Beryllium mirrors are made in a variety of configurations. Beryllium mirrors are normally lighter than other types of mirrors regardless of the type of lightweight design. Weight of beryllium mirrors is given by:

$$W = 26 D^{2.31}$$

TABLE 5.1   Lightweight Mirrors

Mirror	Year	Dia. (M)	Thick. (M)	Weight (KG)	Matl.	Config.	Misc.
IRAS	1983	0.60	0.09	12.6	Beryllium	Openback	annular ribs
Ball Relay	1989	0.60	0.06	9.07	Beryllium	Openback tri. cells	Hip process
P-E 40 inch	1989	1.02	0.05	18.14	Beryllium	Sandwich hex cells	2.265 in cells
P-E Scan	1975	.86 × .81	0.08	14.52	Beryllium	Openback sqr. cells	Flat λ/20
P-E second.	1975	1.65 × 1.02	0.08	53.5	Beryllium	Openback sqr. cells	f/0.67
Thematic Map.	1972	.406 × .508	0.04	1.86	Beryllium	Sandwich sqr. cells	Brazed
P-E 9.5 inch	1984	0.24	0.05	0.98	Beryllium	Sandwich hex cells	HIP process
P-E test	—	0.57	0.04	13.25	Beryllium	Double arch	—
P-E test	—	0.51	0.05	6.51	Beryllium	Double arch	1in circ cores
Hale	1950	5.0	0.60	13158	Pyrex	Openback	—
MMT	1979	1.8	0.30	567	F silica	Sandwich sqr. cells	6 mirrors
RCT	1965	1.3	0.15	200	Aluminum	Single arch	—
Spacelab UV	1979	0.92	0.15	100	Cervit	Double arch	—
Hubble	1990	2.48	0.30	773	ULE	Sandwich sqr. cells	—
Teal Ruby	1980	0.50	0.08	7.3	F silica	Sandwich hex cells	—
OAO-c	1972	0.82	0.13	48	F silica	Sandwich sqr. cells	—
U of Colorado	1979	0.41	0.05	9.98	Cervit	Double arch	f/2.5, 1/4λ
Steward Obs.	1985	1.8	0.36	703	Borosilicate	Sandwich hex cells	f/1.0
LDR test	1985	0.38	0.13	6.24	Borosilicate	Sandwich hex cells	sand-hexing
LDR test	1985	0.15	0.05	0.53	Vycor	Sandwich hex cells	air pressure
UTRC	1985	0.30	0.06	1.1	Glass TSC	Sandwich	Frit bonded
Ft. Apache	1986	3.5	0.46	1893	Borosilicate	Sandwich hex cells	—
NASA	1983	2.48	0.30	771	Glass	Sandwich sqr. cells	—
Los Alamos	1982	1.1 × 1.1	0.20	204	Tempax	Openback sqr. cells	—
SIRTF test	1983	0.51	0.089	16–25	quartz	Single arch	—
SIRTF test	1983	0.51	0.102	19–29	F silica	Double arch	f/4
Landsat-D	1979	0.42	0.07	9	ULE	Sandwich sqr. cells	—
GIRL	1985	0.50	0.074	25	Zerodar	Double taper	—
ISO	1985	0.64	0.075	20	F silica	Sandwich	machined
Hextek	1989	1.0	0.15	73	Borosilicate	Sandwich hex cells	f/0.5, meniscus
Hextek	1989	0.46	0.086	5.17	Borosilicate	Sandwich hex cells	—
Hextek	1989	0.38	0.076	7.71	Borosilicate	Sandwich hex cells	—
Shane 3 M.	1959	3.0	0.406	3856	Pyrex	Openback tri. cells	f/5
NASA 2.4 M.	1981	2.4	0.305	748	ULE	Sandwich sqr. cells	f/2.35
Milan 54 inch	1968	1.37	0.20	907.2	Aluminum	Single arch	—
Steward 68 cm.	—	0.68	0.10	25.4	Pyrex	Sandwich hex cells	—
Soviet test	1977	0.506	0.076	13.7	quartz	Openback hex cells	54 mm cells
Soviet test	1977	0.50	0.065	12.5	quartz	Sandwich hex cells	54 mm cells
Soviet test	1977	0.37	0.052	5.2	F silica	Sandwich hex cells	28 mm cells
Soviet test	1983	0.52	0.053	12.4	F silica	Sandwich	70 mm cells
Soviet test	1983	0.57	0.057	13.2	F silica	Sandwich	71 mm cells
Soviet test	1983	0.42	0.059	11.2	F silica	Sandwich	73 mm cells
Soviet test	1985	0.70	0.10	20	Al alloy	Openback	annular ribs
Schott test	—	1.143	0.159	204.12	F silica	Sandwich	—
OSC 16 in scope	1989	0.406	0.076	6.17	SXA	Single arch	—
OSC 12 in scope	1988	0.305	0.064	2.04	Aluminum	Double concave	Al foam core
OSC 12 in scope	1988	0.305	0.043	1.95	Aluminum	Double concave	Al foam core
AFCRL	1972	1.524	0.165	363	Cervit	Single arch	—

This relationship is shown in Figure 5.5.

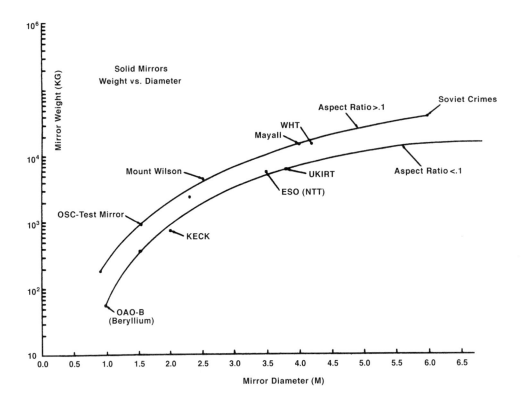

FIGURE 5.1   Weight vs. diameter of solid mirrors. (From Valente, T.M. 1990. *Proc. SPIE 1340*, 47.)

5.3 Mirror Self-Weight Deflection.

A parameter of considerable importance in the design of lightweight mirrors is the self-weight induced deflection. Self-weight deflection is important in terrestrial systems if the direction of the gravity vector changes. An example of an application of a system with a changing gravity vector is an astronomical telescope. As the telescope is pointed at objects at different zenith distances, the direction of the gravity vector acting on the optics changes.

Self-weight deflection in space optical systems is related to the change in optical figure upon gravity release in space and to the fundamental frequency of the mirror. Fundamental frequency is critical in determining the response of the mirror to random vibration during launch. Fundamental frequency and self-weight deflection are related by:[3]

$$f_n \approx \frac{1}{2\pi}\sqrt{\frac{g}{\delta}}$$

where f_n = fundamental frequency (Hz)

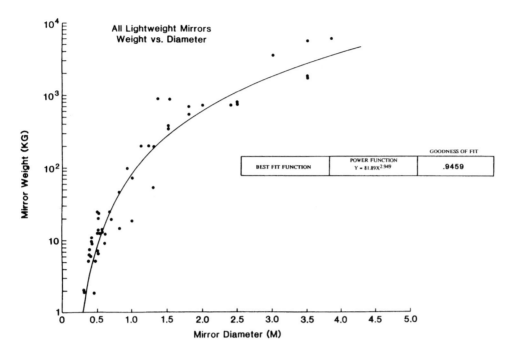

FIGURE 5.2   Weight vs. diameter of all lightweight mirrors. (From Valente, T.M. 1990. *Proc. SPIE 1340*, 47.)

g = acceleration due to Earth's gravity
δ = self-weight deflection of mirror

Self-weight deflection is normally calculated as normal to the mirror surface. For an axisym-metric mirror, the most common self-weight loading condition is the worst case of the gravity vector acting along the axis of symmetry. In this loading case the gravity vector is normal to the plane of the mirror and parallel to the optical axis. This loading condition is called the axial deflection case.

When the gravity vector acts normal to the axis of symmetry, the loading condition is called the radial deflection case. In this case, gravity is acting parallel to the plane of the mirror and normal to the optical axis. Although gravity acts parallel to the mirror surface, deflection normal to the mirror surface is induced by this loading condition.[4]

If the mirror is subjected to a loading condition in which the gravity vector is at an angle to the axis of symmetry, the resulting mirror surface deflections are given by:[5]

$$\delta_\theta \approx \sqrt{\left(\delta_A \cos\theta\right)^2 + \left(\delta_R \sin\theta\right)^2}$$

where δ_θ = mirror self weight deflection when gravity vector is at an angle to mirror axis
δ_A = mirror self-weight deflection in axial deflection case
δ_R = mirror self-weight deflection in radial deflection case
θ = angle between mirror axis and gravity vector

For most lightweight mirrors, the radial deflection is very small and is often ignored in prelim-inary estimates of performance. In some cases, calculation of the radial deflection is important. Such cases include very large mirrors, extremely lightweight systems, and systems used under high accelerations.

Self-weight deflection of mirrors in the axial loading condition is calculated using the classical plate theory. Caution is indicated in applying plate theory to lightweight mirrors. Classical plate

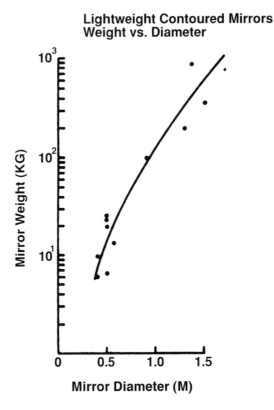

FIGURE 5.3 = axial deflection due to self-weight

		GOODNESS OF FIT
BEST FIT FUNCTION	LINEAR HYPERBOLA Y=-4409+2854X+1439/X	.9967
POWER FUNCTION	$Y=106X^{2.712}$	.9728

FIGURE 5.3 Weight vs. diameter of contoured mirrors. (From Valente, T.M. 1990. *Proc. SPIE 1340*, 47.)

theory assumes axisymmetric plane parallel plates, with a diameter-to-thickness ratio of 10:1 or more. Real lightweight mirrors may depart significantly from these assumptions. Shear deformations may play an important role in self-weight deflection and are ignored in classical plate theory. Shell action may become important if the mirror has significant surface curvature. Classical plate theory is an approximation and is used for preliminary design. More sophisticated design analysis using such techniques as finite element analysis is necessary for final design.

The general equation for axial deflection due to self-weight is[6]

$$\delta_A = C \frac{qr^4}{D}$$

where δ_A = axial deflection due to self-weight
        C = support condition constant
        q = weight per unit area of mirror
        r = mirror radius
        D = flexural rigidity of mirror

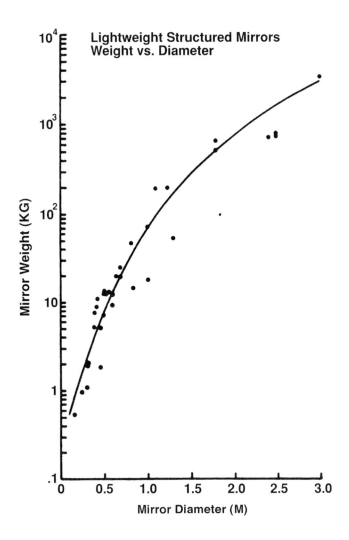

		GOODNESS OF FIT
BEST FIT FUNCTION	LOG NORMAL $Y=.177 \times 10^{-3} \exp\left[(-8.739-\ln X)^{\frac{2}{5.996}}\right]$	.9512
POWER FUNCTION	$Y=106X^{2.712}$	.9499

FIGURE 5.4 Weight vs. diameter of structured mirrors. (From Valente, T.M. 1990. *Proc. SPIE 1340*, 47.)

Axial deflection due to self-weight is reduced by changing the support condition, reducing the weight per unit area of the mirror, or by increasing the flexural rigidity of the mirror. Weight per unit area of the mirror is determined by the mirror material density, the mirror structure, and mirror thickness. When material is removed from the cross section of the mirror, a lightweight structure is produced. Flexural rigidity is determined by the mirror material elastic modulus, mirror thickness, and mirror structure. Removing material from the cross section of the mirror to produce a lightweight structure influences flexural rigidity.

Another form of the general equation for axial deflection due to self-weight is

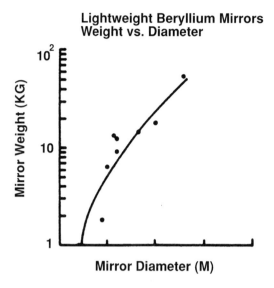

FIGURE 5.5 Weight vs. diameter of beryllium mirrors. (From Valente, T.M. 1990. *Proc. SPIE 1340*, 47.)

$$\delta_A = C\left(\frac{\rho}{E}\right)\frac{V_o}{I_o}r^4\left(1 - v^2\right)$$

where δ_A = axial deflection due to self-weight
           V_o = unit volume of mirror
           ρ = mirror material density
           E = mirror material elastic modulus
           C = support condition constant
           I_o = unit cross-sectional moment of inertia
           r = mirror radius
           v = Poisson's ratio of mirror material

    In the above axial deflection equation, the material parameter determining self-weight deflection is the ratio of mirror material density to elastic modulus. This material properties ratio, ρ/E, is the inverse specific stiffness of the material. This ratio does not change significantly for most common structural materials, and typically has a value of 386×10^{-9} m^{-1}. Significant exceptions to the rule that most materials have about the same specific stiffness are

Material	p/E × 10^{-9} m^{-1}
Unusual Specific Stiffness Materials	
Beryllium	58.8
Silicon carbide	92.1
Metal matrix composite aluminum/silicon carbide	244
Graphite/epoxy composite	188

The ratio of unit volume to unit cross-sectional moment of inertia is the structural efficiency of the mirror cross section. This ratio, V_o/I_o, is a measure of the stiffness-to-weight independent of material properties. The structural efficiency is determined by the distribution of material in the cross section. High structural efficiency is achieved when the material in the cross section is distributed as far as possible from the neutral or bending axis. This condition is satisfied by the sandwich mirror. In the sandwich mirror, most of the mirror material is in the face plate and back plate of the mirror and is distant from the bending axis.

For a solid mirror the following conditions hold:

$$V_o = h$$

$$I_o = \frac{h^3}{12}$$

$$\frac{V_o}{I_o} = \frac{12}{h^2}$$

where V_o = unit volume of mirror
   I_o = unit cross-sectional moment of inertia
   h = mirror thickness

The support condition constant varies with the geometry of mirror support. There are no units associated with the support condition constant. The magnitude of the support condition constant depends on the location and number of the support forces. For a given number of support forces, there is an optimum location geometry to produce minimum axial deflection in the mirror due to self-weight. Increasing the number of support forces usually reduces axial deflection due to self-weight if the supports do not overconstrain the mirror.

A common mirror geometry is a right circular cylinder. One type of axial support for the right circular cylinder-shaped mirror consists of three points on a common diameter. The point supports are equal spaced around the diameter, and the diameter is some fraction of the mirror diameter. There is an optimum location for the three point supports for the right circular cylinder mirror, and that optimum location is a support diameter that is 0.68 of the mirror diameter. When the mirror is supported by three points on the 0.68 diameter, axial deflection due to self-weight is at a minimum for any three-point support. If the mirror is supported on this optimum three-point support, the maximum deflection is given by:

$$\delta_A = 0.02859 \left(\frac{\rho}{E}\right) \frac{V_o}{I_o} r^4 \left(1 - v^2\right)$$

If a right circular cylinder mirror is supported by three points equally spaced at the edge, the maximum deflection is given by:

$$\delta_A = 0.11303 \left(\frac{\rho}{E}\right) \frac{V_o}{I_o} r^4 \left(1 - v^2\right)$$

If a right circular cylinder mirror is supported by six points equally spaced at 0.68 of the mirror diameter, the axial deflection due to self-weight is at a minimum for a six-point support. The maximum deflection for this optimum six-point support is given by:

$$\delta_A = 0.0048973 \left(\frac{\rho}{E}\right)\frac{V_o}{I_o} r^4 \left(1-\nu^2\right)$$

Occasionally rectangular mirrors are used in optical systems. The optimum three-point support for a rectangular mirror consists of one point located at the middle of one long edge, and the other two points located at the corners of the opposite long edge. For this optimum support of a rectangular mirror, axial deflection due to self-weight is given by:[7]

$$\delta_A = \frac{1}{\psi^2}\left(\frac{\rho}{E}\right)\frac{V_o}{I_o} a^4 \left(1-\nu^2\right)$$

$$\psi^2 = \frac{7\left(\dfrac{a}{b}\right)}{\left[1+0.461\left(\dfrac{a}{b}\right)^{13}\right]^{\frac{1}{13}}}$$

where: δ_A = axial deflection of mirror due to self-weight
ρ = mirror material density
E = mirror material elastic modulus
V_o = unit volume of mirror
I_o = unit cross-sectional moment of inertia
a = length of mirror
b = width of mirror
ν = Poisson's ratio for mirror material

These equations are used for rapid estimation of mirror axial self-weight deflection. More sophisticated calculations are typically required for final design. A significant disadvantage of these equations is neglect of shear deformations. A rough correction for shear effects is possible using a method developed by Nelson.[8] The shear correction is

$$\delta_{total} = \delta_{bending}\left(1+\frac{N\pi h^2}{A}\right)$$

where δ_{total} = total axial deflection including shear effects
$\delta_{bending}$ = axial deflection due to bending
N = number of point supports
h = mirror thickness
A = mirror surface area

5.4 Contoured Back Mirrors

Contoured back mirrors are mirrors with a back contour shaped to reduce weight, and in some cases self-weight deflection. Three types of contoured back mirrors are used: symmetric, single arch, and double arch. These three types of mirrors are shown in Figure 5.6. Contoured back mirrors offer reduction in weight up to 25% in comparison with a right circular cylinder, 6:1 aspect ratio solid mirror. Contoured back mirrors are low in fabrication cost and are relatively easy to mount. A significant disadvantage of contoured back mirrors is the variation in mirror thickness. This variation in mirror thickness causes different portions of the mirror to reach thermal equilibrium at different times following a change in temperature. The resulting variation in temperature with mirror thickness induces optical surface distortion. Contoured back mirrors are more sensitive to optical surface distortion due to temperature changes than other types of mirrors.

Axial deflection of contoured back mirrors is difficult to calculate using simple closed-form equations due to the substantial variation in mirror thickness and shell action of the curved contours of the mirror. Cho gives a scaling relation for contoured back mirrors of different sizes.[9] This scaling relation is used for mirrors of similar contours to scale deflections as the mirror size changes. This scaling relation is

$$\delta = \left(\frac{\rho}{\rho_{ref}}\right)\left(\frac{E_{ref}}{E}\right)\left(\frac{A}{A_{ref}}\right)\delta_{ref}$$

where δ_{ref} = axial deflection due to self-weight of reference mirror
δ = axial deflection due to self-weight of new mirror
ρ_{ref} = mirror material density of reference mirror
ρ = mirror material density of new mirror
E_{ref} = elastic modulus of reference mirror material
E = elastic modulus of new mirror material
A_{ref} = cross-sectional area of reference mirror
A = cross-sectional area of new mirror

Symmetric mirror shapes are used to minimize axial deformation due to self-weight when the gravity vector is perpendicular to the mirror optical axis. Bi-metallic bending of plated metal mirrors is reduced through the use of symmetric shapes. The front and back of the symmetric mirror are given equal radii, but of opposite sign. A symmetric mirror is either double concave or double convex. Normally a symmetric mirror is supported either by a small ring near the center or by multiple points at the edge.

Self-weight axial deflection of the symmetric mirror is worse for equal weights than the single arch or double arch shape. Radial deflection of the symmetric mirror is much smaller than that of single arch or double arch mirror shapes. Therefore, the symmetric shape is often used when the operating position of the mirror is such that the gravity vector is perpendicular to the optical axis. The extremely small radial deflection of the symmetric mirror shape makes it attractive as a candidate space mirror. The very small radial deflection of the symmetric mirror minimizes residual optical surface error after gravity release.[10] This advantage is offset by the lower fundamental frequency of the symmetric mirror in comparison with other mirror shapes.

The symmetric shape is used to minimize bi-metallic bending effects in electroless nickel-plated metal mirrors.[11] The thermal coefficient of expansion of electroless nickel is either 12.5 or 15 × 10^{-6} m/m-K depending on whether the nickel is annealed after plating.[12] In comparison, the thermal coefficients of expansion of aluminum and beryllium are 23 × 10^{-6} and 11 × 10^{-6} m/m-K, respectively. Electroless nickel is normally plated onto the mirror surface to a thickness of 75 to 125 μm or more. The difference in thermal coefficient of expansion between substrate material and plating, and the relatively thick plating layer leads to bi-metallic bending effects when the temperature of the plated mirror is changed.

Plating both sides of a symmetric mirror shape with the same thickness of electroless nickel minimizes this bi-metallic bending, since equal and opposite bending forces are produced in the mirror. This technique of suppressing bi-metallic bending reduces bending deflection, but does

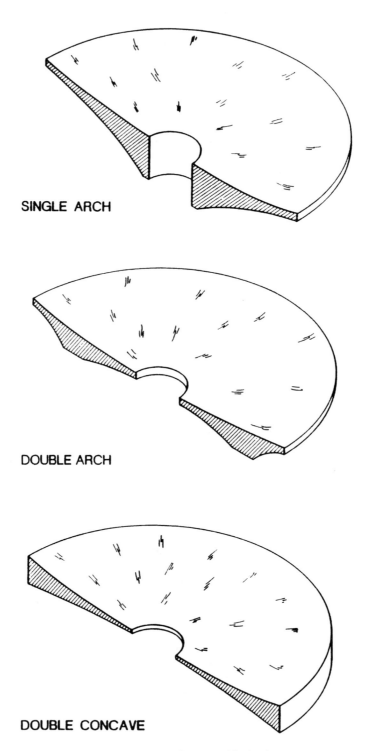

SINGLE ARCH

DOUBLE ARCH

DOUBLE CONCAVE

FIGURE 5.6 Types of contoured back mirrors.

not affect bending stress. Bending stress may still exceed the microyield strength of the material (microyield strength is defined as that amount of stress required to produce a permanent strain of 10^{-6} in the material; for maximum dimensional stability the rule of thumb is to keep all stress in the substrate below one half of the microyield strength of the material). If the microyield strength of the material is approached or exceeded due to bi-metallic bending, thermal hysteresis results, with poor optical figure stability.

Cho's studies include a 40-in.-diameter, 5-in.-thick (at the edge) double concave mirror.[13] The radius of curvature is 160 in., so the center thickness is 2.49 in. Mirror material is an aluminum/silicon carbide reinforced metal matrix composite, SXA™, with a density of 0.10 lb/in.[3] and an elastic modulus of 16×10^6 lb/in.[2]. The mirror support consists of either a continuous edge ring or multipoint supports spaced around the edge. Optical surface deflection in both zenith (axial) and horizontal (radial) for ring and multipoint supports is given by:

Self-Weight Deflection of 40 in. Double Concave Mirror
Surface Deformation (RMS Wave, 1 Wave = 633 nm)

Gravity Load	Ring	12–30°	6–60°	4–90°	3–120°
Zenith	0.282	0.284	0.292	0.402	0.883
Horizon	0.002	0.003	0.008	0.012	0.020

The single arch mirror is a contoured back mirror with a taper from a thick center to a thin edge. Three kinds of back contours or tapers are used. These are a straight taper,[14] producing a conical back, a convex back taper, and a parabolic taper.[15] The vertex of the parabolic taper is located either at the back of the mirror or the edge of the mirror.

Very good stiffness-to-weight is obtained with a single arch mirror shape using a parabolic taper, with the vertex of the parabola located at the mirror edge. Lower weight, but reduced stiffness, is obtained by locating the vertex of the parabola at the back of the mirror. Both the straight taper or conical back and convex back mirrors are inferior in stiffness-to-weight when compared against parabolic back mirrors.

Cho's studies include a series of single arch shapes, of 16 in. diameter, 3 in. thick, with different back tapers. In all cases the mirrors are SXA™, with a 48-in. optical radius of curvature and a concave shape. Typical edge thickness is 0.5 in. The following self-weight deflections of these mirrors are

Self-Weight Deflection of 16 in. Single Arch Mirrors (1 Wave = 633 nm)

Mirror Type	Horizon (RMS Waves)	Zenith (RMS Waves)	Mirror Weight (Lb)
Straight taper	0.003	0.007	27.3
Convex back	0.003	0.015	29.6
Parabola, vertex at edge	0.003	0.004	18.9
Parabola, vertex at back	0.003	0.012	16.2

For typical single arch mirror designs, the mirror center of gravity is either very close to the optical surface vertex or actually outside the mirror, beyond the vertex. The forward location makes center of gravity support in the radial direction very difficult. Since the mirror cannot easily be supported through the plane of its center of gravity, the optical surface develops astigmatism when the optical axis is in the horizontal position. This astigmatism in the axis horizontal position is a serious limiting factor for larger single arch mirrors. For typical single arch mirrors at a diameter of 1.2 m the self-weight-induced astigmatism is about 1 wave (1 wave = 633 nm) peak-to-valley in the axis horizontal position.[16]

The poor radial bending stiffness of the single arch mirror causes problems for both plated metal mirrors and in a dynamic environment. The very thin edge of the single arch mirror is subject to significant distortion due to bi-metallic bending effects when made of a nickel-plated metal. Vibration can excite the thin edge of the mirror, leading to blur in the final image.

Like all contoured back mirrors, the changing thickness of the single arch mirror causes distortion in the optical surface when the mirror is exposed to a rapid change in temperature. This distortion is exploited in some applications. In the single arch mirror used in the primary of the Mars Observer Camera, a radial temperature gradient is used to control focus. The temperature gradient is created by heating elements at center and edge of the mirror.[17]

The single arch mirror is relatively easy to produce. Typically the mirror is generated from a solid; alternately the mirror may be cast. The thin edge and cantilever form of the single arch complicate optical fabrication, driving up the production cost. Special blocking techniques are sometimes used to support the thin edge of the single arch mirror during polishing.

Ease of mounting is an important advantage of the single arch mirror. The single arch is normally mounted by a center hub support. The central hub may be bonded to an axial hole in the mirror. An athermal center hub mount uses a conical hole in the mirror, with the apex of the hole coincident with the back of the mirror. Figure 5.7 shows a bonded athermal single arch mirror mount. A conical mount is installed into the central conical hole and acts to pull the mirror into contact with a rear flange. The conical mount is provided with an axial spring preload.[18]

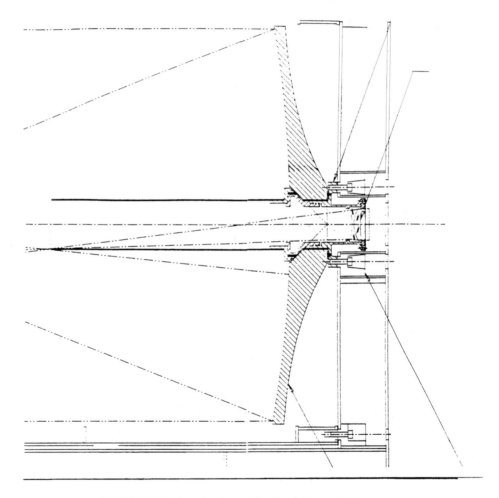

FIGURE 5.7 Single arch mirror with athermal center hub mount.

When compared to other types of lightweight mirror, such as the double arch or sandwich, the single arch mirror is relatively poor in stiffness-to-weight. The extremely low mass and simplicity of the central hub mount for the single arch mirror may make this type of mirror very competitive when the weight of mirror and mount are considered together. In particular, for diameters below 0.5 m the single arch is very competitive in performance and cost to other types of mirrors. An example of the single arch is the primary mirror for the Mars Observer Camera.

The double arch mirror is supported at the back on a ring intermediate in diameter between center and edge. The mirror thickness is reduced away from the support ring, so that both edge and mirror center are thin in comparison with the part of the mirror above the support ring. The cross section of the double arch mirror resembles a bridge with two piers with arches between the piers.[19] This is the source of the "double arch" name. A straight taper, convex taper, or parabolic taper is used on the inner and outer tapered sections of the mirror. The vertex of the parabolic taper may be located at either the edge or back of the mirror.

Stiffness-to-weight of the double arch mirror shape is the best of any contoured back mirror,[20] and is competitive with other types of lightweight mirrors. Optimization of the stiffness-to-weight of the double arch mirror requires selection of an optimum radius of support for the ring and an optimum taper for the mirror back. Cho's studies include four types of back contours: straight taper, convex taper, parabolic taper with vertex at edge, and parabolic taper with vertex at mirror back. All mirrors are made of SXA™, with a 40-in. diameter and 5 in. thick. Typical edge thickness is 0.5 in. Optical surface radius of curvature is 160 in. and the surface shape is concave. The ratio of support ring diameter to mirror diameter varied from 0.5 to 0.65, with both continuous ring and multiple point supports considered. Cho's results are as follows:

Double Arch Mirror Self-Weight Deflection (All Deflections in Units of RMS Waves, 1 Wave = 633 nm)

Mirror Shape	Mirror Weight (lb)	Axis	Ring	Support Location			
				12–30°	6–60°	4–90°	3–120°
Support Ring Ratio = 0.5							
Parabola back vertex	256	Zenith	0.021	0.021	0.023	0.078	0.253
		Horizon	0.027	0.027	0.028	0.046	0.119
Parabola edge vertex	254	Zenith	0.021	0.022	0.023	0.073	0.234
		Horizon	0.027	0.027	0.028	0.041	0.098
Straight taper	324	Zenith	0.069	0.070	0.070	0.092	0.229
		Horizon	0.021	0.021	0.022	0.044	0.136
Support Ring Ratio = 0.55							
Parabola back vertex	256	Zenith	0.004	0.004	0.016	0.090	0.291
		Horizon	0.020	0.020	0.021	0.041	0.097
Parabola edge vertex	254	Zenith	0.013	0.013	0.019	0.085	0.272
		Horizon	0.046	0.046	0.047	0.051	0.065
Straight taper	324	Zenith	0.036	0.036	0.038	0.083	0.260
		Horizon	0.025	0.025	0.026	0.043	0.109
Support Ring Ratio = 0.60							
Parabola back vertex	256	Zenith	0.045	0.046	0.050	0.114	0.352
		Horizon	0.019	0.018	0.021	0.045	0.103
Parabola edge vertex	254	Zenith	0.036	0.037	0.041	0.093	0.266
		Horizon	0.007	0.007	0.011	0.027	0.044
Straight taper	324	Zenith	0.006	0.007	0.017	0.085	0.279
		Horizon	0.008	0.008	0.012	0.038	0.094
Support Ring Ratio = 0.65							
Parabola back vertex	256	Zenith	0.065	0.066	0.072	0.133	0.368
		Horizon	0.006	0.006	0.012	0.033	0.052
Parabola edge vertex	254	Zenith	0.065	0.067	0.072	0.131	0.355
		Horizon	0.010	0.010	0.014	0.032	0.045
Straight taper	324	Zenith	0.043	0.044	0.049	0.114	0.343
		Horizon	0.007	0.007	0.012	0.037	0.187

Optimum support for the double arch depends on the number of supports and the contour of the back. If three supports equally spaced on a common diameter are used, as is normal practice with lightweight space mirrors, the optimum support diameter ratio is about 0.5. This is true regardless of back shape. The best stiffness-to-weight is obtained with a parabolic set of back contours. For the optimum shape, the outer parabola vertex is at the edge, and the inner parabola vertex is at the center.

A set of six individual supports equally spaced on a common diameter provides a support condition which closely approximates a ring support. This suggests that for critical applications the double arch should be supported by a six-point support. For this type of support the optimum support diameter ratio is 0.55. The optimum back contour associated with the six-point support is identical in form to that of the six-point support. The inner and outer portions of the double arch back contour are parabolas; the vertex of the outer parabola is at the edge, and the inner at the center.

Like the single arch, the variable thickness of the double arch causes the mirror to distort following a sudden change in temperature. The double arch mirror usually develops a more complex optical surface distortion than the single arch when a similar temperature gradient is introduced into the mirror. This suggests that the double arch mirror is not well suited for applications in which the temperature is changing rapidly. Owing to better radial stiffness, the double arch mirror is subject to less bi-metallic bending than the single arch mirror shape when used with plated metal.

Unlike the single arch, the center of gravity location of the double arch is usually below the optical surface vertex. In most applications, the plane of the center of gravity is accessible for mounting. Since the double arch is readily supported through the plane of its center of gravity, deflection of the optical surface producing astigmatism is limited when the optical axis is horizontal. The double arch is well suited for use in mirror sizes over 1 m. Double arch mirror designs up to 4 m diameter are discussed in the literature.[21]

Mounting of the double arch is significantly more complex than the single arch. Standard practice is to produce cylindrical pockets in the back of the mirror, at the support ring. These pockets extend axially into the mirror to sufficient depth to reach the plane of the center of gravity. Radial support forces act through the plane of the center of gravity, against the side wall of the pockets. Axial supports are provided either at the same pockets, or spaced in-between the pockets. A key design issue is the athermalization of the mounting hardware in the pockets. If the temperature is limited, a simple Invar ring is bonded into the pocket, coincident with the plane of the center of gravity. For larger temperature changes an athermal socket is used, typically with a single or double conical taper.[22] These athermal sockets are difficult to fabricate in the back of the mirror, and significantly increase the mirror cost. Performance of conical athermal sockets is very good. In a test performed at NASA Ames Research Center, a 0.5-m-diameter fused silica double arch mirror with three athermal conical sockets was taken from room temperature to about 10 K. Total change in figure over this range of temperature was about 0.1 wave RMS (1 wave = 633 nm).[23] Figure 5.8 is a schematic drawing of the athermal mount used in the NASA Ames tests.

The double arch mirror is easy to fabricate.[24] It is generated from a solid or casting. Support during polishing is often provided by a continuous compliant ring. Mirror stiffness is sufficient, and a special support is not required for the inner and outer portions of the mirror. Edge thickness is often reduced in an effort to minimize the weight. The edge thickness of a metal matrix composite double arch mirror may be only 3 mm. Such thin edges pose a significant risk during fabrication and handling of the mirror.

Mounting sockets are machined in the back of glass material double arch mirrors using diamond tools. Fixed abrasive tools with a shape corresponding to the required socket shape are used. Mushroom-shaped holes in the mirror back are produced by rotating the mirror as well as the tool during the socket generating process. A cylindrical hole is first cored out of the back of the mirror. A tool with a corresponding contour is then inserted into the socket, and the tool rotated about its axis. The tool is then de-centered relative to the socket. Once the tool is de-centered, the mirror is rotated about the axis of the socket. This causes the tool to sweep out a circle concentric with the socket axis, producing the mushroom-shaped hole.

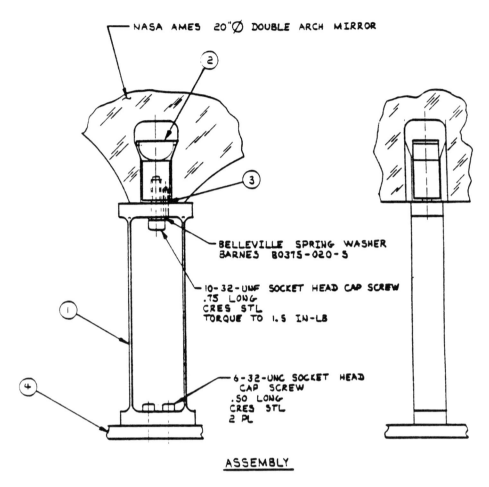

FIGURE 5.8 Athermal mounting socket for double arch mirror.

5.5 Sandwich Mirrors

Sandwich mirrors achieve the highest stiffness-to-weight ratios of any type of lightweight mirrors. Typically a sandwich mirror is from 40 to 20% of the weight of an equivalent solid right circular cylinder 6:1 diameter-to-thickness ratio mirror. Weight ratios below 20% are possible with the sandwich mirror, although cost and fabrication risk are high. The sandwich mirror is relatively expensive and difficult to fabricate. Mounting of the sandwich mirror is technically challenging, particularly when large loads must be accommodated in a dynamic environment. Thermal response of the sandwich mirror is controversial. For terrestrial applications there is the possibility of reduced thermal equilibrium time by ventilating the interior of the sandwich mirror.[25] Sandwich mirrors consist of a thin face sheet, a thin back sheet parallel to the face sheet, and a shear core connecting the two sheets. The shear core normally consists of thin ribs at right angles to the face and back sheets. These ribs intersect to form pockets between face and back sheets. The pocket geometry consists of triangular, square, or hexagonal cells.

Other types of shear cores are used. Cylindrical cell cores are used in machined sandwich mirrors. Tubular cells are employed in blow-molded borosilicate sandwich mirrors.[26] Foam cores are used to make ultralightweight sandwich mirrors.[27] The structural foam in the shear core is aluminum, metal matrix composite, or fused silica glass.

In the discussion of contoured back mirrors, the bridge analogy is used to explain the development of the double arch mirror. A beam analogy is likewise useful to explain the structural efficiency of the sandwich mirror. High stiffness is provided when the mass of a structure is distributed as far as possible from the neutral or bending axis. An I-beam distributes most of its mass in the flanges, which are far from the bending axis. Relatively little mass is placed in the web of the I-beam. In a similar fashion, the sandwich mirror places most of its mass in the face and back sheet, with as little mass as possible in the shear core. This provides a very efficient structure in bending.

Although the structure of a sandwich mirror is complex, the self-weight deflection of this type of mirror is readily calculated through the use of the concept of equivalent flexural rigidity.[28] The equivalent flexural rigidity of a sandwich mirror is the flexural rigidity of a solid plate of equal thickness. The flexural rigidity of a solid plate without a lightweight section is given by:

$$D_{solid} = \frac{Eh^3}{12\left(1-v^2\right)}$$

where D_{solid} = flexural rigidity of plate
           E = elastic modulus of plate material
           h = thickness of plate
           v = Poisson's ratio of plate material

In a similar manner, the flexural rigidity of a lightweight mirror is given by:

$$D_{lightweight} = \frac{Et_b^3}{12\left(1-v^2\right)}$$

where $D_{lightweight}$ = flexural rigidity of lightweight mirror
           E = elastic modulus of mirror material
           t_b = equivalent bending thickness of lightweight mirror
           v = Poisson's ratio of mirror material

The equivalent bending thickness is given by:

$$t_b^3 = \left(2t_f + h_c\right)^3 - \left(1 - \frac{\eta}{2}\right)h_c^3$$

where t_b = equivalent bending thickness of mirror
           t_f = face sheet thickness
           h_c = rib height
           η = rib solidity ratio

The above equations assume that the face and back sheets are of equal thickness. A key parameter is the rib solidity ratio, which is a function of the rib thickness and pocket size. The size of the pockets in the shear core is expressed by the diameter of a circle that is tangent to all walls of the pocket. This circle is the inscribed circle. The rib solidity ratio is given by:

$$\eta = \frac{\left(2B + t_w\right)t_w}{\left(B + t_w\right)^2}$$

where η = rib solidity ratio
           B = inscribed circle diameter
           t_w = rib thickness

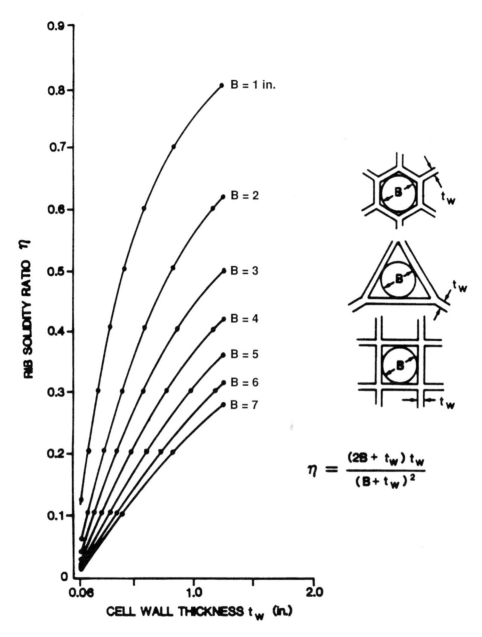

FIGURE 5.9 Rib solidity ratio. (From Valente, T.M. and Vukobratovich, D. 1989. *Proc. SPIE 1167*, 20.)

The rib solidity ratio is shown in Figure 5.9. The inscribed circle diameter for a triangular cell is

$$B_{triangular} = \frac{L}{\sqrt{3}}$$

where $B_{triangular}$ = inscribed circle diameter of triangular pocket
L = length of side of triangular pocket

The inscribed circle diameter for a square cell is

$$B_{square} = L$$

where B_{square} = inscribed circle diameter of square pocket
           L = length of side of square pocket

The inscribed circle diameter for a hexagonal pocket is

$$B_{hexagonal} = \sqrt{3}\,L$$

where $B_{hexagonal}$ = inscribed circle diameter of hexagonal pocket
            L = length of one of six sides of hexagonal pocket

Another parameter useful in analysis of lightweight sandwich mirrors is the cell pitch. This is the spacing of the cells in the shear core, or distance from center of inscribed circle to center of inscribed circle. The cell pitch is given by:

$$P = B + t_w$$

where   P = cell pitch
        B = inscribed circle diameter
        t_w = rib thickness

The weight of a sandwich mirror is given by:

$$W = \rho A \left(2 t_f + \eta h_c \right)$$

where   W = sandwich mirror weight
        ρ = mirror material density
        A = area of the mirror
        t_f = face sheet thickness
        η = rib solidity ratio
        h_c = rib height

Mehta has developed equations which optimize the distribution of mass in the face sheets and core of a sandwich mirror.[29] For a given overall height or weight, the optimum face sheet thickness is found for varying rib solidity which produces a mirror with the greatest possible flexural rigidity. Flexural rigidity is an important measure of stiffness, but is not necessarily a measure of stiffness-to-weight. For an optimum symmetric sandwich section:

$$t_f = \frac{W\left(\sqrt{1 - \dfrac{\eta}{2}} - \sqrt{1 - \eta} \right)}{\rho A \left\{ 2 \left[\sqrt{1 - \dfrac{\eta}{2}} - \sqrt{\left(1 - \eta \right)^3} \right] \right\}}$$

where t_f = optimum face sheet thickness
        W = mirror weight
        η = rib solidity ratio
        ρ = mirror material density
        A = area of mirror

Figure 5.10 shows the relationship between face sheet thickness, rib thickness, mirror thickness, and inscribed circle diameter. In this figure, the minima of the curves represent mirrors with

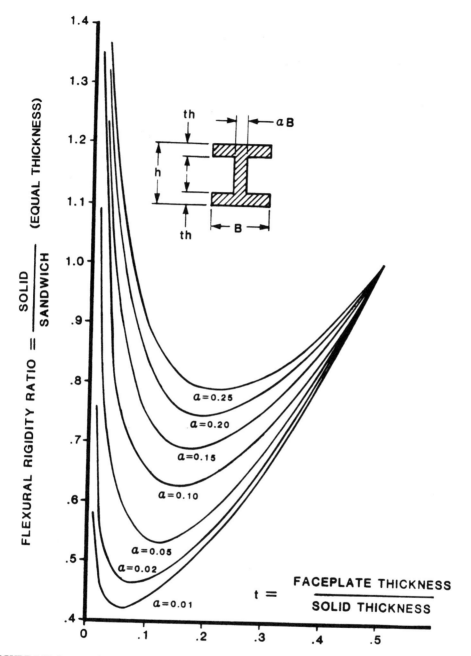

FIGURE 5.10 Symmetric sandwich mirror flexural rigidity. (From Valente, T.M. and Vukobratovich, D. 1989. *Proc. SPIE 1167*, 20.)

optimum stiffness-to-weight. The figure illustrates that sandwich mirrors are capable of better stiffness than solid mirrors of the same thickness.

The structural efficiency of the sandwich mirror is given by the ratio V_o/I_o. This is the ratio of unit volume to unit cross-sectional moment of inertia, and is given for the sandwich mirror by:

$$\frac{V_o}{I_o} = \frac{12(2t_f + \eta h_c)}{t_b^3}$$

where　V_o = unit volume of mirror
　　　　I_o = unit cross-sectional moment of inertia
　　　　t_f = face sheet thickness of mirror
　　　　η = rib solidity ratio
　　　　h_c = rib height
　　　　t_b = equivalent bending thickness of mirror

Shear deflection is an important component of the deflection of sandwich mirrors. Corrections for shear in a sandwich mirror are difficult to implement in simple closed-form equations. One approximation including shear effects for self-weight deflection of circular lightweight mirrors on multiple point supports is[30]

$$\delta \approx 0.025 \left(\frac{W}{A}\right)\left(\frac{r^4}{D}\right)\left(\frac{3}{n}\right)^2 + 0.65\left(\frac{W}{A}\right)\left(\frac{r^2}{S_c G A_o}\right)\left(\frac{3}{n}\right)$$

where　δ = peak-to-peak surface deflection of mirror
　　　　W = mirror weight
　　　　A = area of mirror
　　　　r = radius of mirror
　　　　D = flexural rigidity of mirror
　　　　n = number of support points
　　　　S_C = shear coefficient
　　　　G = shear modulus
　　　　A_o = cross-section area/unit width

The shear relations are

$$S_c = \frac{A_w}{A_w + A_f} = \frac{1}{1 + \dfrac{4t_f}{\eta h_c}}$$

$$A_o = \frac{2Pt_f + h_c t_w}{P}$$

$$G = \frac{E}{2(1+v)}$$

where　S_C = shear coefficient
　　　　A_W = area of rib
　　　　A_f = area of face sheet within pitch
　　　　t_f = face sheet thickness
　　　　η = rib solidity ratio
　　　　h_C = rib height
　　　　t_w = rib thickness
　　　　A_o = cross-section area/unit width
　　　　P = core pitch
　　　　G = shear modulus of mirror material
　　　　E = elastic modulus of mirror material
　　　　v = Poisson's ratio of mirror material

For any given pocket geometry or cell pattern, the shear core has equal shear rigidity if the pitch is held equal. For sandwich mirrors, structural efficiency is independent of cell geometry.[31] The equivalence of different cell or pocket geometries is controversial.[32] Experience with actual mirrors indicates at best a very weak dependence on shear core geometry.

Another controversial area of sandwich mirror design is the use of an edge band. An edge band provides additional tangential stiffness, or stiffness in the direction of the circumference of the mirror. This additional stiffness helps prevent deformation of the mirror edge when the mirror surface changes radius. The edge band provides protection for the thin ribs of the sear core. In some applications the edge band is used to provide an anchor point for the mirror mount. The disadvantage of the edge band is the additional weight of the band at the edge of the mirror.

Contouring the back of a sandwich mirror provides additional weight reduction at a relatively small penalty in stiffness. Such contouring is expensive and may add to the cost of mirror fabrication. Contouring the back may also present mounting problems, and degrade the thermal response of the mirror for the same reasons given for contoured back mirrors. For these reasons, contoured back sandwich mirrors are relatively uncommon.

"Quilting" is an issue that is related to the optimization process in the design of a sandwich mirror. Quilting is a permanent pattern of deformation that is polished into the mirror during optical fabrication. This deformation is due to deflection of the face sheet of the mirror between the ribs under polishing pressure. When the surface of the mirror is viewed by an optical test the resulting deflection pattern resembles the squares of a quilt. This resemblance explains the use of the term "quilting".

Quilting creates surface errors that are periodic and of relatively small amplitude. These periodic surface errors act like a diffraction grating. In a diffraction-limited system, quilting scatters light from the central maximum of the diffraction disk. The reduction in energy due to quilting is given by:

$$\frac{I_1}{I_0} = \frac{4\pi^2 \left(\dfrac{\delta_c}{2\lambda}\right)^2}{\left[1 - 2\pi^2 \left(\dfrac{\delta_c}{2\lambda}\right)^2\right]\left[1 - 4\pi^2 \left(\dfrac{\delta_c}{2\lambda}\right)^2\right]}$$

where I_1 = energy in central maximum with quilting
          I_0 = energy in central maximum without quilting
          δ_C = face sheet deflection due to quilting
          λ = wavelength

The relationship between quilting deflection and reduction of energy in the central maximum of the diffraction disk is shown in Figure 5.11.

The quilting deflection due to polishing pressure is given by:

$$\delta_c = \frac{PB^4}{\psi \left(\dfrac{Et_f^3}{12\left(1 - v^2\right)}\right)}$$

where δ_C = face sheet deflection due to quilting
        P = polishing pressure
        B = inscribed circle diameter
        E = mirror material elastic modulus

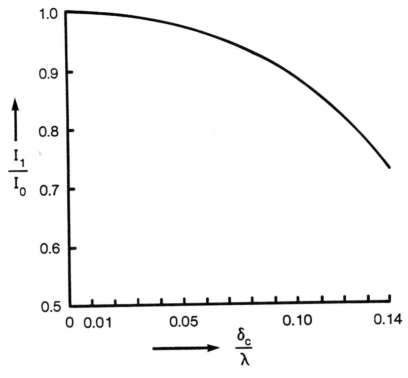

FIGURE 5.11 Reduction in energy of the central maximum of the diffraction disk due to quilting.

t_f = face sheet thickness
v = Poisson's ratio for mirror material
ψ = geometric quilting constant

The geometric quilting constant depends upon the shape of the cell or pocket in the mirror and is given by:

Geometric Quilting Parameters

Cell Geometry	Ψ
Triangular	0.00151
Square	0.00126
Hexagonal	0.00111

Figure 5.12 shows examples of quilting for a sandwich mirror with triangular cells.

Even at modest levels of weight reduction, quilting becomes a significant issue in lightweight mirror design. Quilting is relatively independent of cell shape. As indicated by the above table, quilting varies by a factor of less than 1.5 between hexagonal and triangular cells. This is shown in Figure 5.13. More important is the inscribed circle diameter and face sheet thickness.

Several solutions are suggested for quilting. Reducing polishing pressure is the simplest solution. Quilting is linearly dependent on polishing pressure. Reducing polishing pressure directly reduces quilting. The drawback to this idea is that polishing time is also linearly proportional to polishing pressure. Reducing polishing pressure increases polishing time, and therefore increases polishing cost. Increased polishing time increases the possibility of the mirror developing high spatial frequency errors called "ripple". High spatial frequency errors affect the mirror the same way as quilting.

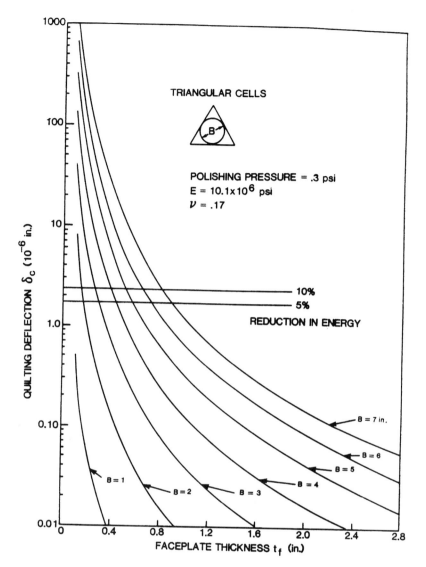

FIGURE 5.12 Example quilting deformation of triangular cell. (From Valente, T.M. and Vukobratovich, D. 1989. *Proc. SPIE 1167*, 20.)

Another possible solution is pressurization of the cells with a gas. This produces a pressure equal and opposite to the polishing pressure. This approach is expensive, and may be difficult to implement in a sandwich mirror with multiple holes in the core and back sheet. Polishing pressure is normally not uniform across the polishing tool, which limits the utility of the deflection compensation due to internal pressure. Part of the quilting deflection is due to poorly understood thermal effects, and pressurization does not reduce these effects.

If there are holes in the back sheet of the mirror, supports can be run through these holes to provide additional stiffness for the face sheet. These supports are called "quilting posts". Use of these supports is dependent on access through the back of the mirror. Adjusting the posts to provide just the right amount of support is difficult. If not properly adjusted the posts may act as hard points to cause another quilting pattern corresponding to the post location.

Filling the shear core with an incompressible fluid to limit quilting is sometimes attempted. This approach has not proven successful. Handling a fragile lightweight sandwich mirror when

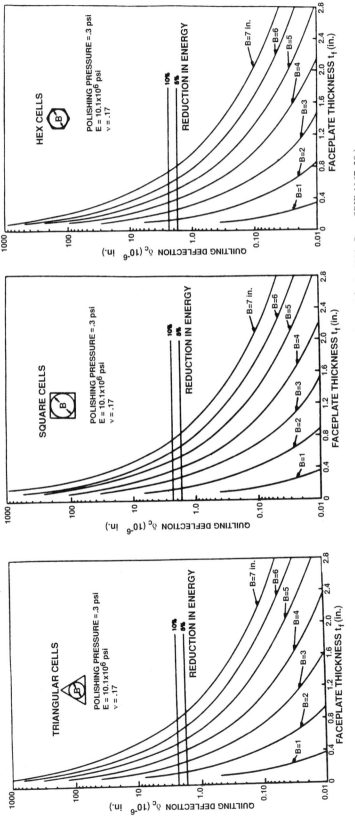

FIGURE 5.13 Quilting compared in triangular, square, and hexagonal cells. (From Valente, T.M. and Vukobratovich, D. 1989. *Proc. SPIE 1167, 20.*)

filled with fluid is very hazardous. The cells must be completely full of fluid and sealed relative to each other for this approach to succeed.

One solution to the quilting problem is local polishing of the face sheet above each cell. Polishing is normally done by hand. This approach is the most common. Although tedious, hand polishing of lightweight sandwich has been successful even on very large and lightweight mirrors.

Quilting posts require holes in the back sheet of the mirror. These holes represent discontinuities in the back sheet. Holes reduce the bending stiffness of the back sheet. As a rule of thumb, if the total area of the holes is less than 10% of the back sheet area, the mirror flexural rigidity is not significantly affected.

Holes in the shear core are also occasionally necessary. Holes located in the ribs between cells should be placed at the midplane of the mirror. In this position the holes have the smallest effect on flexural rigidity. Holes in the shear core should be circular to minimize stress concentration. Holes in the ribs should be less than 1/3 the height of the ribs. As a rule of thumb, the total area of the holes in the ribs should be less than 10% of the total rib surface area.

Lightweight sandwich mirrors are made by assembly from pieces, casting, or machining from a solid. The earliest technique for making a sandwich mirror was assembly from pieces, using a Bakelite adhesive. Ritchey pioneered this method, but was not successful in its use.[33]

Corning and others developed techniques for fusing together fused silica sections to produce a lightweight sandwich mirror. Typically the face and back sheets are made as individual plates. The shear core is made of slotted ribs. The assembly is placed into a furnace and fused together. The ribs of shear core produced by this assembly technique resemble the wall of an egg carton. This resemblance leads to the name "egg crate mirror".[34] Richard and Malvick demonstrated that the flexural behavior of egg crate mirrors is highly isotropic despite the apparent poor adhesion in the slotted sections of the core.

A similar fusion technique is used by Hextek to produce "blow molded" sandwich mirrors. In this technique the shear core is produced from tubes of glass close packed between face and back sheet. During the fusing process an inert gas is blown into the tubes. This gas causes the tubes to expand and adhere to each other. The cells produced in the shear core by this process are roughly hexagonal.[35]

Shear cores are sometimes produced by machining cells or pockets in a monolithic blank. The blank is placed between face and back sheet, and the assembly is fused in an oven. This technique is often used to produce hexagonal pockets. The shear core is normally machined using classic glass fabrication methods.[36] Recently Eastman Kodak developed a technique for machining the shear core by water jet cutting.[37]

Frit bonding is another technique for assembling lightweight sandwich mirrors. Frits are special glass materials that act as cements when heated.[38] A torch is used to heat the frits during assembly. Bulk heating often creates an undesirable sag of the face sheet between the ribs during fusing. Fritting avoids the bulk heating of the mirror. Extremely lightweight sandwich mirrors are often assembled using frits to avoid this bulk heating.

Lightweight sandwich mirrors are sometimes machined from a single solid blank. This approach is extremely expensive and involves substantial risk. The possibility of damage to the mirror during the machining operations is very high. The main advantage of this method is the high degree of uniformity of the resulting mirror, since it is produced from a single piece of material. This technique was pioneered in the 1960s. Mirrors up to 1.8 m diameter were produced in materials such as Cer-Vit using this method. Today REOSC in France is the main proponent of this technique. An example of such a mirror is the primary mirror of the ISO (Infrared Satellite Observatory).

Glasses which melt at relatively low temperatures are used to cast lightweight mirrors.[39] Casting was used to produce the 5-m primary mirror for the Hale telescope at Mt. Palomar. Casting is often combined with the use of spinning furnace to produce a near-net optical surface shape.[40]

There are a number of significant problems with the casting process. Hydrostatic pressures on the molds for the shear cores are substantial. The core molds may break loose, as occurred during the Mt. Palomar casting operation. Pressure may cause the walls of the shear core to deform. This may produce ribs of uneven thickness, which reduces the flexural rigidity of the mirror. At high casting temperatures borosilicate glass is very chemically active and may react with the mold material. This chemical reaction leads to undesirable properties in the glass. These properties include compositional inhomogeneity, staining of the surface, and de-vitrification of the glass.

Bubbles rise to the surface of the mirror during the casting process. The very high viscosity of the molten glass causes the bubbles to remain, creating voids in the surface of the mirror. Removal of the molds following casting is difficult. One technique is the use of water-soluble mold materials. Molds made of such materials are flushed out of the core using high pressure water jets.

Beryllium sandwich mirrors are produced using a particle metallurgy process. This process is high isostatic pressing or the HIP process. The HIP process is somewhat similar to casting in that molds are used. Copper molds are the most common. The copper molds are placed in a steel canister containing beryllium powder. Heat and pressure are used to consolidate the powder. Acid is then used to remove the copper molds from the mirror.[41] Thermal coefficient of expansion differences between the steel canister, copper mold, and beryllium power may cause cracking of the mirror during this process.

Other metals such as aluminum are used to make lightweight sandwich mirrors by casting, welding, or brazing. Brazing is the most common assembly method for producing metal sandwich mirrors, and is used with aluminum, metal matrix composite (SXA™), and beryllium. Adhesive bonding is not generally used due to the extreme difference in thermal coefficient of expansion between adhesive and metal mirror material.

Mounting lightweight sandwich mirrors requires incorporation of special mounting features into the mirror. Such features consist of solid or near-solid cells. Alternately local regions of high density are placed at the perimeter of the mirror or on the back sheet. On blow-molded and cast mirrors it is common to incorporate mounting surfaces in open bottom cells.[42] These mounting surfaces are coincident with the neutral or bending axis of the mirror. Such mounting surfaces are used to carry loads in the plane of the center of gravity of the mirror.

Frit bonding is used to attach mounting features to the surface of the mirror. Pads are attached to the edge band or the back sheet of the mirror. An alternate method uses conventional adhesives to attach the mounting pads. The use of adhesives requires great caution since the thermal coefficient of expansion of most adhesives is much higher than that of common mirror materials. In an extreme case, a large change in temperature may induce failure in the bond between mirror and pad.[43]

Most mounting geometries are kinematic. Principal concerns are differences in the thermal expansion coefficient between mirror and mount, and lack of co-planarity in the mounting pads. Flexures are often used to isolate the mirror from expansion or contraction of the mount. If the mounting surfaces are not in the same plane, moments are introduced into the mirror. These moments are reduced by adding additional degrees of rotational compliance in the mirror mounts.

5.6 Open-Back Mirrors

Open-back mirrors consist of a thin face sheet with an array of ribs on the back side of the face sheet. These ribs intersect to form pockets in the back of the mirror. Unlike the sandwich mirror these pockets are completely open in the back. Open-back mirrors are a traditional means of producing lightweight mirrors. Normally open-back mirrors are comparable in weight to sandwich mirrors, with a weight reduction of 30 to 40% of the same diameter 6:1 diameter-to-thickness ratio right circular cylinder mirror. Extremely lightweight mirrors are produced using the open back geometry, with weight reductions in some case below 20%. Stiffness-to-weight ratio of the open-back mirror is poor, and is inferior to both sandwich and contoured back mirrors. Thermal behavior of the open-back mirror is very good, due to the favorable ratio of volume-to-surface area. In addition, all portions of the mirror are relatively thin, producing short thermal time constants. The open-back mirror is normally lower in cost than the sandwich mirror, but higher in cost than the contoured back mirror. Mounting of open-back mirrors is relatively easy.

The cells or pockets of the shear core of a open-back mirror are open in the back. This open geometry is responsible for the term "open-back" mirror. Normally, cell or pocket geometry is similar to that found in sandwich mirrors. Triangular, square, and hexagonal cells or pockets are used in the shear core of open-back mirrors. Circular pockets are produced in mirrors machined from a solid blank. Other cell geometries are produced by combinations of radial and concentric circular ribs. This type of geometry is not common.

A beam analogy is useful in understanding the bending behavior of the open-back mirror. The open-back mirror is comparable to a T-shaped beam. A T-shaped beam lacks symmetry about its neutral or bending axis. Such a beam is poor in structural efficiency in comparison with an I-shaped beam. This analogy is extended to explain the poor bending stiffness of the open back in comparison with a sandwich mirror.

The stiffness of an open-back mirror is determined using an approach very similar to that used in finding the stiffness of a sandwich mirror. Many of the equations used in calculating the bending of an open-back mirror are identical to those used for the sandwich mirror. Only those equations which are unique to the open-back mirror are presented here. Like the sandwich mirror, the stiffness of an open-back mirror is determined by calculating the flexural rigidity of a solid mirror of equivalent stiffness. The flexural rigidity of a lightweight open-back mirror is given by:

$$D_{lightweight} = \frac{Et_b^3}{12(1-v^2)}$$

where $D_{lightweight}$ = flexural rigidity of lightweight mirror
    E = elastic modulus of mirror material
    t_b = equivalent bending thickness of lightweight mirror
    v = Poisson's ratio of mirror material

The equivalent bending thickness of a lightweight open-back mirror is given by:

$$t_b^3 = \frac{\left[\left(1-\frac{\eta}{2}\right)\left(t_f^4 - \frac{\eta h_c^4}{2}\right) + (t_f+h_c)^4 \frac{\eta}{2}\right]}{\left(t_f + \frac{\eta h_c}{2}\right)}$$

where t_b = equivalent bending thickness of lightweight mirror
    η = rib solidity ratio
    t_f = face sheet thickness
    h_c = rib height, measured from mirror back to back of face sheet

The same rib solidity ratio relationships are used for the open-back mirror as are used for the sandwich mirror. The rib solidity ratio is combined with the face sheet thickness and rib height to find the mirror weight. The weight of an open-back mirror is given by:

$$W = \rho A(t_f + \eta h_c)$$

where W = sandwich mirror weight
    ρ = mirror material density
    A = area of the mirror

t_f = face sheet thickness
η = rib solidity ratio
h_c = rib height

Mehta developed a relationship which is used to optimize the flexural rigidity of an open-back mirror. The lack of symmetry in an open-back mirror results in a significantly more complex relationship for optimization than is used for the sandwich mirror. This complex relationship is normally solved numerically. The relationship for an optimum open-back mirror is

$$4\left(t_f + \frac{\eta h_c}{2}\right)\left[\left(1 - \frac{\eta}{2}\right)\left(t_f^3 - \frac{h_c^3}{2}\right) + \frac{(\eta - 1)(t_f + h_c)^3}{2}\right]$$

$$-\left(\frac{1}{2}\right)\left[\left(1 - \frac{\eta}{2}\right)\left(t_f^4 - \frac{\eta h_c^4}{2}\right) + \frac{\eta(t_f + h_c)^4}{2}\right] = 0$$

where t_f = face sheet thickness
η = rib solidity ratio
h_c = rib height

Figure 5.14 shows the relationship between face sheet thickness, rib thickness, inscribed circle diameter, and mirror thickness. In this figure the optimum mirror designs for the best stiffness-to-weight are found to the left, at the bottom of the curves. Unlike the sandwich mirror, the open-back mirror never exceeds the flexural rigidity of a solid mirror of equal thickness.

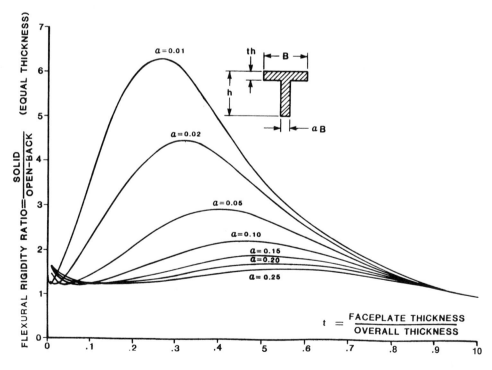

FIGURE 5.14 Open-back mirror flexural rigidity. (From Valente, T.M. and Vukobratovich, D. 1989. *Proc. SPIE 1167*, 20.)

The structural efficiency of an open-back mirror is given by:

$$\frac{V_o}{I_o} = \frac{12\left(t_f + \eta h_c\right)\left(t_f + \frac{\eta h_c}{2}\right)}{\left[\left(1 - \frac{\eta}{2}\right)\left(t_f^4 - \frac{\eta h_c^4}{2}\right) + \left(t_f + h_c\right)\frac{\eta}{2}\right]}$$

where V_o = unit volume of mirror
   I_o = unit cross-sectional moment of inertia
   t_f = face sheet thickness of mirror
   η = rib solidity ratio
   h_c = rib height

Shear deformation is an important component of the bending behavior of an open-back mirror. Self-weight deflection of an open-back mirror is computed using the same equation employed for sandwich mirrors. The lack of symmetry in the open-back mirror leads to a more complex set of equations for the shear coefficient in the deflection equation. The shear coefficient is given by:

$$S_c = \frac{u}{D_1 D_2 D_3 D_4}$$

$$u = 10\left(1 + v\right)\left(1 + 4m\right)^2$$

$$D_1 = 12 + 96m + 276m^2 + 192m^3$$

$$D_2 = v\left(11 + 88m + 248m^2 + 216m^3\right)$$

$$D_3 = 30n^2\left(m + m^2\right)$$

$$D_4 = 10vn^2\left(4m + 5m^2 + m^3\right)$$

$$m = \frac{Bt_f}{ht_w}$$

$$n = \frac{B}{h}$$

where S_C = shear coefficient
   v = Poisson's ratio for mirror material
   B = inscribed circle diameter
   t_f = face sheet thickness
   h = total mirror thickness
   t_w = rib thickness

There is a consensus in the U.S. optical engineering community that the optimum cell or pocket geometry for the open-back mirror is triangular. Other geometries provide less torsional resistance than the triangular geometry. Square cells are considered acceptable, and hexagonal cells are thought to provide inferior shear stiffness.

Open-back mirrors in the past were produced with a radial rib pattern. Concentric ribs (sometimes called intercostal ribs) joined the radial ribs to produce semicircular pockets or cells. This type of construction is used in the 0.6-m-diameter beryllium primary mirror for the IRAS.[44] This type of cell geometry is relatively easy to produce using older machine tools that are not numerically controlled. There is agreement in the U.S. optical community that this type of shear core is inferior in stiffness at comparable weight to shear cores using straight ribs and conventional cell shapes.

The open-back mirror is inferior in stiffness at comparable weights to the sandwich mirrors. Simple analysis of flexural rigidity without a shear correction sometimes indicates that superior stiffness is obtained in open-back mirrors by increasing the depth of the shear core or ribs. When a correction for shear is included for such very deep structures, shear effects are found to increase deflection. Very deep open back structures are not as efficient as comparable weight sandwich structures.

The stiffness of open-back mirrors is comparable with conventional solid mirrors of equal weight or equal thickness. For most applications the open back does not offer any significant advantage in stiffness when compared with the much lower-in-cost solid mirror. Open-back mirrors are used to provide shorter thermal equilibrium times than solid mirrors. If stiffness is not important, as is often the case for space applications, the open back may provide a reduction in weight in comparison with other mirror types.

Open-back mirrors are sometimes used to provide extremely lightweight mirrors. Two modifications used in open-back structures to further reduce weight are tapered backs and cylindrical holes located in the junction of the ribs. Both modifications are undesirable.

Open-back mirrors are low in bending stiffness. Tapering the ribs in the vertical direction tends to further reduce bending stiffness. This reduction is at the mirror edge, which is subject to the great deflection. If a reduction in weight is considered important, a better solution is redesign of the ribs or face sheet. One option is the use of thinner ribs.

Cylindrical holes located at the junction of the ribs interrupt the continuity of the ribs. This interruption significantly reduces the stiffness of the ribs. Any reduction in weight is offset by a decrease in the overall stiffness of the mirror.[45] Such cylindrical holes are sometimes used to improve the thermal equilibrium time of the mirror. Normally the rib junctions are the thickest portion of the shear core. Cylindrical holes reduce the effective thickness of the junction. This practice is questionable, since comparable thickness areas exist in the junction of rib and face sheet. The face sheet and rib junction is likely to be more critical to thermal response time of the mirror than the rib junctions.

Quilting effects in open-back mirrors are identical to those experienced in sandwich mirrors. Open-back mirror designs sometimes feature a pattern of "subribs" on the back of the face sheet between the ribs of the shear core. These subribs are intended to provide additional stiffness to the face sheet to help minimize quilting under polishing loads. Such relatively shallow ribs provide little additional stiffness. The weight of such subribs is better applied to increasing the thickness of the face sheet.

Open-back mirrors are often polished face down on the polishing lap. The pockets or cells of the shear core are provided with weight to offset the deflection of the face sheet under polishing pressure. Lead shot, for example, is used to load the cells. This method of reducing quilting effects appears to work for small, relatively stiff mirrors about 0.5 m in diameter.[46] The efficiency of this method for larger mirrors is controversial.

Open-back mirrors are produced by casting or machining from a solid blank. Casting is the oldest approach and was successfully employed for the primary mirror of the 5-m Hale telescope at Mt. Palomar. Machining is used to produce both metal and glass mirrors. Welding of metal mirrors to produce an open-back section is still largely experimental,[47,48] although large, low precision solar simulator mirrors have been produced this way.

Casting of open-back mirrors requires the use of a relatively low melting temperature glass, such as a borosilicate, or the use of a metal. Cast open-back mirrors are vulnerable to the same problems as discussed for sandwich mirrors. One advantage of the casting process for open-back mirrors is suppression of surface bubbles. Bubbles are suppressed by casting the mirror upside down, that is, with the face sheet down. The core mold is suspended above the mirror and then plunged into the molten material. This casting method is expensive and requires handling of the mold inside the furnace.

Machining of open-back mirrors from a solid is now common in the U.S. optical industry.[49] This approach allows the use of materials that cannot be cast. Machining from a solid is sometimes used as a way of minimizing quilting effects. The mirror optical surface is produced before the mirror is machined into a lightweight configuration. The mirror is then machined into an open-back geometry. There are two very serious difficulties with this approach: residual stress in the mirror and breakage during machining. Residual stress in the mirror is released during the machining of the mirror into the open-back configuration. This residual stress may produce mirror optical surface errors larger than the expected quilting errors from ordinary polishing. Any polishing to remove errors due to residual stress introduces the possibility of quilting. Breakage of the mirror during machining is a significant possibility. Such breakage occurs at the worst possible time, which is after the optical surface figure is produced. Some machining techniques break as many as one out of three mirrors.

Beryllium lightweight open-back mirrors are produced by machining from solid billets.[50] Older vacuum hot-pressed or VHP beryllium billets often are flawed, with internal voids. Such voids interrupt the continuity of the ribs and greatly reduce the stiffness of the mirror. This is an expensive procedure, since as much as 80% or more of the billet is removed during machining. Massive machining puts considerable stress into the beryllium mirror. Very rigorous heat treatment is necessary to remove this residual stress.

Open-back mirrors are straightforward to mount. Open-back mirrors are mounted by attachments at the edge of the mirror, or through the use of the interior of the cells or pockets. Mounting features are attached to either the rib sides or bottom of the pockets. The thickness of the face sheet is sometimes increased in the area of the cells used for mounting. The center of gravity of the open-back mirror is normally close to the bottom of the cell. A relatively small increase in face sheet thickness in the cells used for mounting brings the bottom of the cell into coincidence with the plane of the center of gravity of the mirror. This provides a very favorable location for mounting.

Although the open-back mirror mounting geometry is very favorable, the low stiffness requires attention in the design of the mount. Open-back mirrors are more sensitive to applied forces and moments than sandwich or contoured back mirrors. Particular care is necessary to minimize moments induced in the mirror due to alignment errors between mounts. One approach is to provide a universal joint between the point of attachment to the mirror and the mount at each mounting point. This universal joint consists of an ordinary ball and socket or a multiple degree of freedom flexure assembly.

5.7 Comparison of Mirror Performance

Lightweight mirrors are selected on the basis of the following criteria identified by Valente and Vukobratovich:[51]

1. Self-weight induced deflection
2. Efficiency of mirrors of equivalent weight, where efficiency is defined as a function of self-weight induced deflection and mirror thickness
3. Ease of fabrication

In the study performed by Valente and Vukobratovich, 1-m fused silica lightweight mirrors of single arch, double arch, sandwich, and open-back geometries were compared. Mirror thickness

NOTE: ALL DIMENSIONS
IN INCHES

SINGLE ARCH MIRROR

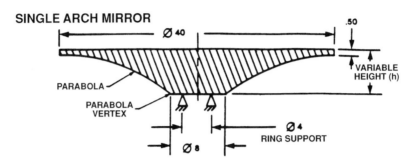

DOUBLE ARCH MIRROR

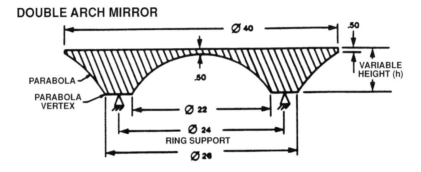

SYMMETRIC SANDWICH MIRROR

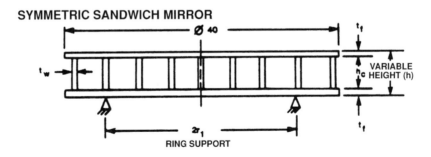

OPEN-BACK MIRROR

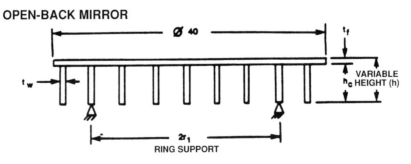

FIGURE 5.15   Mirror geometries used in mirror performance comparison. (From Valente, T.M. and Vukobratovich, D. 1989. *Proc. SPIE 1167*, 20.

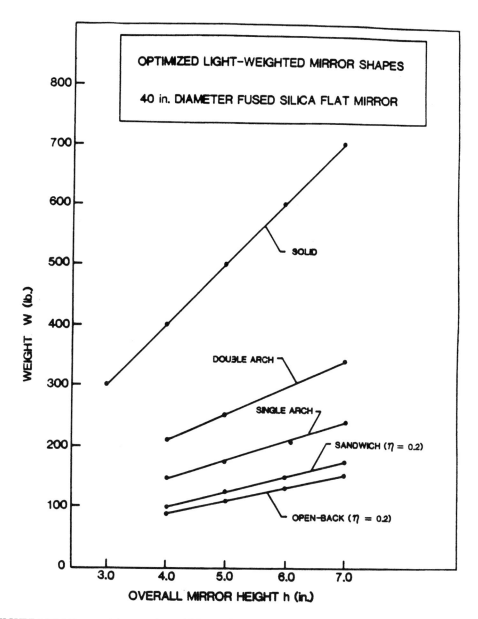

FIGURE 5.16 Mirror weight vs. mirror thickness. (From Valente, T.M. and Vukobratovich, D. 1989. *Proc. SPIE 1167*, 20.)

and weight were varied, with self-weight deflection computed for each variation in parameter. Figure 5.15 shows the mirror geometries used in this study. Figure 5.16 gives the mirror weight vs. mirror thickness or height. Certain reasonable assumptions were made in this study about detailed mirror parameters such as the rib solidity ratio, face sheet thickness, and so forth.

Figure 5.17 is a plot of the mirror height vs. self-weight deflection for the mirrors in the study. The worst deflection, and therefore the worst performance for a given height, is provided by the single arch mirror. The next best performance is obtained by the double arch. Significantly better at constant height than the double arch mirror is the solid mirror. Comparable deflection to the solid mirror is provided by the open-back mirror. Finally, the minimum deflection for a given height is provided by the sandwich mirror.

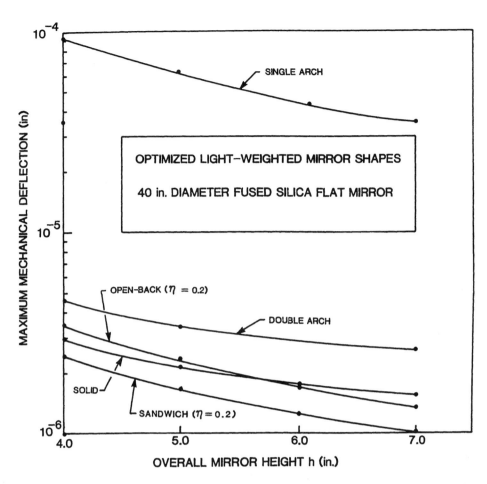

FIGURE 5.17 Mirror self-weight deflection vs. mirror height. (From Valente, T.M. and Vukobratovich, D. 1989. *Proc. SPIE 1167*, 20.)

Mirrors are sometimes produced by machining from solid blanks. For minimum self-weight deflection at a constant height, the best mirrors (excluding the sandwich mirror, which is not easily machined from a solid) are the open-back and solid mirrors. For comparable height, the open-back will be lighter than the solid mirror.

Different rankings are produced when mirror weight is plotted against self-weight deflection in Figure 5.18. Maximum deflection, and therefore the worst performance, is obtained with the single arch design. At comparable weights the solid mirror and open-back are the next best in performance. These two types of mirrors are virtually identical in deflection at comparable weights. Next best is the double arch mirror. Minimum deflection is provided by the sandwich mirror.

Both the deflection vs. height and deflection vs. weight charts indicate that best performance, in the sense of minimum self-weight deflection, is provided by the sandwich mirror. Use of a sandwich mirror may not always be possible, due to fabrication or cost concerns. Next best in minimizing deflection for a given weight is the double arch. At comparable weights, the solid and open-back mirrors provide the same self-weight deflection. If weight is an issue, use of an open-back mirror provides no stiffness advantage over a solid mirror of identical weight. Selection of an open-back mirror is often based on other criteria than stiffness and weight. Worst in performance is the single arch mirror. Mirror efficiency, defined as the total mirror height divided by the mirror self-weight deflection, is given in Figure 5.19.

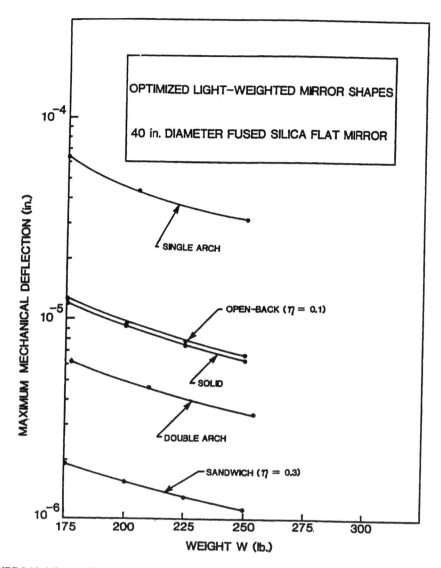

FIGURE 5.18 Mirror self-weight deflection vs. mirror weight. (From Valente, T.M. and Vukobratovich, D. 1989. *Proc. SPIE 1167*, 20.)

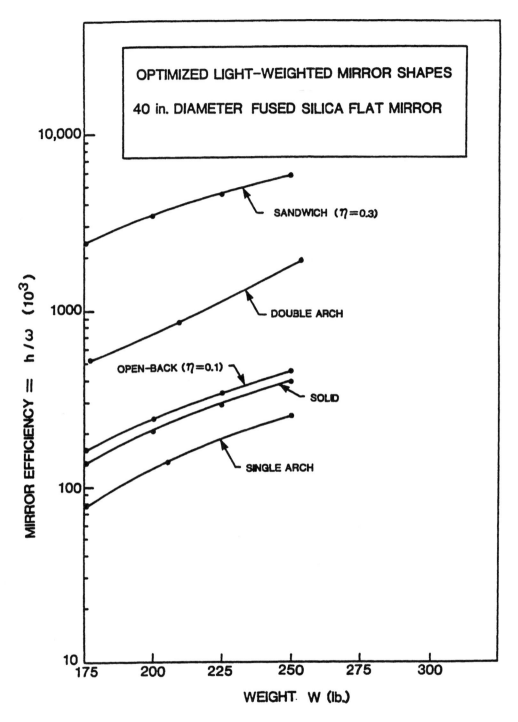

FIGURE 5.19 Mirror efficiency (total mirror height divided by mirror self-weight deflection) vs. mirror weight. (From Valente, T.M. and Vukobratovich, D. 1989. *Proc. SPIE 1167*, 20.)

References

1. Schwesinger, G. 1969. Elastostatischer Vergleich von Teleskopspiegeln in Leichtbauweise, *Messtechnik*, 9, 229.
2. Valente, T.M. 1990. Scaling laws for light-weight optics, *Proc. SPIE 1340*, 47.
3. Belvins, R.D. 1979. *Formulas for Natural Frequency and Mode Shape.* Van Nostrand Reinhold, New York.
4. Schwesinger, G. 1954. Optical effect of flexture in vertically mounted precision mirrors, *J. Optic. Soc. Am.*, 44, 417.
5. Pepi, J.W. 1987. Analytical predictions for lightweight optics in a gravitational and thermal environment, *Proc. SPIE 748*, 172.
6. Williams, R. and Brinson, H.F. 1974. Circular plate on multipoint supports, *J. Franklin Inst.*, 297.
7. Nowak, W.J. 1983. A parametric approach to mirror natural frequency calculations, *Proc. SPIE 450*, 164.
8. Nelson, J.E., Lubliner, J., and Mast, T.S. 1982. Telescope mirror supports: plate deflections on point supports, *Proc. SPIE 332*, 212.
9. Cho, M.K. 1989. Structural Deflections and Optical Performances of Light Weight Mirrors, Ph.D. dissertation. University of Arizona, Tucson.
10. Pepi, J.W. and Wollensak, R.J. 1979. Ultra-lightweight fused silica mirrors for a cryogenic space optical system, *Proc. SPIE 183*, 131.
11. Vukobratovich, D. 1989. Lightweight laser communications mirrors made with metal foam cores, *Proc. SPIE 1044*, 216.
12. Hibbard, D.L. 1990. Dimensional stability of electroless nickel coatings, *Proc. SPIE 1335*, 180.
13. Cho, M.K., Richard, R.M., and Vukobratovich, D. 1989. Optimum mirror shapes and supports for lightweight mirrors subjected to self weight, *Proc. SPIE 1167*, 2.
14. Prevenslik, T.V. 1968. The deflection of circular mirros of linearly varying thickness supported along a central hole and free along the outer edge, *Appl. Opt.*, 12, 1220.
15. Buchroeder, R.A., Elmore, L.H., Shack, R.V., and Slater, P.N. 1972. The Design, Construction and Testing of the Optics for a 147 cm Aperture Telescope, Optical Sciences Center Technical Report 79, University of Arizona, Tucson.
16. Huss, C.E. 1970. Axisymmetric Shells Under Arbitrary Loading, Ph.D. dissertation. University of Arizona, Tucson.
17. Applewhite, R.W. and Telkamp, A.R. 1992. Effects of thermal gradients on the Mars observer camera primary mirror, *Proc. SPIE 1690*, 376.
18. Sarver, G., Maa, G., and Chang, I. 1990. SIRTF primary mirror design, analysis and testing, *Proc. SPIE 1340*, 35.
19. Vukobratovich, D., Iraninejad, B., Richard, R.M., Hansen, Q.M., and Melugin, R. 1982. Optimum shapes for lightweighted mirrors, *Proc. SPIE 332*, 419.
20. Talapatra, D.C. 1975. On the self-weight sag of arch-like structures in the context of lightweight mirror design, *Opt. Acta*, 22, 745.
21. Meinel, A.B., Meinel, M.P., Hu, N., Hu, Q., and Pan, C. 1980. Minimum-cost 4-m telescope developed at October 1979 Nanjing Study of Telescope Design and Construction, *Appl. Opt.*, 4, 1674.
22. Iraninejad, B., Vukobratovich, D., Richard, R.M., and Melugin, R. 1983. A mirror mount for cryogenic environments, *Proc. SPIE 450*, 34.
23. Miller, J.H., Melugin, R.K., and Augason, G.C. 1988. Ames Research Center cryogenic mirror testing program. A comparison of the cryogenic performance of metal and glass mirrors with different types of mounts, *Proc. SPIE 973*, 62.
24. Anderson, D. and Parks, R.E. 1982. Gravity deflections of lightweighted mirrors, *Proc. SPIE 332*, 424.

25. Angel, J.R.P., Cheng, A.Y.S., and Woolf, N.J. 1985. Steps toward 8 m honeybomb mirrors. VI. Thermal control, *Proc. SPIE 571*, 123.

26. Parks, R.E., Wortley, R.W., and Cannon, J.E. 1990. Engineering with lightweight mirrors, *Proc. SPIE 1236*, 735.

27. Pollard, W., Vukobratovich, D., and Richard, R. 1987. The structural analysis of a lightweight aluminum foam core mirror, *Proc. SPIE 748*, 180.

28. Barnes, W.P., Jr. 1969. Optimal design of cored mirror structures, *Appl. Opt.*, 8, 1191.

29. Mehta, P.K. 1987. Flexural rigidity characteristics of light-weighted mirrors, *Proc. SPIE 748*, 158.

30. Seibert, G.E. 1990. Design of lightweight mirrors, *SPIE Short Course Notes*.

31. Richard, R.M. and Malvick, A.J. 1973. Elastic deformation of lightweight mirrors, *Appl. Opt.*, 12, 1220.

32. Sheng, S.C.F. 1988. Lightweight mirror structures best core shapes: a reversal of historical belief, *Appl. Opt.*, 27, 354.

33. Osterbrook, D.E. 1993. *Pauper and Prince: Ritchey, Hale and Big American Telescopes*. University of Arizona Press, Tucson.

34. Lewis, W.C. and Shirkey, W.D. 1982. Mirror blank manufacturing for the emerging market, *Proc. SPIE 332*, 307.

35. Melugin, R.K., Miller, J.H., Angel, J.R.P., Wangsness, P.A.A., Parks, R.E., and Ketelsen, D.A. 1985. Development of lightweight, glass mirror segments for the Large Depolyable Reflector, *Proc. SPIE 571*, 101.

36. Ruch, E. 1991. The manufacture of ISO mirrors, *Proc. SPIE 1494*, 265.

37. DeRock, J.W. and Wilson, T.J. 1991. Large, ultralightweight optic fabrication: a manufacturing technology for advanced optical requirements, *Proc. SPIE 1618*, 71.

38. Spangenberg-Jolley, J. and Hobbs, T. 1988. Mirror substrate fabrication techniques of low expansion glasses, *Proc. SPIE 1013*, 198.

39. Goble, L.W., Ford, R.M., and Kenagy, K.L. 1988. Large honeycomb mirror molding methods, *Proc. SPIE 966*, 291.

40. Goble, L.W., Angel, J.R.P., and Hill, J.M. 1988. Spincasting of a 3.5-m diameter f/1.75 mirror blank in borosilicate glass, *Proc. SPIE 966*, 300.

41. Paquin, R.A. 1985. Hot isostatic pressed beryllium for large optics, *Proc. SPIE 571*, 259.

42. Cannon, J. and Wortley, R. 1988. Gas fusion center-plane-mounted secondary mirror, *Proc. SPIE 966*, 309.

43. Huang, E.W. 1990. Thermal stress in a glass/metal bond with PR 1578 adhesive, *Proc. SPIE 1303*, 59.

44. Young, P. and Schreibman, M. 1980. Alignment design for a cryogenic telescope, *Proc. SPIE 251*, 171.

45. Schwesinger, G. 1968. General characteristics of elastic mirror flexure in theory and applications. In *Support and Testing of Large Astronomical Mirrors*, Crawford, D.L., Meinel, A.B., and Stockton, M.W., eds., p. 11. Kitt Peak National Observatory, Tucson.

46. Ulph, E. 1988. Fabrication of a metal-matrix composite mirror, *Proc. SPIE 966*, 166.

47. Forbes, F.L. July 1969. A 40-cm welded-segment lightweight aluminum alloy telescope mirror, *Appl. Opt.*, 8, 1361.

48. Rozelot, J.P. and Leblanc, J.M. 1991. Metallic alternative to glass mirrors (active mirrors in aluminum). A review, *Proc. SPIE 1494*, 481.

49. Mastandrea, A.A., Benoit, R.T., and Glasheen, R.R. 1989. Cryogenic testing of reflective optical components and telescope systems, *Proc. SPIE 1113*, 249.

50. Altenhof, R.R. 1975. The design and manufacture of large beryllium optics, *Proc. SPIE 65*, 20.

51. Valente, T.M. and Vukobratovich, D. 1989. A comparison of the merits of open-back, symmetric sandwich, and contoured back mirrors as light-weighted optics, *Proc. SPIE 1167*, 20.

6

Optical Mounts: Lenses, Windows, Small Mirrors, and Prisms

Paul R. Yoder, Jr.

6.1 Introduction and Summary

A variety of common techniques for mounting optical components such as individual and multiple lenses, windows, domes, filters, small mirrors, and prisms are discussed here. Numerous examples from the literature illustrate these techniques. Analytical relationships are given for estimating selected important attributes of the designs such as contact stress due to forces imposed during assembly or due to temperature changes or acceleration. Principles of optomechanical design and material selection intended to minimize the adverse effects of these imposed forces while retaining component function, location, and alignment are explained.

0-8493-0133-5/97/$0.00+$.50

6.2 Mounting Lenses

Low Precision Mounts

In this section, configurations for relatively low cost, low precision mounts for lenses are considered. In each case, the parts are premachined to specified dimensions and assembled without adjustment. Although the examples show single lens elements, some of these concepts are applicable to multiple element designs. For simplicity, it is assumed in most cases that the lenses are glass and the mount is a simple, cylindrical metal cell.

Spring Suspension

In applications involving large temperature changes with loose centration, tilt, and/or axial positioning tolerances, lenses might be supported by springs.[1,2] One such mounting, typically used to support condenser lenses or filters made of heat-absorbing glass in projector illuminators, is illustrated in Figure 6.1. Three flat springs spaced at 120° intervals around the lens rim are shaped to interface with that rim. Symmetry of the cantilevered springs tends to keep the lens centered. Free circulation of air around the lens is allowed. This type of mount also offers some protection against shock and vibration.

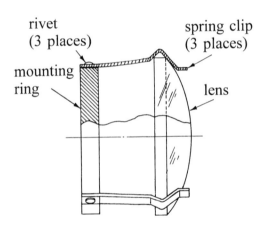

FIGURE 6.1 Typical configuration of a spring-mounted lens element. (From Yoder, P.R., Jr. 1993. *Opto-Mechanical Systems Design*, 2nd ed. Marcel Dekker, New York.)

Interference-Fit Ring

A lens can be held against a shoulder or spacer in a cell by an axial force exerted against the lens near its rim by a continuous ring as shown in Figure 6.2. The outside diameter (OD) of the ring is made slightly oversize with respect to the inside diameter (ID) of the cell. After installing the lens, the ring can be pressed into place or (preferably) the ring shrunk by cooling and inserted into a cell expanded by heating. The cell and ring materials should have similar thermal expansion coefficients to prevent loosening at extreme temperatures.

It is difficult to determine exactly when the ring touches the lens surface during assembly so achievement of a particular axial force on the lens is difficult.[2] Assembly by this technique is essentially permanent since it is virtually impossible to remove the ring without damaging either it or the lens.

Snap Ring

A discontinuous ring that drops into a groove machined into the inside surface of a cell is commonly termed a "snap" ring.[1-3] This ring, which acts as a spring, usually has a circular cross section as shown in Figure 6.3. Rectangular cross-section rings are less frequently used. The opening or slot

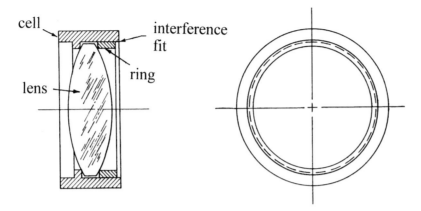

FIGURE 6.2 Typical configuration of a lens held in place by a pressed-in-place continuous ring. (Adapted from Yoder, P.R., Jr. 1993. *Opto-Mechanical Systems Design*, 2nd ed. Marcel Dekker, New York.)

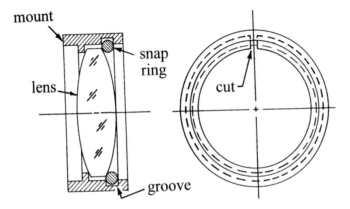

FIGURE 6.3 Typical configuration of a lens held in place by a discontinuous snap ring with circular cross section. (From Yoder, P.R., Jr. 1993. *Opto-Mechanical Systems Design*, 2nd ed. Marcel Dekker, New York.)

in the ring allows it to be compressed slightly while sliding into alignment with the groove. The groove cross section is usually rectangular.

Ensuring contact between the lens surface and the ring using this technique is difficult since thickness, diameter, and surface radius of the lens as well as ring dimensions, groove location, dimensions, and temperature changes all affect the degree of mechanical interference, if any, existing between the lens and ring. For this reason, this technique is used only where the location and orientation of the lens is not critical. Provision of a specific axial restraining force to the lens with this type mount is virtually impossible.

If the cell is designed without a groove, a snap ring can be inserted against the lens and constraint offered by friction between the ring and cell wall.[1] A rectangular ring is preferred in this case. Disassembly is possible. This design is sensitive to shock and vibration.

Burnished Cell

If the cell is made of malleable material such as brass or certain aluminum alloys, it can be designed to be mechanically deformed around the rim of a lens at assembly so as to secure that lens against an internal cell shoulder or spacer.[1,2,4] Figure 6.4 illustrates a typical example. At left is shown the cell prior to assembly. The chucking thread allows the cell to be installed onto a lathe spindle. In some designs, the cell lip is tapered to facilitate intimate contact with the lens bevel.

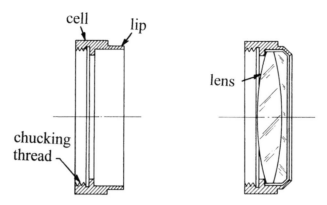

FIGURE 6.4 Typical configuration of a lens held in place by burnishing the cell rim. (From Yoder, P.R., Jr. 1993. *Opto-Mechanical Systems Design*, 2nd ed. Marcel Dekker, New York.)

Deformation of the cell lip is accomplished by bringing one, or preferably several, hardened tools or rollers against the lip at an oblique angle while the cell is rotated slowly. The lens should be held axially against the cell shoulder by external means (not shown in the figure) during the burnishing procedure to help keep it centered. If the radial fit between the lens and cell wall is close and the lens rim is accurately ground, this technique results in a well-centered subassembly. Once completed (see right view of Figure 6.4), the subassembly is essentially permanent.

This technique is most frequently used for mounting small lenses such as those for microscope or camera objectives where space constraints restrict use of separate retainers.

In some designs, a thin, narrow "washer" of resilient material such as a plastic or a thin rubber O-ring is inserted between the lens and the shoulder to soften the interface and provide some measure of sealing. Other designs may incorporate a coil spring between the lens and shoulder to offer some preload and flexibility against shock and vibration loads.[2,5]

Retaining Ring Mounts

The most frequently used technique for mounting lenses is to clamp the lens near its rim between a shoulder (or spacer in multiple component designs) and a retaining ring. The ring may be threaded loosely (Class 1 or 2 fit per ANSI Publication B1.1-1982) into the cell ID or held by screws as if it were a flange. The axial force exerted by the ring onto the lens is termed axial preload. The magnitude of this preload is determined at assembly and generally varies with temperature due to differences in thermal expansion coefficients of the materials involved. One reason for providing axial preload is to hold the lens in place under acceleration due to shock and/or vibration. The magnitude of preload, P_{ACC}, required for this purpose may be approximated by the expression:

$$P_{ACC} = W A F_S \tag{1}$$

where W is the weight of the lens, A is the maximum acceleration expected, and F_S is a safety factor (typically at least 2).

Axial preload induces axial stress into the lens and cell as discussed in the Section "Axial Stress at Single Element Interfaces". Manufacturing variations in axial dimensions of lenses and cells can be compensated with this type of mounting. It is compatible with environmental sealing with a cured-in-place elastomer or O-ring. Retaining ring designs also accommodate multiple component lens systems that are separated by spacers as discussed in Section 6.3.

Threaded Ring

Figure 6.5 illustrates a typical threaded retaining ring mount design for a biconvex lens. Contact between the lens and the mechanical parts occurs on the polished glass surfaces as recommended for precise centering of the optical axis to the mechanical axis of the cell and to minimize the need for precise edging or close tolerances on diameter of the lens.[2] This contact usually occurs slightly outside the clear aperture of the optical surface. To minimize bending of the lens, contact should occur approximately at the same height from the axis on both sides of the lens.[6] Since the spherical surface (either convex or concave) is more or less tilted with respect to the axis at the contact region due to its curvature, an axial preload applied at any point around the lens rim develops a force component directed toward the axis that tends to center the lens. When the lens is centered, these radial components balance each other and tend to hold the lens in the aligned condition. Hopkins[7] reported that a net difference in inclination of the front and back lens surfaces at the contact height of at least $17°$ is needed to achieve centering by means of axial preload.

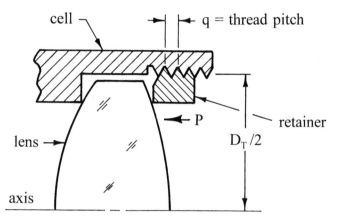

FIGURE 6.5 Typical configuration of a lens held in place by axial preload, P, from a threaded retaining ring. (From Yoder, P.R., Jr. 1993. *Opto-Mechanical Systems Design*, 2nd ed. Marcel Dekker, New York.)

The magnitude of the total preload (P) developed by a specific threaded retainer lens mount design with a specific torque (Q) applied to the ring at a fixed temperature can be estimated by the following equation:

$$P = 5\,Q/D_T \tag{2}$$

where D_T is the pitch diameter of the thread as shown in Figure 6.5.[2,8]

Clamping (Flange) Ring

A typical design for a lens mount involving a clamped (flange-type) retaining ring is shown in Figure 6.6. This type constraint is most frequently used with large aperture lenses where manufacture and assembly of a threaded retainer would be difficult. The retainer is usually configured with a sufficient tangential stiffness so that it does not warp significantly between the clamping bolts, thereby ensuring approximately uniform pressure against the lens surface around its rim. With preload applied symmetrically, this type of mount functions essentially like the threaded ring mount described above.

The magnitude of the preload produced by a given axial deflection of the flange can be approximated by considering it to be a perforated circular plate with outer edge fixed and uniform axially

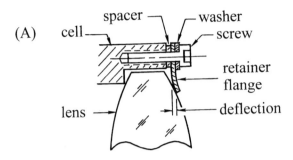

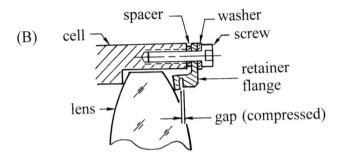

FIGURE 6.6 Typical configurations of lenses held in place by clamped flange-type retainers.

directed load applied along the inner edge to deflect that edge. Applicable equations are given by Rourk.[9] The spacer under the flange can be ground at assembly to the particular axial thickness that produces the predetermined flange deflection when firm metal-to-metal contact is achieved by tightening the clamping bolts. Variations in as-manufactured lens thicknesses, are easily accommodated with this technique. The flange material, thickness, and annular width are the prime design variables. The change in gap upon tightening the mounting screws can be measured to determine the flexure deflection in configuration (B) of Figure 6.6.

Techniques for Distributing Preload

Mounting designs using stiff flanges or retainers tend to contact the lens at the three highest points at low preload and at many points at higher preload. Stress concentrations and surface deformations may result in the latter case.[6]

Preload can be distributed more evenly with flexure designs such as shown in Figure 6.6 or 6.7. Registry for lens alignment purposes occurs at the lens-to-shoulder interface in all cases. Each type of threaded retainer provides some measure of resiliency in an attempt to distribute the force uniformly around the lens rim. In view (A), multiple flexures are built into a separate ring.[8] In view (B), an O-ring of about 70 durometer is compressed to 50 to 70% of nominal deflection.[11] In the design shown in view (C), the dimension "x" is machined at assembly to cause a predetermined amount of bending of the flexure when the retainer is firmly seated.[2] It is used with a convex surface. The configuration of view (D) serves the same function for a concave surface.

Sealing Techniques

Lenses mechanically clamped with threaded or flange-type retainers can be sealed to their cells by injecting elastomeric sealants into annular grooves machined into the retainer or cell. O-rings can be incorporated into some designs for this purpose.[10] Figure 6.8 illustrates each of these techniques. The lens should register against the cell for alignment purposes in both cases. Injected elastomeric sealant is usually inserted after all adjustments between lens and cell have been completed. Note that the elastomer must touch the lens all around its rim. If a retainer is used, it is advisable to

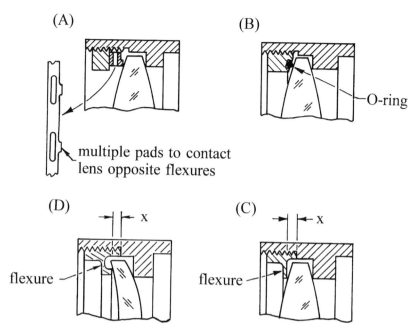

FIGURE 6.7 Some concepts for creating axially resilient interfaces between a lens and its mount to distribute preload more uniformly. (Views (A), (B), and (C) from Yoder, P.R., Jr. 1993. *Opto-Mechanical Systems Design*, 2nd ed. Marcel Dekker, New York.)

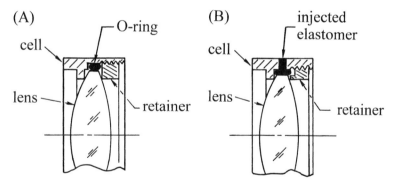

FIGURE 6.8 Typical means for sealing a lens into its cell with (A) an O-ring and (B) cured-in-place elastomer.

provide an annular airspace for the elastomer to expand into at high temperature. This is hard to do with the sealant injected after the retainer is in place. In designs with O-rings, the ring should have a durometer as large as 70 and preferably would be located around the periphery of the lens. Interfaces should be dimensioned so the ring is nominally compressed about 50 to 70% of full recommended compression at assembly. More or less compression can then take place as the temperature changes without losing sealing capability or causing undue stress.[11]

Axial Stress at Single Element Interfaces

General Considerations

The axial stress developed within the lens due to the applied axial preload depends upon the magnitude of that preload, the geometry of the interface, and the physical properties of the materials involved. The preload generally varies with temperature and this causes related changes in axial stress.

The axial stress is maximum within the narrow annular area of contact between the metal and glass. It therefore is frequently called "contact stress". The stress is generally lower at points within the lens more remotely located from the contact area.

The axial contact stress (S_A) in a lens preloaded at a height y from the axis is estimated from the following equation adapted by Yoder[2] from Roark:[9]

$$S_A = 0.798 \left(K_1 \, p / K_2 \right)^{1/2} \tag{3}$$

where K_1 depends upon the optomechanical interface design and the lens surface radius, K_2 depends upon the elastic properties of the glass and metal materials, and p is the linear preload as determined from the total preload (P) by:

$$p = P/2\pi y \tag{4}$$

The term K_1 will be discussed later in conjunction with the various interface types.

For all interface types, the term K_2 is given by:

$$K_2 = K_G + K_M = \left[\left(1 - \upsilon_G^2 \right) / E_G \right] + \left[\left(1 - \upsilon_m^2 \right) / E_M \right] \tag{5}$$

where υ_G, E_G, υ_M, and E_M are Poisson's ratio and Young's modulus values for the contacting glass and metal, respectively.

The size of the contact area depends upon the same parameters as the stress. Under light preload, the contact is essentially a "line" of length $2\pi y$. As the preload increases, the line contact widens and the resulting area is computed as $2\pi y \Delta y$ where Δy is the annular width of the elastically deformed area. The equation for Δy as adapted by Yoder[2] from Roark[9] is

$$\Delta y = 1.6 \left(K_2 \, p / K_1 \right)^{1/2} \tag{6}$$

where all terms except K_1 are as defined above.

The "Sharp Corner" Interface

The "sharp corner" interface was first defined by Delgado and Hallinan[12] as one in which the nominally 90° intersection of the machined surfaces on the metal part has been burnished in accordance with good shop practice to a radius of the order of 0.002 in. (0.05 mm). This small-radius surface contacts the glass at a height y. Figure 6.9(A) illustrates a typical design for a biconvex lens. The "hole" in the retainer referred to in the figure accepts a pin on a wrench used to tighten the retainer. A diametrical slot is frequently used for this purpose.

Hopkins[7] indicated that the machinist is more likely to achieve a smooth edge on a "sharp corner" and the chance of damage to that edge during assembly is minimized if the angle between intersecting surfaces is greater than 90°. Figure 6.9(B) shows such a design with 135° included angles as applied to a biconcave lens.

Again applying equations from Roark,[9] Yoder[2] showed that the value of K_1 in Equation 3 for any optomechanical interface is given by:

$$K_1 = \left(D_1 \pm D_2 \right) / D_1 D_2 \tag{7}$$

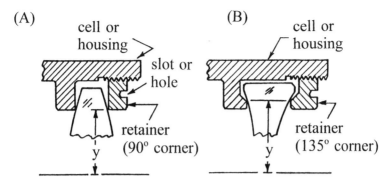

FIGURE 6.9 Schematics of "sharp corner" interfaces on (A) convex lens surfaces and (B) concave lens surfaces.

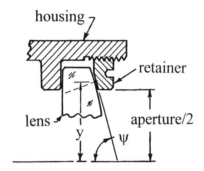

FIGURE 6.10 Schematic of a tangential interface on a convex lens surface.

where $D_1 = 2$(surface radius) and $D_2 = 2$(corner radius). The "+" sign is used for convex surfaces and the "−" sign is used for concave surfaces. K_1 is always assigned a positive sign.

In the case of the "sharp corner" interface, D_2 is typically 0.004 in. (0.1 mm).[12] For surface radii larger than about 0.2 in. (5.1 mm), D_2 can be ignored and the value of K_1 is constant at 250/in. (10/mm).[2]

The Tangential Interface

An interface design in which the lens surface contacts a conical surface in the mount is called a tangential interface. Figure 6.10 illustrates such a design. Note, this type interface is not feasible with a concave lens surface. The cone half-angle ψ is determined by the following equation:

$$\psi = 90° - \arcsin(y/R) \tag{8}$$

where R is the surface radius.

It is common practice to define the contact height y as the midpoint between the clear aperture and the edge of the polished surface. The tolerance on ψ in a given design depends primarily on the radial width of the conical annulus on the metal part and the allowable error in axial location of the lens vertex. Typically, this tolerance is about ±1°.

Since D_2 of Equation 7 is infinite for a tangential interface, the value of K_1 reduces to $1/D_1 = 0.5/R$ where R is the surface radius.[2] The axial stress developed in a lens of given surface radius by a given preload with a tangential interface is smaller by a factor of $(250\,D_1)^{1/2}$ than that with a "sharp corner" interface.

The Toroidal Interface

Figure 6.11 shows toroidal (or donut-shaped) mechanical surfaces contacting convex and concave lens surfaces. Yoder[13] demonstrated that the axial stress developed in a given lens with surface radius R at given preload with a toroidal interface is essentially the same as that of the tangential interface if the cross section radii of the toroids are at least −10R for a convex lens surface and 0.5R for a concave lens surface. The corresponding values for K_1 are −0.55/R and 0.5/R for the convex and concave cases, respectively. The axial stresses developed in these lenses with these preferred toroidal radii are significantly reduced from those that would prevail with "sharp corner" interfaces.

Achievement of accurate cross-sectional radii on toroidal interfaces is not essential since stresses vary slowly with these parameters in the regions of the preferred values.[13] Tolerances of +100% are common. Figure 6.12 shows typical concepts for toroidal spacers to be used between concave or convex lens surfaces.

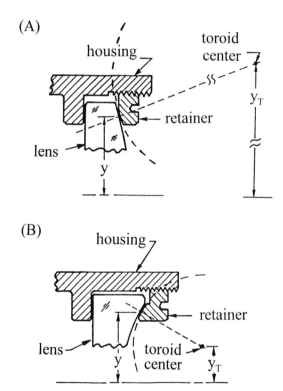

FIGURE 6.11 Schematics of toroidal interfaces on (A) a convex lens surface and (B) a concave lens surface.

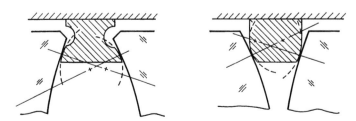

FIGURE 6.12 Schematics of spacers with toroidal interfaces. (From Yoder, P.R., Jr. 1991. *Optomechanics and Dimensional Stability*, SPIE Proc. Vol. 1533, Paquin, R.A. and Vukobratovich, D., eds., p. 2.)

The Spherical Interface

Figure 6.13 illustrates typical spherical contact lens-to-cell interfaces for convex and concave surfaces. Such designs have the advantage of distributing axial preloads over large annular areas and hence are virtually stress-free. If the surfaces match closely in both curvature and optical figure, the contact stress equals the total preload divided by the annular area of contact. As indicated by the dashed lines in both views, the design must provide access for lapping in order to produce accurate spherical interfaces on the mount. Surface matching requires very careful manufacture and increases cost. For this reason, the spherical interface is not frequently used.

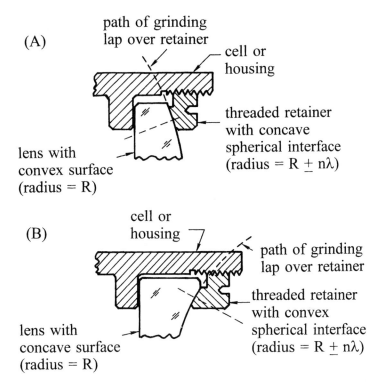

FIGURE 6.13 Schematics of spherical interfaces on (A) a convex lens surface and (B) a concave lens surface.

Flat and Step Bevel Interfaces

In Figure 6.14 a lens mount configuration involving flat bevels on concave and convex surfaces is shown. To facilitate alignment, a flat bevel should be accurately perpendicular to the optical axis of the lens. If used on a convex surface, a step should be ground into the rim of that surface. If intimate contact with the mount occurs over a flat bevel uniformly around the rim of the lens, the contact stress at the interface equals the total preload divided by the annular area of contact.

If flat bevels are used on both sides of a lens having optical power in both surfaces, as frequently is the case with biconcave lenses, both should be perpendicular to the lens optical axis. If this is not the case, it is impossible to accurately align the optical axis to the mechanical axis of the subassembly. Self-centering by applying axial preload also is impossible with flat bevels applied to both lens surfaces. A preferred interface for a concave surface is the toroidal one illustrated in Figure 6.11(B).

Parametric Comparisons of Interface Types

Figure 6.15 shows the nature of the variation of axial stress with radius of the contacting corner for a particular design having a given convex lens surface radius and a given mechanical preload.

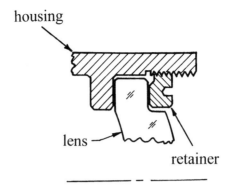

FIGURE 6.14 Schematics of flat bevel interfaces on (left) a concave lens surface and (right) a convex lens surface.

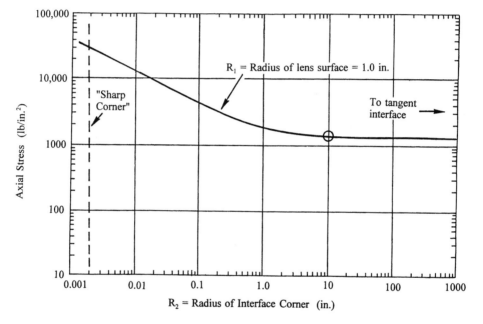

FIGURE 6.15 Variation of axial stress in a typical lens at constant preload as the radius of the mechanical surface contacting its convex surface is changed. (From Yoder, P.R., Jr. 1993. *Optomechanical Design, SPIE Proc. Vol. 1998*, Vukobratovich, D., Yoder, P.R., Jr., and Genberg, V., eds., p. 8.)

Both stress and corner radius are plotted logarithmically to cover large ranges of variability. At the left is the short corner radius characteristic of the "sharp corner" interface, while at the right, the tangential interface case is approached asymptotically. Between these extremes are an infinite number of toroidal interface designs. The "preferred" toroidal radius (equal to −10R) for which the stress is within 5% of the value for a tangential interface is indicated by the circle.[13]

Figure 6.16 shows a similar relationship for a concave lens surface example. The "sharp corner" case is again at the left. As the toroidal radius increases toward the matching radius (spherical interface) limit, the stress decreases. The circle represents the "preferred" toroidal radius of 0.5R for which the stress approximates that which would prevail at the same preload on a convex surface of the same radius using a −10R toroidal interface.[13]

The last two figures show conclusively that the axial contact stress is always significantly higher with a "sharp corner" interface than with any other type. It has been recommended that whenever

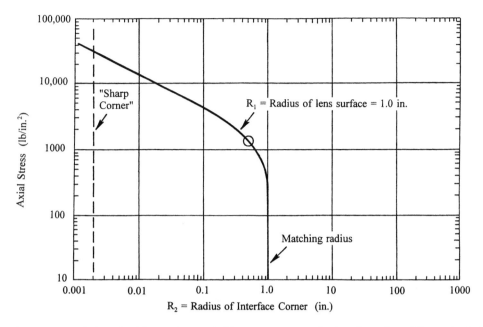

FIGURE 6.16 Variation of axial stress in a typical lens at constant preload as the radius of the mechanical surface contacting its concave surface is changed. (From Yoder, P.R., Jr. 1993. *Optomechanical Design, SPIE Proc. Vol. 1998*, Vukobratovich, D., Yoder, P.R., Jr., and Genberg, V., eds., p. 8.)

slightly higher manufacturing cost can be tolerated, tangential interfaces be used on all convex lens surfaces and toroidal interfaces of radius approximately 0.5R be used on all concave surfaces.[13]

Stress Variation with Surface Radius

Figures 6.17 and 6.18 show graphically how the axial stress varies as the surface radius is changed by successive factors of 10 for convex and concave surface cases, respectively. The preload is held constant. The stress is seen to be independent of surface radius or its algebraic sign for a "sharp corner" interface (left side of each graph). The greatest changes occur for long-radii toroids on either type surface, for the "tangential interface on a convex surface", and the "matching radii on a concave surface" cases. It has been shown[13] that, for the toroids indicated by the circles on each curve (toroid radius = −10R for convex and 0.5R for concave), if the surface radius changes from R_1 to R_2 with all other parameters unchanged, the corresponding stress changes by $(R_1/R_2)^{1/2}$. Hence, for the 10:1 increases in surface radius depicted in Figures 6.17 and 6.18, the stress decreases by a factor of $0.1^{1/2} = 0.316$.

Stress Variation with Preload

If the total preload, P, on a lens with any type interface and any surface radius increases from P_1 to P_2 while all other parameters remain fixed, the resulting axial contact stress changes by a factor of $(P_2/P_1)^{1/2}$. A tenfold increase in preload therefore increases the stress by a factor of 3.162.[13]

Effects of Changing Materials

The first (K_G) and second (K_M) terms of Equation 5 apply independently to the two materials in contact at the lens-to-mount interface.[14] Although lenses are commonly made of glass, crystals, or plastic, considerations here are limited to optical glass materials.

Walker[15] selected 62 basic types of optical glass offered by various manufacturers that "span the most common range of index and dispersion and have the most desirable characteristics in terms of price, bubble content, staining characteristics and resistance to adverse environmental conditions." The factor K_G of Equation 5 has been calculated for each of the 68 Schott varieties included

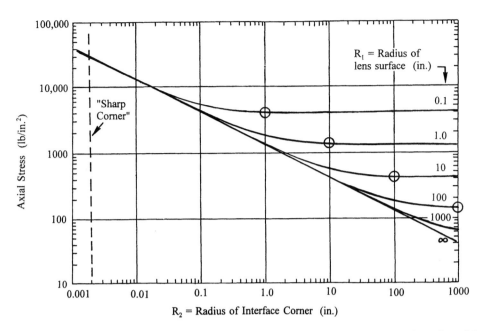

FIGURE 6.17 Variations of axial stress in a family of typical lenses at constant preload as the radius of the mechanical surface contacting a convex surface and the radius of that surface are changed. (From Yoder, P.R., Jr. 1991. *Optomechanics and Dimensional Stability, SPIE Proc. Vol. 1533*, Paquin, R.A. and Vukobratovich, D., eds., p. 2.)

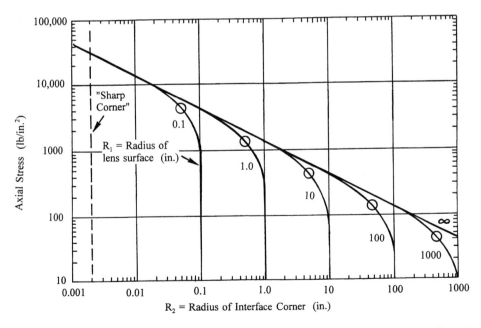

FIGURE 6.18 Variations of axial stress in a family of typical lenses at constant preload as the radius of the mechanical surface contacting a concave surface and the radius of that surface are changed. (From Yoder, P.R., Jr. 1991. *Optomechanics and Dimensional Stability, SPIE Proc. Vol. 1533*, Paquin, R.A. and Vukobratovich, D., eds., p. 2.)

in Walker's list. Figure 6.19 shows, in bar-graph form, how the magnitude of K_G varies for this family. The sequence is by increasing glass-type designation and hence by increasing index of refraction. There is no apparent correlation between K_G and index of refraction. The glasses with

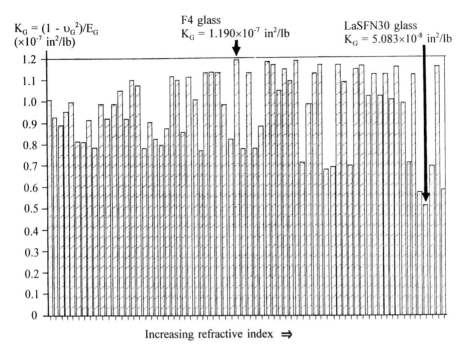

FIGURE 6.19 Variation of magnitude of K_G for 68 Schott glasses selected by Walker.[15] The left-to-right sequence is by increasing glass type code and, hence, by refractive index. (From Yoder, P.R., Jr. 1993. *Optomechanical Design, SPIE Proc. Vol. 1998*, Vukobratovich, D., Yoder, P.R., Jr., and Genberg, V., eds., p. 8.)

Table 6.1   Selected Mechanical Properties of the Schott Glasses Included in Walker's[14] List of Preferred Glass Types that have the Lowest and Highest Value for K_G

Glass Name	Glass Type	$K_G = (1 - \upsilon_G^2)/E_G$ in.²/lb (m²/N)	Young's Modulus E_G lb/in.² (N/m²)	Poisson's Ratio υ_G	Thermal Expansion Coefficient α_G/°F (/°C)
F4	617366	1.190×10^{-7} (1.726×10^{-11})	7.98×10^6 (5.50×10^{10})	0.225	4.6×10^{-6} (8.3×10^{-6})
LaSFN30	803464	5.083×10^{-8} (7.372×10^{-12})	1.80×10^7 (1.24×10^{11})	0.293	3.4×10^{-6} (6.2×10^{-6})

Adapted from Yoder, P.R., Jr. 1991. *Optomechanics and Dimensional Stability, SPIE Proc. Vol. 1533*, Paquin, R.A. and Vukobratovich, D., eds., p. 2.

the highest (F4) and lowest (LaSFN30) values of K_G are indicated in the figure by arrows. The ratio of K_G values for these extreme glasses is 2.34. Table 6.1 gives the pertinent mechanical properties of these two glasses. Properties data were obtained from the Schott catalog.[16]

Table 6.2 lists key mechanical properties of six types of metals selected for consideration here. K_M was calculated from Equation 5 and varies from 2.366×10^{-8} in.²/lb (for beryllium) to 1.350×10^{-7} in.²/lb (for magnesium). The ratio of these extreme values is 5.70. Figure 6.20 shows graphically how K_M varies for these metals.

It should be noted that low values for either K_G or K_M tend to increase lens stress since these factors appear in the denominator of Equation 3.

Yoder[14] analyzed combinations of the metals from Table 6.2 with the two glasses of Table 6.1 in a typical glass-to-metal design with preload and interface-type constant. Figure 6.21 shows plots of variations of axial stress with material type. The vertical scale is normalized to the stress level of a BK7 lens in an aluminum mount (triangle). The horizontal scale is Young's modulus for the metals. The horizontal spacings of the vertical dashed lines representing the selected metals give a sense of the variation of this important parameter from one metal to another. The two curved

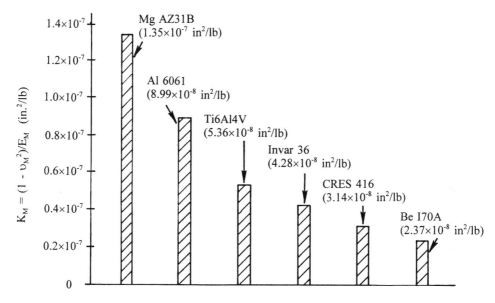

FIGURE 6.20 Variation of K_M for six metals typically used in lens mounts in optical instruments. (From Yoder, P.R., Jr. 1993. *Optomechanical Design, SPIE Proc. Vol. 1998*, Vukobratovich, D., Yoder, P.R., Jr., and Genberg, V., eds., p. 8.)

Table 6.2 Selected Mechanical Properties of Some Metals Used in Lens Mounts. Sequence is by Increasing K_M

Metal Type	$K_M = (1 - v_M^2)/E_M$ in.²/lb (m²/N)	Young's Modulus E_M lb/in.² (N/m²)	Poisson's Ratio v_M	Thermal Expansion Coefficient α_M °F (/°C)
Be I70A	2.366×10^{-8} (3.438×10^{-12})	4.2×10^7 (2.89×10^{11})	0.080	1.6×10^{-5} (1.13×10^{-5})
CRES 416	3.138×10^{-8} (4.55×10^{-12})	2.90×10^7 (2.00×10^{11})	0.300	5.5×10^{-6} (9.9×10^{-6})
Invar 36	4.28×10^{-8} (6.231×10^{-12})	2.14×10^7 (1.47×10^{11})	0.290	7.0×10^{-7} (1.26×10^{-6})
Ti6Al4V	5.36×10^{-8} (7.758×10^{-12})	1.65×10^7 (1.14×10^{11})	0.340	4.9×10^{-6} (8.8×10^{-6})
Al 6061	8.988×10^{-8} (1.305×10^{-11})	9.9×10^6 (6.82×10^{10})	0.332	1.3×10^{-5} (2.36×10^{-5})
Mg AZ31B	1.350×10^{-7} (1.959×10^{-11})	6.50×10^6 (4.48×10^{10})	0.350	1.4×10^{-5} (2.52×10^{-5})

Adapted from Yoder, P.R., Jr. 1991. *Optomechanics and Dimensional Stability, SPIE Proc. Vol. 1533*, Paquin, R.A. and Vukobratovich, D., eds., p. 2.

lines connect discrete points representing particular combinations of glasses and metals and are not really continuous functions. The fact that the curves diverge toward the right indicates the greater significance of differing glass characteristic K_G for the stiffer metals having smaller K_M values.

Rate of Change of Preload with Temperature

If, as is usually the case, the lens and mount materials have dissimilar thermal expansion coefficients; temperature changes, ΔT, cause changes in total axial preload, P, exerted onto the lens. The following equation[2] quantifies this relationship:

$$\Delta P = K_3 \Delta T \qquad (9)$$

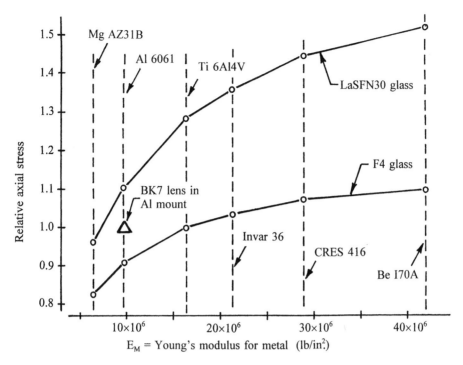

FIGURE 6.21 Variations of normalized axial stress in lenses made of two types of glass when mounted in different metals. (From Yoder, P.R., Jr. 1993. *Optomechanical Design, SPIE Proc. Vol. 1998*, Vukobratovich, D., Yoder, P.R., Jr., and Genberg, V., eds., p. 8.)

where

$$K_3 = \frac{-E_G A_G E_M A_M (\alpha_M - \alpha_G)}{E_G A_G + 2 E_M A_M} \quad (10)$$

Here, υ_G, E_G, υ_M, and E_M are as defined above and α_M and α_G are thermal expansion coefficients of the two materials. The terms A_G and A_M represent cross-sectional areas of the annular stressed regions within the lens and within the mount. These geometric parameters are shown in Figures 6.22 and 6.23.

Equations for A_G and A_M follow:

$$\text{If } (2y + t_E) < D_G, \text{ then } A_G = 2\pi y t_E \quad (11)$$

$$\text{If } (2y + t_E) \geq D_G, \text{ then } A_G = (\pi/4)(D_G - t_E + 2y)(D_G + t_E - 2y) \quad (12)$$

$$A_M = 2\pi t_C \left[(D_M/2) + (t_C/2) \right] \quad (13)$$

where t_E is the edge thickness of the lens at the contact height y, t_C is the radial wall thickness of the mount at the lens rim, D_M is the ID of the mount at the lens rim, and D_G is the OD of the lens.

The relationship between preload change and temperature change (Equation 9) is linear for any combination of lens and mount materials. The factor K_3 is the slope of this line. A negative value for K_3 means that a drop in temperature (negative ΔT) increases preload. As shown by Yoder,[14] it depends upon geometry of the design as well as the material properties. With lens and mount

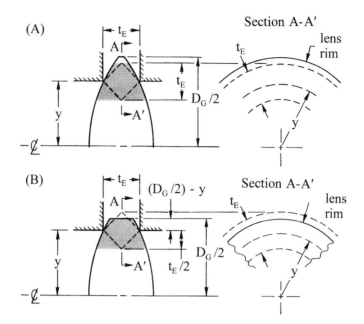

FIGURE 6.22 Geometric relationships used to determine the cross-sectional areas of the stressed regions within a clamped lens, (A) when the region lies within the lens rim and (B) when the stressed region is truncated by the rim. (From Yoder, P.R., Jr. 1992. *Optomechanical Design, SPIE Proc. CR43*, Yoder, P.R., Jr., ed., p. 305.)

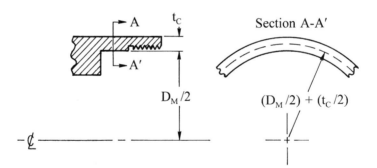

FIGURE 6.23 Geometric relationships used to determine the cross-sectional area of the axially stressed lens cell. (From Yoder, P.R., Jr. 1992. *Optomechanical Design, SPIE Proc. CR43*, Yoder, P.R., Jr., ed., p. 305.)

materials having a small difference between their expansion coefficients, the values of K_3 and the significance of postassembly temperature changes would be expected to be smaller than if this difference is large.

Growth of Axial Clearance at Increased Temperature

If α_M exceeds α_G (as is usually the case), the metal of the mount will expand more than the glass component as the temperature rises. Any axial preload existing at assembly temperature, T_A (typically 20°C [68°F]), will then be reduced. If the temperature rises sufficiently, that preload will disappear and, if not otherwise constrained (as by an elastomeric sealant), the optic is free to move within the mount due to external forces. In nearly all applications, some changes in position and orientation of the lens within small axial and radial gaps created by differential expansion are allowable. The mount should maintain contact with the lens to some elevated temperature, T_C, defined so that *further* temperature increase to the specified maximum survival temperature would not cause the axial gap between mount and lens to exceed the design tolerance.

The axial gap developed between the cell and the lens as temperature rises above T_C can be approximated as:

$$Gap_A = (\alpha_M - \alpha_G)(t_E)(T_{MAX} - T_C) \tag{14}$$

where all terms are as defined above. If Gap_A is equated to the tolerable value, a unique value for T_C of a given design can be calculated. Knowing the rate of change of preload with temperature, K_3, from Equations 10 through 13, the required preload at assembly can be adjusted so that it is just reduced to zero at T_C. Defining the temperature change from T_A to T_C as $-\Delta T$, Equation 9 can then be used to estimate the assembly preload.[2]

Preload and Stress at Low Temperature

In designs where α_M exceeds α_G, the magnitude of the assembly preload is increased whenever the temperature drops below T_A. The total preload at any low temperature, T, can be estimated from Equation 9 by setting ΔT equal to $(T - T_C)$. The axial contact stress created by that preload can then be calculated with the aid of Equations 4 and 3. When T equals the specified minimum survival temperature, T_{MIN}, the stress should not exceed the tolerable compressive value for the glass.

Bending Stress Due to Preload

If the annular areas of contact between the mount and the lens are not directly opposite (i.e., at the same height from the axis on both sides), a bending moment is created within the glass. This moment causes the lens to bend so one side becomes more convex and the other side becomes more concave. The surface that becomes more convex is placed in tension while the other surface is compressed. Since glass breaks much more easily in tension than in compression, especially if the surface is damaged by scratches or has subsurface cracks, catastrophic failure may occur.

An analytical model based upon a thin plane-parallel plate and using an equation from Roark[9] applies also to simple lenses.[6] This is illustrated in Figure 6.24. The tensile stress due to bending of the lens is given by:

$$S_T = \frac{3P}{2\pi mt_E^2}\left[0.5(m-1)+(m+1)\ln\frac{y_2}{y_1}-(m-1)\frac{y_1^2}{2y_2^2}\right] \tag{15}$$

where   P = total applied preload

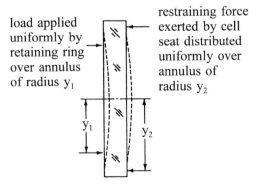

load applied uniformly by retaining ring over annulus of radius y_1

restraining force exerted by cell seat distributed uniformly over annulus of radius y_2

y_1 y_2

FIGURE 6.24 Simplified representation of optical element (plane-parallel plate) bent by clamping between interfaces at different heights. (From Yoder, P.R., Jr. 1993. *Opto-Mechanical Systems Design*, 2nd ed. Marcel Dekker, New York.)

   m = 1/Poisson's ratio for the element
   t_E = element edge thickness
   y_i = contact height on surface "i"

To decrease the probability of breakage from this cause, the contact heights should be made equal within a few percent. Increasing the lens thickness also tends to reduce this danger.

Axial Stress at Multiple Element Interfaces

In the above section ("Axial Stress at Single Element Interfaces"), techniques for estimating axial contact stresses in single element lenses were discussed. That theory was extended by Yoder[17] to include multiple lens designs such as cemented doublets and optomechanical designs with spacers or equivalent cell shoulders between separated lenses. The applicable equations and the procedures for use thereof are summarized here for the convenience of the reader.

The Cemented Doublet

Figure 6.25 shows a typical cemented doublet clamped between a cell shoulder and a threaded retainer. For simplicity, the contact heights are assumed to be the same at both interfaces. The stressed region in the glass is the annulus of radial width $(t_{E1} + t_{E2})$ as indicated by the dashed diamond. Equation 11 or 12, as appropriate, is used to calculate A_G as if the lens were a homogeneous single element. Equation 13 is used to determine A_M. These areas, pertinent component dimensions and the applicable material properties, can then be substituted into the following equation to determine the temperature sensitivity factor K_3:

$$K_3 = \frac{-(\alpha_M - \alpha_{G1})t_{E1} - (\alpha_M - \alpha_{G2})t_{E2}}{\dfrac{2t_{E1}}{E_{G1}A_G} + \dfrac{2t_{E2}}{E_{G2}A_G} + \dfrac{(t_{E1} + t_{E2})}{E_M A_M}} \quad (16)$$

Given the total preload, P, at assembly, the linear preload, p, at any temperature can be calculated using Equations 9 and 4. Note that this preload is the same at both the first and third surfaces of the lens. Then the applicable value of K_2 at either of these surfaces can be estimated by Equation 5 using the material properties prevailing at that interface. Finally, knowing the type of interface and surface radius at each surface, the value for K_1 can be calculated and the contact stress at that surface can be estimated through use of Equation 3. In general, the stresses at the two surfaces will

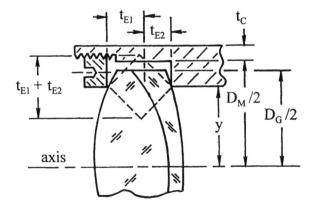

FIGURE 6.25 Schematic of a cemented doublet clamped axially in a cell. The stressed region in the lens is indicated by the dashed diamond. (From Yoder, P.R., Jr. 1994. *Current Developments in Optical Design and Engineering*, *SPIE Proc. Vol. 2263*, Fischer, R.E. and Smith, W.J., eds.)

differ because the glasses have different elastic and thermal properties. The interface types and surface radii also may differ, thereby affecting the values of K_1.

If the temperature rises sufficiently to dissipate assembly preload, an axial gap between the doublet and the mount develops for additional temperature increases in accordance with the following equation:

$$\Delta x = \left[\left(\alpha_M - \alpha_{G1} \right) t_{E1} + \left(\alpha_M - \alpha_{G2} \right) t_{E2} \right] \Delta T \tag{17}$$

The Air-Spaced Doublet

A simple mounting for an air-spaced doublet comprising two unequal diameter elements with differing edge thicknesses is illustrated in Figure 6.26. The spacer material may be different from that of the cell. The glasses also may be different. The contact heights at both surfaces of a given lens are assumed equal and the cell wall thickness is assumed to be constant in this example. If the contact heights at the individual lenses are the same, a cylindrical spacer with parallel OD and ID is used. In the figure, the spacer has a cylindrical OD and a tapered ID. The preload, P, is the same at all lens surfaces.

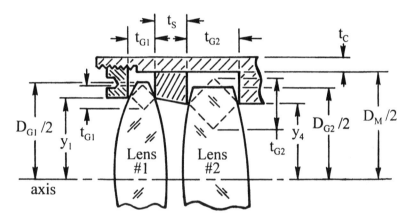

FIGURE 6.26 Schematic representation of two air-spaced lens elements clamped axially in a cell. The stressed regions in the lenses are indicated by the dashed diamonds. (From Yoder, P.R., Jr. 1994. *Current Developments in Optical Design and Engineering, SPIE Proc. Vol. 2263*, Fischer, R.E. and Smith, W.J., eds.)

The following equations give the applicable values of K_3 and the axial gap, Δx, for temperature increases above that for which the preload reaches zero:

$$K_3 = \frac{-\left(\alpha_M - \alpha_{G1} \right) t_{G1} - \left(\alpha_M - \alpha_S \right) t_S - \left(\alpha_M - \alpha_{G2} \right) t_{G2}}{\dfrac{2 t_{G1}}{E_{G1} A_{G1}} + \dfrac{t_S}{E_S A_S} + \dfrac{2 t_{G2}}{E_{G2} A_{G2}} + \dfrac{\left(t_{G1} + t_S + t_{G2} \right)}{E_M A_M}} \tag{18}$$

$$\Delta x = \left[\left(\alpha_M - \alpha_{G1} \right) t_{G1} + \left(\alpha_M - \alpha_S \right) t_S + \left(\alpha_M - \alpha_{G2} \right) t_{G2} \right] \Delta T \tag{19}$$

where all terms are as defined above.

Sectional views of two types of simple lens spacers are shown in Figure 6.27. Both are solid cylinders fitting closely into the ID of the lens cell. The version shown in view (B) has a tapered ID to accommodate different heights of contact with the lenses. It also shows tangential interfaces. Equations 20 through 25 allow the annular areas, A_S, to be calculated for each of these spacers.

(A)

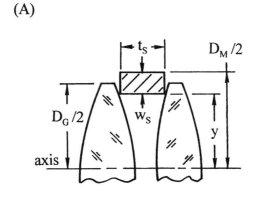

(B)

FIGURE 6.27 Sectional views of two types of lens spacers, (A) solid cylindrical-type with "sharp corner" interfaces at equal heights and (B) solid tapered-type with tangential interfaces at different heights. (From Yoder, P.R., Jr. 1994. *Current Developments in Optical Design and Engineering, SPIE Proc. Vol. 2263*, Fischer, R.E. and Smith, W.J., eds.)

For spacer version (A):

$$w_S = \left(D_M/2\right) - y \tag{20}$$

For spacer version (B), the wall thickness of the tapered spacer is taken as its average annular thickness calculated as follows:

$$\Delta y_i = \left(D_{Gi}/2\right) - y_i \tag{21}$$

$$y_i' = y_i - \Delta y_i \tag{22}$$

$$w_S = \left(D_M/2\right) - \left[\left(y_1' + y_2'\right)/2\right] \tag{23}$$

In both cases:

$$r_S = \left(D_M/2\right) - \left(w_S/2\right) \tag{24}$$

$$A_S = 2\pi r_S w_S \qquad (25)$$

By following the same sequence of calculations as described for the cemented doublet, the contact stresses at the four air–glass interfaces can be estimated.

General Formulation for Multiple Elements

Figure 6.28 provides a schematic example for stress estimation in a more complex multiple element design. Here, different cemented doublets, "A" and "B", are separated by a spacer of uniform annular thickness, t_C, in a cell with constant ID adjacent to the lens rims. A single retainer applies axial preload. The interfaces are all "sharp corners". The cross-sectional areas of the lenses are A_A and A_B, while those for the cell wall and spacer are A_M and A_S. These areas are calculated with the aid of Equations 11 or 12, 13, 20, 24, and 25.

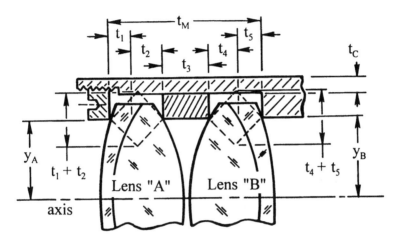

FIGURE 6.28 Schematic of two air-spaced cemented doublet lenses clamped axially in a cell. The stressed regions in the lenses are indicated by the dashed diamonds. (From Yoder, P.R., Jr. 1994. *Current Developments in Optical Design and Engineering, SPIE Proc. Vol. 2263*, Fischer, R.E. and Smith, W.J., eds.)

The applicable equation for K_3 of this design is

$$K_3 = \frac{-(\alpha_M - \alpha_1)t_1 - (\alpha_M - \alpha_2)t_2 - (\alpha_M - \alpha_3)t_3 - (\alpha_M - \alpha_4)t_4 - (\alpha_M - \alpha_5)t_5}{\dfrac{2t_1}{E_1 A_A} + \dfrac{2t_2}{E_2 A_A} + \dfrac{2t_3}{E_3 A_3} + \dfrac{2t_4}{E_4 A_B} + \dfrac{2t_5}{E_5 A_B} + \dfrac{t_M}{E_M A_M}} \qquad (26)$$

This equation has, in its numerator, the sum of negative terms comprising the axial thicknesses of each lens element and of the spacer at the applicable height of contact multiplied by the pertinent differences in thermal expansion coefficients for those parts relative to that of the cell. In the denominator is found the sum of reciprocals of the spring constants for each part of the subassembly. The first five terms in the denominator represent parts in compression and the last three terms represent segments of the cell wall in tension.

The calculations leading to estimation of the axial contact stress at each interface involve first the application of Equations 9 and 4 to determine the linear preload, p, for all air–glass interfaces at any temperature given the total applied preload, P. Then the applicable value of K_2 at each interface is calculated by Equation 5 using the material properties prevailing at that interface. Finally, knowing the type of interface and surface radius at each surface, the value for K_1 can be calculated and the contact stress at each surface can be estimated through use of Equation 3.

The axial gap, Δx, for temperature increases above that for which the preload reaches zero can be calculated for this design by the following equation:

$$\Delta x = \left[(\alpha_M - \alpha_1)\,t_1 + (\alpha_M - \alpha_2)\,t_2 + (\alpha_M - \alpha_3)\,t_3 \right.$$
$$\left. + (\alpha_M - \alpha_4)\,t_4 + (\alpha_M - \alpha_5)\,t_5 \right] \Delta T \tag{27}$$

With understanding of the general formats of Equations 26 and 27, their extension to even more complex multiple element designs is facilitated. The procedure explained earlier can then be applied to determine the axial contact stress for those designs at any temperature.

Radial Stress

Radial Stress in Single Elements

In all the designs considered above, radial clearance was assumed to exist between the lens and the mount. In some designs, this clearance is the minimum allowing assembly so, at some reduced temperature, the metal touches the rim of the optic and, at still lower temperatures, a radially-directed force and resultant radial stress develops. The magnitude of this stress, S_R, for a given temperature drop, ΔT, can be estimated as[2]

$$S_R = K_4 K_5 \, \Delta T \tag{28}$$

Here:

$$K_4 = (\alpha_M - \alpha_G) \big/ \left[(1/E_G) + (D_G/2E_M t_C) \right] \tag{29}$$

and

$$K_5 = 1 - \left\{ (2\,\Delta r) \big/ \left[D_G\,\Delta T (\alpha_M - \alpha_G) \right] \right\} \tag{30}$$

where
$\quad D_G$ = lens OD
$\quad t_c$ = mount wall thickness outside the rim of the lens
$\quad \Delta r$ = radial clearance

If Δr exceeds $D_G\,\Delta T(\alpha_M - \alpha_G)/2$, the lens will not be constrained by the cell ID and radial stress will not develop within the temperature range ΔT due to rim contact.

Tangential Hoop Stress within the Cell Wall

As another consequence of differential contraction of the cell relative to the lens, stress is built up within the metal in accordance with the equation:

$$S_M = S_R D_G \big/ 2t_c \tag{31}$$

where all terms are as defined above.[6] With this expression, one can determine if the cell is strong enough to withstand the force exerted upon the lens without exceeding its elastic limit. If the yield strength of the metal exceeds S_M, a safety factor exists.

Radial Stress within Multiple Elements

In designs involving multiple separated lenses, the radial stress in any element is determined on an individual basis as discussed in the above section "Radial Stress within the Cell Wall". Cemented doublets of uniform OD made of glasses with different coefficients of thermal expansion are usually treated by considering only the element with greatest difference in α as compared to that of the mount. Doublets made up of elements with significantly unequal ODs are treated by considering only the largest element.

Radial Forces Resulting from Axial Preload

Axial preload, P, applied symmetrically to a curved lens surface of radius, R, at some height, y, from the axis produces an inwardly directed radial force component at all contact points. This force tends to compress radially that portion of the glass within the contact zone. The magnitude of this radial force equals $(P \sin \theta \cos \theta)$, where θ is the angular inclination of the surface normal at the contact height relative to the axial direction. For designs with $\theta = \arctan (y/R)$ no larger than about 6°, this radial force is no larger than P/10. It reaches P/3 at about 21°. Only with large axial preloads and/or short surface radii does this factor become a significant contributor to radial stress.

Growth of Radial Clearance at Increased Temperature

The increase in radial clearance, ΔGap_R, between the optic and the mount due to a temperature increase of ΔT from that at assembly can be estimated by the equation:

$$\Delta Gap_R = (\alpha_M - \alpha_G) D_G \, \Delta T / 2 \tag{32}$$

where all terms are as previously defined.

Elastomeric Suspension Interfaces

A Typical Configuration

Figure 6.29 shows a typical design for a lens suspended by an annular ring of resilient elastomeric material (typically epoxy, urethane, or room temperature vulcanizing rubber) within a cell.[6] One side of the elastomer ring is unconstrained so as to allow the material to deform under compression or tension due to temperature changes and maintain a constant volume.[18] Registration of one optical surface against a machined surface of the cell helps align the lens.[3] Centration can be established prior to curing and maintained throughout the cure cycle with shims or external fixturing.

First-Order Thermal Effects

If the resilient layer has a particular radial thickness, the assembly will be athermal to first-order approximation in the radial direction. Stress buildup within the optomechanical components due to differential expansion or contraction is then resisted. This thickness is

$$t_E = (D_G / 2)(\alpha_M - \alpha_G)/(\alpha_E - \alpha_M) \tag{33}$$

where α_E is the thermal expansion coefficient of the elastomer and all other terms are as defined above.

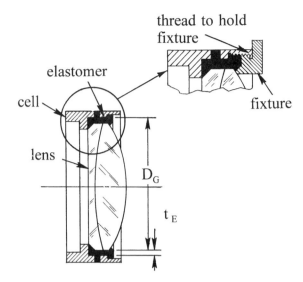

FIGURE 6.29 Schematic of a lens component supported within a cell by an annular layer of cured-in-place elastomer. The detail view shows one means for retaining the injected elastomer during cure. (From Yoder, P.R., Jr. 1993. *Opto-Mechanical Systems Design*, 2nd ed. Marcel Dekker, New York.)

Gravity and Acceleration Effects

Valente and Richard[19] reported an analytical technique for estimating the decentration, Δ, of a lens mounted in a ring of elastomer when subjected to radial gravitational loading. Their method was extended[20] to include more general radial acceleration forces resulting in the following equation:

$$\Delta = A W t_E \Big/ \left[\pi R d \left\{ \left[E_E \big/ \left(1 - \upsilon_E^2 \right) \right] + E_S \right\} \right] \tag{34}$$

where

$\quad$ A $\;$ = acceleration factor

$\quad$ W $\;$ = weight of optical component

$\quad$ t_E $\;$ = thickness of elastomer layer

$\quad$ R $\;$ = optical component OD/2

$\quad$ d $\;$ = optical component thickness

$\quad$ E_E $\;$ = Young's modulus of elastomer

$\quad$ E_S $\;$ = Shear modulus of elastomer

and υ_E $\;$ = Poisson's ratio of elastomer

The decentrations of modest-sized optics corresponding to normal gravity loading are generally quite small, but tend to grow under shock and vibration loading. Fortunately, the resilient material will tend to restore the lens to its unstressed location and orientation when the acceleration loading dissipates.

6.3 Lens Assemblies

"Drop-In" Assembly

Designs in which the lens(es) and the features of the mount that interface therewith are manufactured to specified dimensions within specified tolerances and assembled without further machining and with a minimum of adjustment are called "drop-in" assemblies. Low cost, ease of assembly, and simple maintenance are prime criteria for these designs. Typically, relative apertures are f/4.5 or slower and performance requirements are not particularly high.

An example is shown in Figure 6.30. This is a fixed-focus eyepiece for a military telescope.[2] Both lenses (identical doublets back to back) and a spacer fit into the ID of the cell with typically 0.003 in. (0.075 mm) diametric clearance. The threaded retainer holds these parts in place. "Sharp corner" interfaces are used throughout. Accuracy of centration depends primarily upon the accuracy of lens edging and the ability of the axial preload to "squeeze out" differences in edge thickness before the rims of the lenses touch the cell ID. The axial air space between the lenses depends upon the spacer dimensions which are typically held to design values within 0.010 in. (0.25 mm).

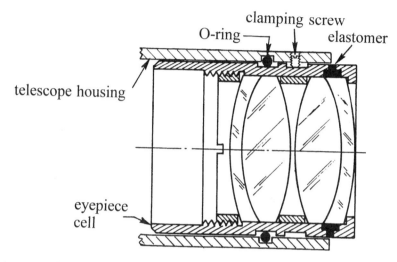

FIGURE 6.30 Example of a fixed-focus eyepiece for a military telescope with lenses and mechanical parts premachined and assembly by the "drop-in" technique. (Adapted from Yoder, P.R., Jr. 1993. *Optical Systems Engineering III, SPIE Proc. Vol. 389*, Taylor, W.H., ed., pp. 2–11.)

Lens assemblies for many commercial applications traditionally follow the "drop-in" design concept. Most involve high volume production and many are intended for assembly by "pick and place" robots. Thorough tolerancing guided by knowledge of normal optical and mechanical shop practices is essential since parts are usually selected from stock at random, and few, if any, adjustments at assembly are feasible.[1] It is expected that a small percentage of the end items will not meet performance requirements. Those that fail are usually discarded — that action being more cost effective than troubleshooting and fixing the problem.

An example of a commercial lens assembly is shown in Figure 6.31. This is an objective for large-screen projection television.[21] The three lenses are injection-molded polymethyl methacrylate. The molded plastic mount is constructed as two symmetrical half-cylinders that are joined longitudinally with adhesive, tape, and/or self-tapping screws after insertion of the lenses. Shoulders that locate the lenses axially are molded in place as are radially oriented mounting pads that center the lens rims. Molded-in pressure tabs are designed to flex slightly as the lenses are inserted so as to constrain the lenses even with minor axial thickness variations.

"Lathe" Assembly

A "lathe-assembled" lens is one in which the lens seats in the mount are custom machined on a lathe or similar machine to fit closely to the measured ODs of a specific lens or specific set of lenses.[2] Axial position of each seat is usually determined during this operation. For this to be successful, the lenses should be precision edged to a high degree of roundness. The tolerances on lens ODs can be relatively loose if sufficient material is provided at the corresponding seat IDs to

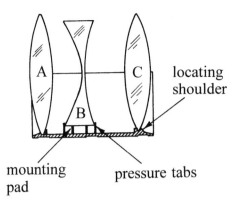

FIGURE 6.31 Optomechanical schematic of a plastic triplet mounted in a plastic mount. (From Betinsky and Welham. 1979. *Optical Systems Engineering, SPIE Proc. Vol. 193*, Yoder, P.R., Jr., ed., p. 78.)

ensure significant material removal during the fitting process. Radial clearances between lens and mount of 0.0002 in. (5.1 µm) are common, while clearances as small as 50×10^{-6} in. (1.3 µm) are feasible. With such small clearances, these lens mountings are frequently referred to as "hard mounts".

An example of the measurement/machining sequence is illustrated in Figure 6.32. View (A) shows the complete optical subassembly comprising an air-spaced doublet in a cell. Required measurements of the lenses are indicated in view (B). Surface radii also are known from test plate or interferometric measurement during manufacture. The mechanical surfaces designated by letters "A" through "E" are machined to suit this specific set of lens measurements and to position the lenses axially and radially within specified tolerances. Machining of surface "D" which provides a tangential interface for lens no. 1 is an iterative process with trial insertions of the lens and measurement of its vertex location relative to flange surface "B" to ensure achievement of the specified 57.150 mm dimension within the 10-µm tolerance. The spacer thickness also is machined iteratively with trial assembly and measurement of overall axial thickness to ensure meeting the design tolerance on this dimension.

This technique is often used in the assembly of lenses for high performance aerial reconnaissance and space science payloads.[6,22,23] For example, Figure 6.33 shows a 24-in. (61-cm) focal length, f/3.5 aerial camera objective lens designed for this method of assembly.[6] The titanium barrel is made in two parts so a shutter and iris can be inserted between lenses 5 and 6 following optical alignment. The machining of lens seats to fit measured lens ODs and to provide proper air spaces begins with the smaller diameter components and progresses toward the larger ones. Each lens is held with its own retainer so no spacers are required. The lenses are fitted into the front and back barrel components in single lathe setups to maximize centration. These optomechanical subassemblies are mechanically piloted together so their mechanical (and optical) axes coincide. An O-ring is used to seal this interface with metal-to-metal contact between the flanges. Tangent contacts are used as the convex surface interfaces. Flat bevels on concave surfaces are made with accurate perpendicularity to the lens' optical axes to facilitate centration. Because of space constraints between lenses 2–3 and 3–4, deep step recesses are ground into the rims to provide space for the retainers. Injected elastomer rings (not shown) seal lenses 1, 5, 6, and 7 to the barrel and all internal air spaces (interconnected) are purged with dry nitrogen to minimize moisture condensation at low temperatures.

Extreme care is required when inserting lenses with small radial clearances to prevent damage. As shown in Figure 6.34, the rims of thick lenses are sometimes edged spherical to minimize the risk of jamming during assembly.[22] The centers of curvature of these rims are located within close tolerances at the optical axis of the respective lenses to maximize centration accuracy. The lens

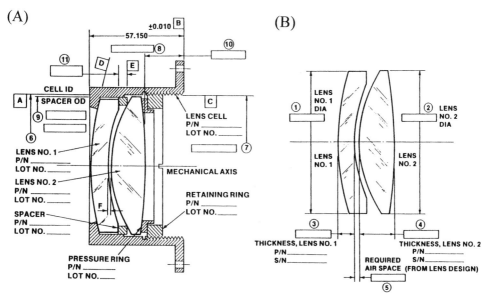

FIGURE 6.32 (A) Example of an optomechanical lens subassembly custom machined by the "lathe assembly" technique to fit a specific set of lens dimensions per (B). (From Yoder, P.R., Jr. 1993. *Optical Systems Engineering III, SPIE Proc. Vol. 389*, Taylor, W.H., ed., pp. 2–11.)

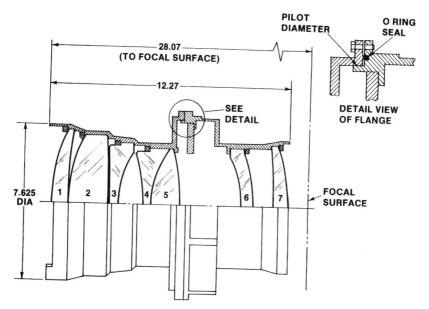

FIGURE 6.33 Sectional view of a 24-in. (61-cm) focal length, f/3.5 aerial camera objective lens designed for the lathe assembly method of assembly. (From Bayar, M. 1981. *Opt. Eng.*, 20, 181.)

assembly shown in this figure is a 9-in. (23-cm) focal length, f/1.5 objective with coaxial laser channel designed for a military night vision periscope application.[22,24]

Vukobratovich[11] described a technique for custom fitting shims between the lens rim and cell ID that essentially achieves a hard mounting with very small radial clearance or, in some cases, radial compression of the shims. Full contact around the lens rim is provided in some designs, while, in others three shorter shims are inserted symmetrically to give more kinematic support. To lock the latter shim segments in place, adhesive can be inserted through radial holes in the cell

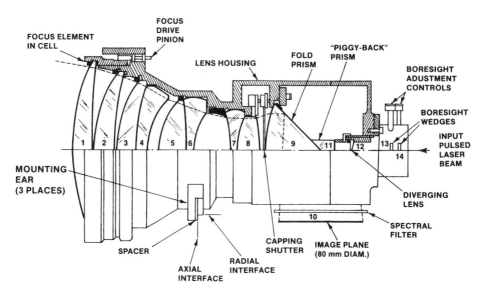

FIGURE 6.34 Sectional view of a lens assembly featuring several lenses with spherical rims assembled with small radial clearance by the "lathe assembly" technique. Shaded components are retainers. (From Yoder, P.R., Jr. 1986. *Contemporary Optical Instrument Design, Fabrication, Assembly and Testing*, SPIE Proc. Vol. 656, Beckmann, L.H.J.F., Briers, J.D., and Yoder, P.R., Jr., eds., p. 225.)

and shim walls. The segmented shim technique has been found especially useful in mounting large diameter lenses.[23]

Subcell Assembly

Optomechanical subassemblies with the lenses mounted and aligned precisely within individual subcells and those subcells inserted in sequence into precisely machined IDs of outer barrels have been described by several authors.[2,6,7,10,25,26] One recent design is illustrated in Figure 6.35.[27] The lenses of this low-distortion, telecentric projection lens were aligned within their respective stainless steel cells to tolerances as small as 0.0005-in. (12.7-μm) decentration, 0.0001-in. (2.5-μm) edge thickness runout due to wedge, and 0.0001-in. (2.5-μm) surface edge runout due to tilt. They then were potted in place with 0.015-in. (0.381-mm)-thick annular rings of 3M 2216 epoxy adhesive injected through radial holes in the subcells to secure the lenses in place. The subcell thicknesses were machined such that the air spaces between lenses were within design tolerances without adjustment. After curing, the subcells were inserted into the stainless steel barrel and secured with retainers.

Vukobratovich described an alternate technique for mounting the lenses within the subcells. Here, each lens is burnished into a subcell and then the outer surfaces of that cell machined concentric with the lens' optical axis and to proper OD for insertion along with similarly machined subcells into a barrel. In other designs, the prealigned subcells were press-fitted with radial mechanical interference into the barrel.[11,28]

Modular Assembly

Optical instrument design, assembly, and maintenance are all simplified if groups of related optical and mechanical components are constructed as prealigned and interchangeable modules. In some cases, the individual modules are nonmaintainable and repair of the instrument is accomplished by replacement of defective modules; sometimes without subsequent system alignment.

A classic example of this type design is shown in Figure 6.36. This military 7 × 50 binocular has prealigned and parfocalized objective and eyepiece assemblies as well as left and right housings with prealigned Porro-type erecting prisms.[2,29,30] Manufacture of such subassemblies is somewhat

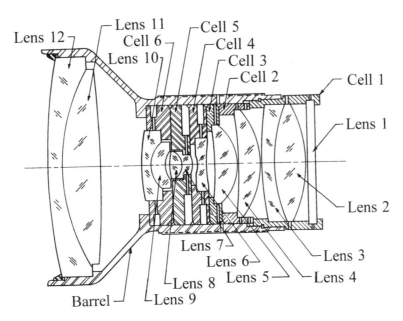

FIGURE 6.35 Sectional view of a lens system comprising several lenses mounted in and aligned to individual subcells that all fit closely within the ID of an external barrel. (Adapted from Fischer, R.E. 1991. *Optomechanics and Dimensional Stability, SPIE Proc. Vol. 1533*, Paquin, R.A. and Vukobratovich, D., eds., p. 27.)

more complex than for the equivalent nonmodular subassemblies due to the requirement for complete interchangeability. In some cases, adjustments are made within the module during assembly, while in other cases, mounting surfaces are machined to specific orientations and/or locations with respect to optical axes and focal planes. Achievement of performance goals is greatly facilitated by the design and fabrication of optomechanical fixtures specifically intended for manufacture and alignment of the modules.[30]

Many photographic and video camera lenses, microscope objectives, and telescope eyepieces are optomechanical modules. Lenses of different focal lengths, relative apertures, and physical sizes have identical mounting features so they can be installed on different instruments. In the photographic application, a variety of lenses can be interchanged on a single camera body or moved from one camera to another of similar type. These lens modules are parfocalized so their calibrated infinity focal planes automatically coincide with the camera's film plane. In some cases, adapters are available to allow lenses from one manufacturer to interface correctly with cameras made by another manufacturer.

Use of advanced injection molding techniques allows complex optomechanical subassemblies to be fabricated from plastic materials in modular form. Figure 6.37 shows such a module designed for use in an automatic coin-changer mechanism.[31] It comprises two acrylic lens elements (one aspheric) molded integral with a mechanical housing having prealigned mounting provisions and interfaces for attaching two detectors. When manufactured in large quantities, this type module is inexpensive. Since it requires no adjustments, it is easy to install and virtually maintenance free.

Many instrument designs utilize single-point diamond machining fabrication techniques to create interchangeable optomechanical modules involving precisely located and contoured reflecting surfaces. This topic is considered in Section 6.5 ("Single-Point Diamond-Turned Mirrors and Mounts"). The methods and machines described there are, in some cases, also applicable to fabricating mounts for lenses.

Elastomeric Assembly

The basic techniques for elastomeric mounting of lenses has been described in Section 6.2 ("Elastomeric Suspension Interfaces"). The advantages of resilient mounting, thermal isolation, ease of

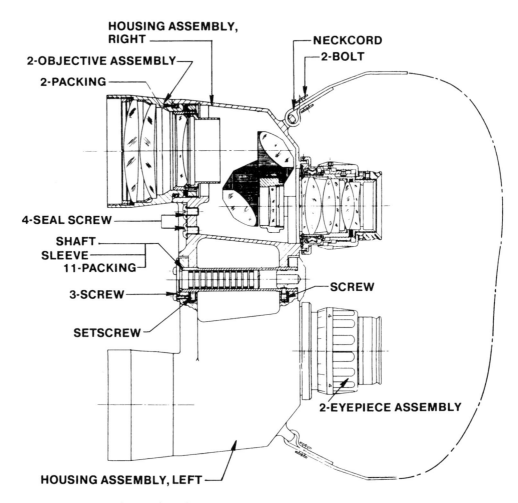

FIGURE 6.36 Sectional view of a military 7 × 50 binocular featuring interchangeable modular objective, eyepiece, and prism housing assemblies. (From Yoder, P.R., Jr. 1993. *Opto-Mechanical Systems Design*, 2nd ed. Marcel Dekker, New York.)

assembly, and inherent sealing can also be achieved in multiple element optical subassemblies. For example, Figure 6.38 shows an aerial camera objective lens in which all optical elements are suspended in rings of elastomeric material.[6] Sufficient radial clearances are provided at each lens seat so that the lenses can be centered and squared-on with respect to pilot diameters and mounting flanges on the barrel halves. Usually these features of the barrels are prealigned to the axis of rotation of a precision spindle so errors in lens alignment can be detected during rotation. In this particular design, the lenses are clamped by threaded retainers after alignment and then the elastomer is injected through several radial holes in the barrel walls to fill the annular space between the lens rim and the cell ID.

In this type design, the thickness of the resilient layer is frequently determined from Equation 33 so the assembly is approximately athermalized in the radial direction. The design is not, however, athermalized in the axial direction since the length of the elastomer layer essentially equals the edge thickness of the lens as well as the applicable length of barrel wall. Since the elastomer is here completely encapsulated and it tends to maintain constant volume with temperature change,[18] the lenses may be stressed at extreme temperatures.

The last mentioned problem can be avoided if, as shown in Figure 6.29, at least one surface of the elastomer ring is not constrained, but is free to deform (i.e., indent or bulge) with temperature changes.

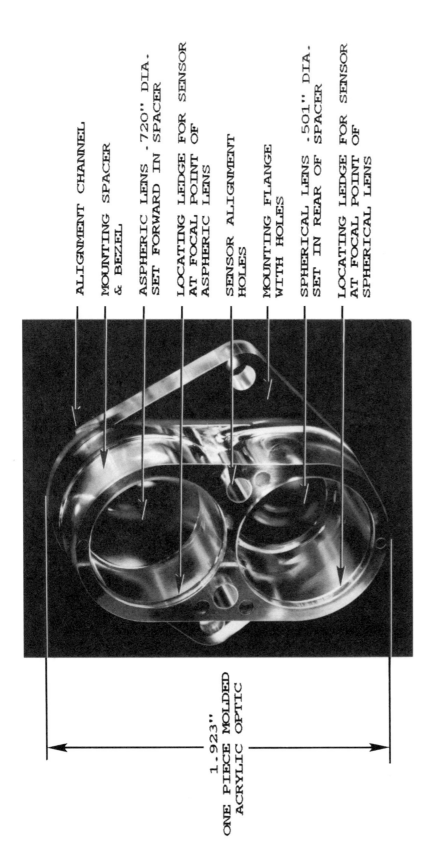

ALIGNMENT CHANNEL

MOUNTING SPACER & BEZEL

ASPHERIC LENS .720" DIA. SET FORWARD IN SPACER

LOCATING LEDGE FOR SENSOR AT FOCAL POINT OF ASPHERIC LENS

SENSOR ALIGNMENT HOLES

MOUNTING FLANGE WITH HOLES

SPHERICAL LENS .501" DIA. SET IN REAR OF SPACER

LOCATING LEDGE FOR SENSOR AT FOCAL POINT OF SPHERICAL LENS

1.923"
ONE PIECE MOLDED ACRYLIC OPTIC

FIGURE 6.37 One-piece molded plastic assembly with two integral lenses, interfaces for sensors, and a mounting flange. (From 1983. *The Handbook of Plastic Optics*, 2nd ed. U.S. Precision Lens, Inc., Cincinnati, OH.)

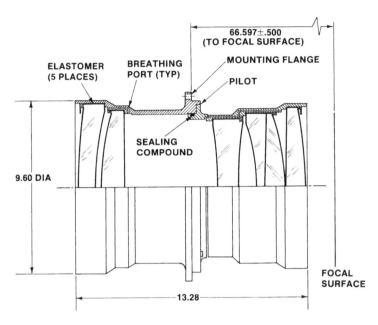

FIGURE 6.38 Partial sectional view of a photographic objective subassembly with elastomerically suspended lenses. (From Bayar, M. 1981. *Opt. Eng.*, 20, 181.)

Operational Motions of Lenses

In many optical instruments, internal adjustments are required during normal operation as, for example, to focus a camera or binocular on objects at different distances, to change focal length (and hence magnification) of a zoom lens, or to adjust focus of a microscope eyepiece to suit the observer's eye. Most of these adjustments involve axial motions of certain lenses or groups of lenses. A few applications, such as the range compensator of a camera rangefinder or rectification of converging images of parallel lines in architectural photography, may involve decentration and/or tilting of lenses.

Focus changes in a camera are generally achieved by moving the entire objective system relative to the film or by moving one or more lens elements within the objective relative to the rest of the lenses while the latter remain fixed with respect to the film. The required motions may be small or large depending upon the lens focal length and object distance, but these motions always must be made precisely and with minimum decentration of the moving elements.

Figure 6.39 shows schematically a typical mechanism used in a camera objective module to couple rotation of an external focus ring through a differential thread to move all the lens elements axially as a group. The differential thread comprises a coarse pitch thread and a slightly finer one on outer and inner surfaces of the intermediate cylinder. They act together to move the lenses as if they were driven by a fine pitch thread, but without the problems normally associated with manufacture, assembly, and possibly reduced lifetime of such a fine thread. The pitch of the equivalent fine thread equals the product of the actual pitches divided by the differences of those pitches.

Since they are used to observe objects at great distances, the optics of military telescopes, binoculars, and periscopes traditionally cannot be refocused for nearby objects. Calibration of reticle patterns used for weapon fire control purposes then remains constant. Whenever the magnification of such an instrument is greater than about 3 power, the eyepiece(s) is(are) individually focusable to suit the user's eye.

Many nonmilitary telescopes and binoculars utilize different means for focusing on objects at different distances. Since there is no reticle pattern to keep in focus, either the eyepiece or the objective can be moved for this purpose. The classical design for focusable binoculars, exemplified

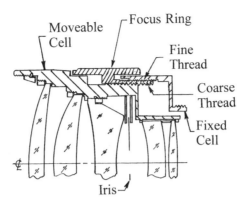

FIGURE 6.39 Simplified schematic sectional view of a camera objective featuring a differential thread focusing movement. Means for preventing rotation of the movable cell (a pin riding in an axial slot) is not shown.

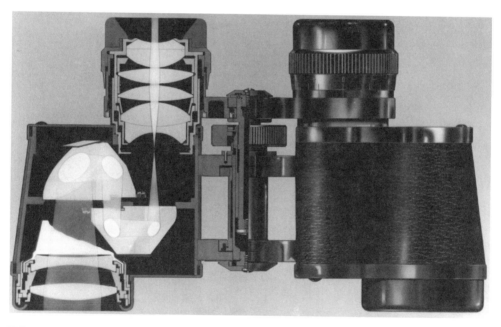

FIGURE 6.40 Partial section view of a commercial 8 × 30 binocular in which both eyepieces are moved axially to focus on objects at different distances. (Courtesy of Carl Zeiss, Inc., Aalen, Germany.)

by Figure 6.40, moves both eyepieces simultaneously along the axis as the knurled ring on the central hinge is rotated. One eyepiece has individual focus capability to allow accommodation errors between right and left eyes to be compensated in what is called the "diopter adjustment". The eyepieces in this design slide in and out of holes in the back cover plates on the prism housings. In low-cost binoculars, no attempt is made to seal the gaps between the eyepieces and these plates.

A more elegant approach for focusing a binocular is illustrated in Figure 6.41. Here, rotation of the focus ring on the central hinge moves internal lens elements of both objectives axially so as to adjust focus. Rotation of another knurled ring adjacent to the focus ring biases the position of the focusable lens of one objective so as to provide required diopter adjustment. Improved sealing is provided with this design since all external lenses can be sealed to the instrument housings.

Figure 6.42 shows an eyepiece for a low-cost commercial binocular in which the entire internal lens cell rotates on a coarse thread to move axially for diopter adjustment.[4] Figure 6.43 shows an eyepiece for a military binocular in which the entire internal lens cell slides axially without turning.

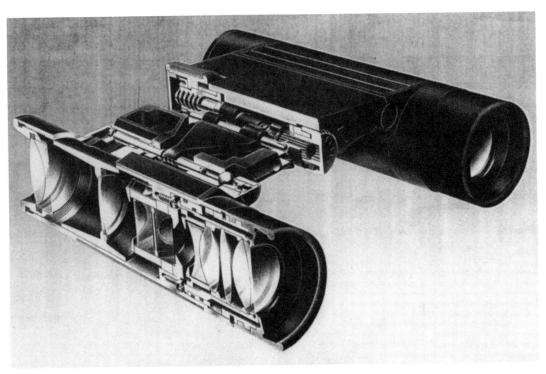

FIGURE 6.41 Partial section view of a commercial 8 × 20 binocular in which internal lens elements are moved axially to focus on objects at different distances. (Copyright: Swarovski Optik KG, Hall in Tirol, Austria. Used with permission.)

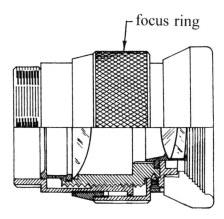

FIGURE 6.42 Simplified sectional view of an eyepiece for a commercial binocular in which the inner cell and lenses rotate on a coarse thread to focus. (Adapted from Horne, D.F., 1972. *Optical Production Technology*, Adam Hilger, Ltd., Bristol. *SPIE Proc. Vol. 163*, p. 92.)

The latter configuration has the performance advantage of maintaining better lens centration, but is more complex and so more expensive.

Most camera zoom lenses have two lens groups that move axially in accordance with specified mathematical relationships to vary the focal length and maintain image focus. Movement of a third lens group may adjust focus for different object distances. These motions usually are controlled by mechanical cams and driven manually by the operator. Smoothness of motion is very critical and lost motion (backlash) in the mechanism should be minimized.

Figure 6.44 shows a sectional view of a representative zoom mechanism. This lens was designed for use in the infrared and has four concentric cylinders machined as matched sets for straight

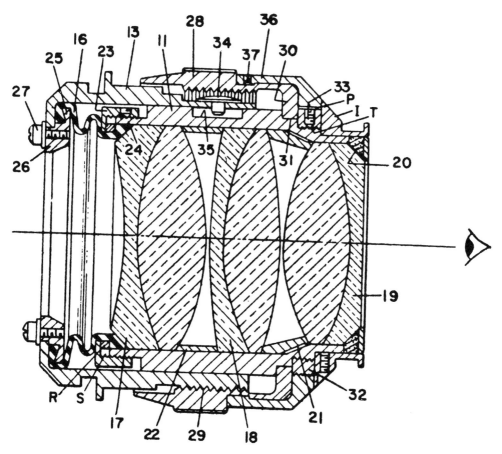

FIGURE 6.43 Simplified sectional view of an eyepiece for a military binocular in which the inner cell and the lenses are constrained by a pin (34) and slot (35) to move axially to focus without rotating. (From Quammen, M.L., Cassidy, P.J., Jordan, F.J., and Yoder, P.R., Jr. 1966. Telescope Eyepiece Assembly with Static and Dynamic Bellows-Type Seal, U.S. Patent No. 3, 246, 563.)

line motion of the movable lenses. It has independent cams to drive two lens groups. Close fits between the cam followers and cam slots are essential in all zoom lenses in order that the active lens motions agree adequately with their design relationships.[32,33] Particular care is taken to maintain contact between the followers and slots in cases where reversal of direction of motion occurs. Otherwise, perceptible image degradation and/or image displacement may occur at those points in the zoom motion.

Sealing Considerations

An important consideration in the design of optical instruments is keeping moisture, dust, and other contaminants from entering and depositing on optical surfaces, electronics, or delicate mechanisms. The need for protection from adverse environments depends upon the intended use. Military optical equipment is subject to very severe environmental exposures, whereas the optics used in scientific or clinical laboratories and in commercial and consumer applications (such as in interferometers, spectrographs, microscopes, cameras, surveyor's transits, binoculars, laser copiers and printers, compact disc players, etc.) usually experience a much more benign environment. The latter types of instruments generally have few, if any, provisions for sealing.

Static protection from the environment at normal temperatures can be provided by sealing exposed lenses and windows with cured-in-place elastomeric gaskets or O-rings. See Figures 6.8,

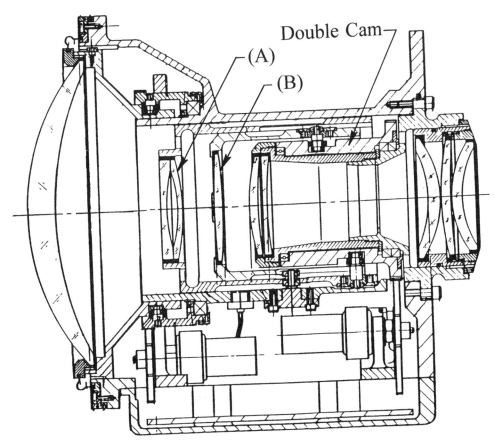

FIGURE 6.44 Simplified exploded view of a 10:1 zoom IR camera objective with two movable lens groups (A and B) each driven by a double-track cam to change focal length and maintain focus. (Adapted from Parr-Burman, P. and Gardam, A. 1985. *Infrared Technology and Applications, SPIE Proc. Vol. 590*, Baker, L.R. and Masson, A., eds., p. 11.)

6.30, and 6.36. Hermetic seals, such as may be required for hard-vacuum applications, are frequently created with gaskets made of indium, lead, or similar soft metals. An example of the latter is given in Section 6.4 ("Examples of Simple Window and Filter Mounts"). High temperature applications may require the use of seals made of formed resilient metal (such as gold-plated Inconel C-rings).[34] Nonporous materials are preferred for housings and lens barrels. Castings may need plating or impregnation with plastic resins to seal pores.

Exposed sliding and rotating parts are frequently sealed with dynamic seals such as O-rings, glands with formed lips, or a flexible bellows made of rubber or metal. In Figure 6.43, a rubber bellows seals the moving lens cell to the fixed housing at left while the outermost lens at right of the cell is sealed statically with an elastomeric seal. An O-ring inserted into the groove at the mounting flange seals the entire eyepiece to the instrument.[35]

Many sealed instruments are purged with dry gas such as nitrogen or helium as part of the assembly process. Positive pressure differential above ambient of perhaps 5 lb/in.^{2} (3.4×10^4 N/m^2) is sometimes generated within the instrument to help prevent intrusion of contaminants. Access through the instrument walls is, in this case, provided by a spring-loaded valve similar in function to those used on automobile tires.

Access for flushing of nonpressurized instruments can be by means of threading through holes into which seal screws are inserted after flushing. For example, two seal screws can be seen in each housing of the binocular shown in Figure 6.36.

Internal cavities of sealed instruments, such as those between lens elements, should be interconnected by leakage paths (bored holes, grooves, etc.) to the main cavity in order for the flushing process to work properly. Removal of moisture and/or products of outgassing from these ancillary cavities is facilitated if the instrument is evacuated and backfilled with the dry gas. Baking the instrument at elevated temperature for several hours also tends to eliminate these materials. To prevent potentially harmful pressure changes due to temperature changes, sealed instruments can be allowed to "breathe" through desiccators.[2]

Protection of outermost optical surfaces on cameras and electro-optical systems while not in use may be afforded by removable covers. Examples would include caps on personal binocular or camera lenses or ones placed over the apertures of airborne electro-optical sensors during takeoff and nonoperational phases of a military mission. The latter caps would be ejected before the instrument is put in use. Similarly, protective doors or covers are placed over exposed optics in space payloads prior to launch. These are opened mechanically or blown away upon reaching the orbit. In each case, the covers protect the optics from exposure to wind-driven dust, debris, and other abrasives as well as high velocity impact with rain, ice, and snow.

6.4 Mounts for Windows, Filters, Shells, and Domes

General Considerations

Generically, a window serves the very important function of isolating the interior of an optical instrument from physical damage and from adverse environmental conditions that may exist outside while allowing useful radiation to pass through. In its simplest form, the window is a plane parallel plate of optical glass, fused silica, crystals, or optical plastic. Shells and domes are special types of windows that have deep meniscus spherical or aspherical surfaces. Sometimes these components have zero net optical power even though the surfaces may be curved. In some applications, a window must mechanically support a positive or negative pressure differential between the outer and inner atmospheres. It may also constrain the internal environment as in the case of a gas laser tube, a spectrometer sample cell, or the observation port of a wind tunnel. An optical filter is another special type of window that serves to modify the spectral character of the transmitted radiation through selective absorption, reflection, or scatter.

The location of a window in the optical system determines to large extent the critical aspects of its design.[2,11] If located near an image plane, dirt, moisture, surface, and coating defects (scratches, digs, sleeks, or inadequate polish) may appear superimposed upon the image. The optical figure of the surfaces, wedge angle, and refractive index homogeneity of the material is less important in this case. If the window is located close to a pupil of the system, the relative importance of these characteristics is reversed, i.e., refractive properties are more important than cosmetic defects or surface contamination. Since, in most cases, windows and filters are relatively thin, their optical aberration contributions are relatively small, especially if located in collimated beams. If thick plane parallel windows are used in beams with large convergence or divergence they may contribute significant aberrations in the same manner as prisms.

Windows that are intentionally wedged to control spurious surface reflections or to deviate the beam in a specific direction may require special mounting arrangements to ensure proper orientation of the wedge apex. They may also cause spectral dispersion if the wedge angle is significant.

Environmental conditions surrounding a window affect its performance. For example, a window exposed to intense thermal radiation may develop a temperature gradient from side to side, front to back, or radially. These gradients tend to change the refractive properties and may change the physical shape of the element as well. Pressure differentials through the window also tend to change the shape of the element. For example, a window on an aerial camera exposed to rapidly moving air flowing over aircraft's skin may heat due to aerodynamic friction and bow into a meniscus shape. These effects might cause the camera image to go out of focus or otherwise deteriorate.

Table 6.3 lists important parameters to be considered in the design of optical windows. Only rarely would all these factors apply to a given case.

Table 6.3   Parameters of Importance in Optical Window and Filter Design

Transmission
   Intensity loss throughout applicable spectral range
   Blocking requirements for undesired radiation
Dimensions
   Optical aperture (instantaneous and total)
   Diameter or width and height
   Thickness
   Wedge angle and orientation
   Special shape and/or bevel requirements
Optical properties
   Optical power contribution
   Transmitted wavefront quality requirements (or surface flatness/irregularity and index of refraction
  homogeneity)
   Transmitted wavefront relative aperture (f/no)
   Surface and bulk scatter characteristics
   Coating requirements (reflectance, thermal emissivity, electrical)
   Bubbles, inclusions, and striae
   Polarization characteristics
Environment
   Temperature extremes and exposure profiles (storage and operational)
   Pressure (including ram air and turbulance effects)
   Exposure to humidity, rain erosion, and particulate matter
   Radiation (thermal, cosmic, nuclear)
   Vibration (amplitude and frequency power spectral density)
   Shock (amplitude, duration, and direction)
Mounting configuration
   Orientation relative to optical beam(s) and vehicle motion
   Mechanical stresses induced (operation and storage)
   Thermal properties of materials
   Heat transfer mechanisms and paths
   Mechanical interface (mounting hole pattern)
   Sealing requirements

Adapted from Yoder, P.R., Jr. 1985. Geometrical Optics, SPIE Proc. Vol. 531, Fischer, R.E., Price, W., and Smith, W.J., eds., p. 206.

Rather than to review these items individually, a few representative window configurations are described from the optomechanical viewpoint. The interrelationships between different applications and technical requirements are stressed. Although important, windows and filters for use in high energy laser systems are intentionally omitted from consideration because space limitations preclude adequate treatment of those topics. The windows fused in place for applications such as gas discharge laser tubes are also omitted.

Examples of Simple Window and Filter Mounts

A large variety of plane-parallel plates made of glass, crystals, etc. are available as catalog items from optical component manufacturers to meet many needs for windows to be used in optical instruments, laboratory experiments, and for other purposes. They can be purchased in standard sizes and uncoated or antireflection coated. Needs for components not available directly from a catalog can frequently be met by modification of standard parts to special order. If this approach is unsuccessful, the needed parts can be custom fabricated to print.

Figure 6.45 shows a plane-parallel window intended to cover the aperture of a visual telescope, to protect the significantly more expensive nearby objective lens from damage, and to prevent entrance of moisture and other contaminants from the military environment.[36] Its diameter is nominally 52 mm (2.05 in.), its aperture is 48 mm (1.89 in.), and its thickness 8.8 mm (0.346 in.). It is mounted in a stainless steel cell designed for flange mounting to the telescope housing. The window is clamped in the cell by a threaded stainless steel retainer and subsequently sealed with injected elastomer per the referenced military specification. The cell, in turn, is bolted to the telescope housing and sealed with an O-ring that fits into the groove shown on the flange. The subassembly is intended to maintain a positive pressure within the telescope of at least 5 lb/in.² $(3.45 \times 10^4 \, N/m^2)$ over external ambient pressure.

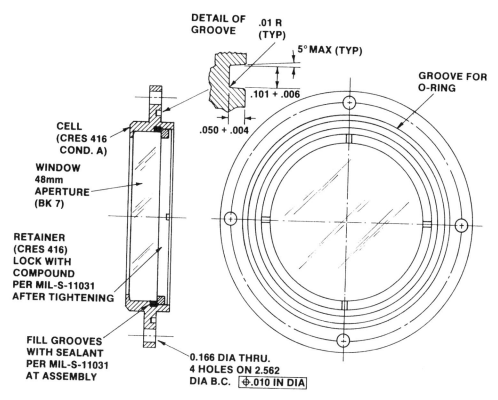

FIGURE 6.45 Instrument window subassembly with window held in place by a retaining ring and sealed with an elastomer. (From Yoder, P.R., Jr. 1985. *Geometrical Optics, SPIE Proc. Vol. 531*, Fischer, R.E., Price, W., and Smith, W.J., eds., p. 206.)

Because the window is located in a collimated beam near the telescope's pupil (which is at the objective) and its aperture is always filled, the maximum transmitted wavefront error is specified as ±5 waves optical power and 0.05 wave peak-to-valley irregularity in 633-nm laser light. Maximum wedge angle is specified as 30 arc seconds. By choosing high quality optical glass for the window (in this case BK7 borosilicate crown) the designer is ensured a high degree of refractive index homogeneity, freedom from striae, bubbles, and inclusions, adequate climatic resistance of the substrate, ease of fabrication, and reasonable material cost. Antireflection coatings of magnesium fluoride provide high durability.

One solution to the problem of providing a vacuum-tight, chemically inert mounting for an infrared crystal window in a stainless steel instrument housing for multiple-photon laser-induced chemistry in gaseous media[37] is illustrated in Figure 6.46. The window is a 7.6-cm (3-in.)-diameter disk of single crystal or polycrystalline sodium chloride approximately 9 mm (0.35 in.) thick. During long-term use, this window was intended to hold internal vacuum of a few millitorr pressure at temperatures of 200 to 275°C with helium leak rates of the order of 3×10^{-10} atm-cm³/sec while

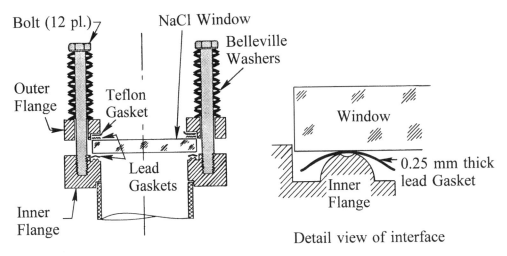

FIGURE 6.46 Diagrams of a high-temperature, vacuum-sealed infrared window subassembly. (Adapted from Manuccia, T.J., Peele, J.R., and Geosling, C.E. 1981. *Rev. Sci. Instrum.*, 52, 1857.)

the sample was being irradiated with pulsed laser radiation. Resistance to thermal shock from 1°C/min temperature changes also were required.

The window surfaces were clamped between two thick stainless steel flanges by 12 Belleville spring-loaded bolts to provide uniform axial pressure and flexibility to accommodate thermal expansion mismatch between the materials. The window was not constrained transversely other than by friction. The inner window surface was sealed to the inner flange through a 0.25-mm (0.010-in.)-thick gasket of lead. The flange had the unique shape described in the detail view. A convex annular toroidal interface 1.9 mm wide projecting from the flange surface was provided with a concave annular groove of 0.19-mm (0.008-in.)-depth. Under compression at high temperature, the edges of the groove cut through the gasket and trap a ring of lead under high hydrostatic pressure inside the groove, where it extrudes into microscale irregularities of the window and of the flange surfaces forming a vacuum seal. The interface between the outer surface of the window and the flange had a 0.25-mm (0.010-in.)-thick lead gasket and a 0.125-mm (0.005-in.)-thick Teflon™ gasket. The surfaces of that lead gasket were roughened so it would deform under preload to distribute the axial force over a large area of the window.

Figure 6.47 shows a much simpler mounting configuration; this for a set of four glass spectral filters each with 25.4 mm (1.00 in.) clear aperture and 3 mm (0.12 in.) thickness located in a multiple-aperture filter wheel inside a laboratory optical system. Since there is no need to seal the filter or even to precisely control its location and/or orientation relative to the optical axis for the intended application, each optical element is held in place by a spring-type snap ring (see Section 6.2 ["Low Precision Mounts" — "Snap Ring"] and Figure 6.3). The wheel is driven manually from one location to another with positioning at 90° intervals determined by a spring-loaded ball (not shown) dropping into the "V" detents shown on the wheel rim. The laboratory environment to which the instrument is exposed is relatively benign and the glass-to-metal interfaces are loose so there are no significant mounting stresses. If it were expected that the filters would not need to be removed during the lifetime of the assembly, they might well be sealed in place with an elastomer.

Example of A Larger Window Mount

Cameras and electro-optical sensors used in high-performance military aircraft are usually mounted on stabilized mounts within environmentally controlled equipment bays in the fuselage or in externally mounted pods. Windows are needed to seal the bay or pod and to provide

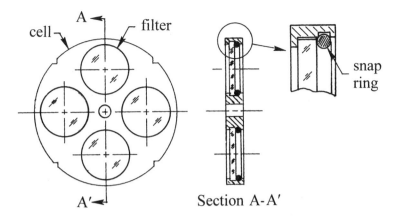

FIGURE 6.47 Schematics of a simple multiple-filter wheel subassembly.

aerodynamic continuity of the enclosure. In this section, an example of such a special design is considered.

The multi-aperture window assembly shown in Figure 6.48 is designed for use in a military aircraft.[36] The larger window is used by a forward-looking infrared (FLIR) sensor and is made of antireflection-coated chemical-vapor-deposited (CVD) zinc sulfide (ZnS). It measures approximately 30 × 43 cm (11.8 × 16.9 in.) and is 1.6 cm (0.63 in.) thick. The irregular aperture conforms with the footprint of the 25.4-cm (10-in.)-diameter beam entering the sensor's optical aperture at the nominal angle of incidence of 47°. The smaller windows are used by a laser range finder/target designator system operating a 1.06-μm wavelength. These windows are identical, have apertures of 9 × 17 cm (3.5 × 6.7 in.), have thicknesses of 1.6 cm (0.63 in.), and are fabricated of BK7 glass. They also are antireflection coated. Specifications for transmitted wavefront error (including mechanical deformations caused by mounting) are 0.1 wave peak-to-valley at 10.6 μm over the beam diameter for the ZnS window and 0.2 wave peak-to-valley power plus 0.1 wave irregularity at 632.8 nm over the full aperture for the BK7 windows.

For maximum strength under environmental and operational stress, all window surfaces (including the rims) are carefully fabricated by the "controlled grinding" process[38, 39] in which progressively finer abrasives are used to remove all traces of subsurface damage caused by the preceding grinding operation. Each window is elastomerically bonded into close-fitting recesses in a contoured anodized 6061-T651 aluminum plate. This plate is bolted to a matching machined interface on the aircraft structure to minimize bending that could stress the refracting materials and distort their optical surfaces.

Examples of Shell and Dome Mounts

Meniscus-shaped optical elements are frequently used as windows for electro-optical sensors having wide fields of view or those with smaller instantaneous fields that are scanned over wider conical fields. Generically they are shells; very deep shells are called domes. Domes subtending >180° from their centers are called hyperhemispheres. Because of their shapes, shells are stiffer than flat windows of the same thickness.

Many shells and domes are made of optical glass and function at visible wavelengths. Others are used in the infrared so must be made of crystalline materials such as zinc selenide, zinc sulfide, germanium, or silicon. Some infrared-transmitting materials are soft and so are difficult to polish and not very resistant to erosion due to impact with dust, rain, ice, or snow at high velocities.[40] Composite substrates such as soft zinc selenide coated with a layer of harder zinc sulfide are sometimes used to combat the latter problem. Shallow shells are frequently used as aberration correctors in objective systems such as Maksutov telescopes.

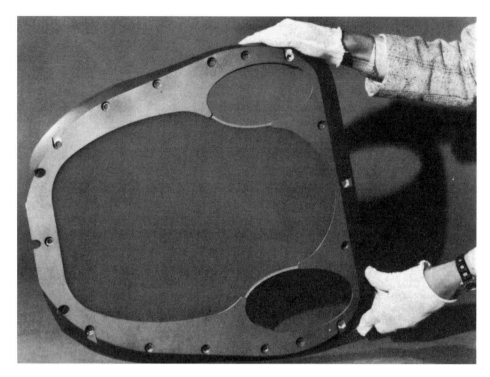

FIGURE 6.48 A multi-aperture window subassembly for a military multiwavelength electro-optical sensor. (From Yoder, P.R., Jr. 1985. *Geometrical Optics, SPIE Proc. Vol. 531*, Fischer, R.E., Price, W., and Smith, W.J., eds., p. 206.)

Mountings for shallow and deep shells typically involve elastomeric potting techniques or mechanical clamping with flange-type retaining rings.[2,11,36] Three examples are shown in Figure 6.49. View (A) shows a crown glass hyperhemisphere with its shaped rim potted into a flange that is, in turn, attached to structure with several bolts. An O-ring seals the flange to the structure. The pilot diameter indicated in the figure serves to center the optical element to the axis of the system. View (B) shows a shallow dome clamped with a Delrin ring to a housing with a series of Nylon screws. A molded Neoprene gasket seals the optic to the mount. View (C) shows a zinc sulfide dome bonded with epoxy to a bezel and supported axially by a threaded retaining ring. This design was successfully used on a projectile fired from a mortar with approximately 11,000-G accelerations.[41]

Pressure Differential Effects

Pressure differentials through flat and curved windows tend to change the shapes of the optical surfaces; these can adversely affect the performance of the optical systems using the windows by distorting the transmitted wavefront. These shape changes also introduce tensile and compressive stresses into the refracting materials. As discussed in Section 6.2 ("Axial Stress at Single Element Interfaces" — "Bending Stress Due to Preload"), tensile stresses are more serious than compressive stresses and may lead to failure.

Vukobratovich[11] outlined techniques for estimating (1) the optical path difference introduced into a wavefront passing through a flat window when that window is deformed by a given pressure differential and (2) the stress introduced into the window by that deformation. He indicated that, for a given window diameter/thickness aspect ratio, in visible light systems, pressure-induced wavefront distortion may outweigh pressure-induced stress effects, while in infrared systems, the reverse may be true because of the less stringent requirements for optical surface figure in the latter

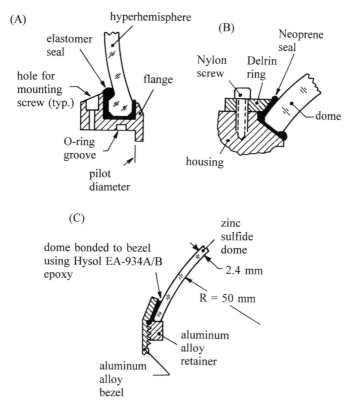

FIGURE 6.49 Three configurations for dome mounting interfaces. (A) Hyperhemisphere potted into a flange. (From Yoder, P.R., Jr. 1993. *Opto-Mechanical Systems Design*, 2nd ed. Marcel Dekker, New York.) (B) Dome clamped by an external ring. (From Vakobratovich, D. 1986. *Introduction to Optomechanical Design, SPIE Short Course Notes.*) (C) Dome bonded in place and clamped by an internal retainer. (From Speare, J. and Belloli, A. 1983. *Structural Mechanics of Optical Systems, SPIE Proc. Vol. 450,* Cohen, L.M., ed., p. 182.)

systems. He also provided a good summary of means for estimating the probability of window failure under a given stress.

A typical application involving large pressure differentials is a window for a deep submergence vehicle. Usually these windows are made of polymethyl methacrolate (acrylic), are quite thick, and have limited aperture. Their rims usually are tapered and closely matched in angle and surface roughness to the interfacing mechanical surface. This allows the window to move axially and avoids hoop stress effects from thermal expansion mismatch under temperature changes. Figure 6.50 shows an example of a window with a 90° taper.[42] The Neoprene gasket seals the window in place at normal pressures; at large underwater depths, the window is driven solidly against and sealed into the mount. Under extreme pressure differentials, the window tends to extrude through the inner mount aperture of diameter D. It is not uncommon for window thickness, d, to be 50% of D.[43]

Domes subjected to large pressure differentials develop stress and may collapse due to elastic buckling.[11]

Thermal Effects

Temperature stabilization of high performance optical systems is important to prevent deterioration of image quality. The segmented window shown in partial section view in Figure 6.51 is an example of design for this purpose.[2] It is used on a high speed military aircraft as part of a high resolution panoramic photographic reconnaissance system capable of scanning from horizon to horizon through nadir transverse to the flight direction. In this example, dual 1-cm (0.39-in.)-

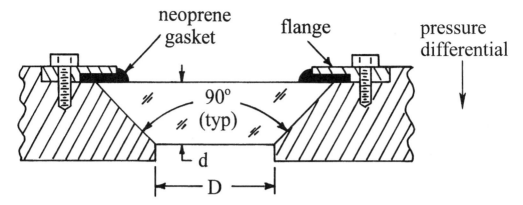

FIGURE 6.50 A tapered-rim acrylic window subassembly intended for a deep-submergence application involving a large hydrostatic pressure differential. (From Vukobratovich, D. 1986. *Introduction to Optomechanical Design, SPIE Short Course Notes.*)

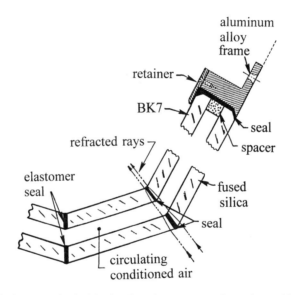

FIGURE 6.51 Partial schematic of a double-glazed, multisegment window subassembly intended for a horizon-to-horizon panoramic aerial camera application. (Adapted from Yoder, P.R., Jr. 1993. *Opto-Mechanical Systems Design*, 2nd ed. Marcel Dekker, New York.)

thick glazings in "thermopane" configuration serve to stabilize the temperature of the camera's environment. The outer glazings are made of fused silica. They carry a low-emissivity (gold) coating on their interior surfaces to minimize transfer of heat to the camera from the windows' outer surfaces as the latter are heated by boundary layer friction during high speed flight. The inner glazings are made of BK7 glass. They are coated for maximum transmission in the spectral region surrounding the peak sensitivity of the film. Further thermal stabilization is afforded to the window, and hence to the camera optics, by passing conditioned air from the aircraft's environmental control system through the air space between the glazings.

The square glazings at the center of each window measure approximately 32 × 33 cm (12.6 × 13.0 in.), while the side glazings are slightly smaller. All window surfaces are processed by the "controlled grinding" method to maximize strength.[38,39] These glazings are sealed with elastomer into recesses machined into an aluminum frame. The frame is contoured to fit closely to a matching machined interface on the aircraft's camera pod and is secured with several bolts.

Under natural (such as solar irradiation) or artificial heating (as in an attempt to stabilize temperature), the window itself may become distorted by axial or transverse thermal gradients, thereby causing the transmitted wavefront to be defocused and deformed. Barnes explained how to estimate the optical path differences introduced by these gradients in circular aperture windows used in high-acuity spaceborne systems.[44] He indicated that, in general, windows should be as thin as possible and the physical aperture should be significantly larger than the optical aperture. The equations given by Barnes were summarized in a more recent publication by Vukobratovich.[11] Rectangular aperture windows with temperature gradients will, in general, deform the transmitted wavefront asymmetrically and introduce astigmatic focus errors.

An example of a mounting arrangement that has a designed-in radial thermal gradient is shown in Figure 6.52. Here, a circular aperture segmented optical interference filter is heated a few degrees above the highest specified ambient by a thermostatically controlled heater coil mounted within the cell wall. The cell is fabricated from a phenolic insulating material (G10) and mounted to an aluminum structure.[2]

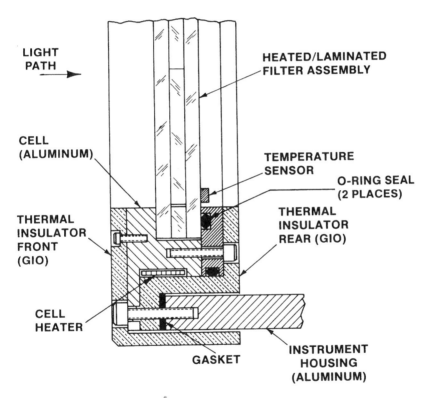

FIGURE 6.52 Schematic of a window (interference filter) mount with active temperature stabilization means. (From Yoder, P.R., Jr. 1993. *Opto-Mechanical Systems Design*, 2nd ed. Marcel Dekker, New York.)

6.5 Mounts for Small Mirrors

Clamped Mounts

A kinematic support for a body has six independent constraints, one each for the six degrees of freedom in inertial space (three rotations about mutually perpendicular axes and three translations along those axes). Ideally, contact with the body occurs at points so moments cannot be exerted thereon. When restraining forces are delivered over very small areas (points), the loads per unit area become large, stresses develop, and elastic bodies can be deformed. To prevent this from becoming a problem in mounting optical elements, finite-sized contact areas are used. The mount is then called semikinematic.

To illustrate a semikinematic mount for a small flat mirror, consider the concept shown schematically in Figure 6.53. Here a circular mirror is clamped between three spring-loaded pads and three coplanar fixed pads located directly opposite through the mirror. The pads are located at 120° intervals on the mirror's surfaces. The contact areas on both sides of the mirror are small and the force vectors pass centrally through those contact areas and perpendicular to the mirror surfaces. These forces constrain the mirror against one translation and two rotations. Two spring-loaded pads press perpendicularly against the rim of the mirror to hold it against a single opposing rigid pad. The rim-contacting pads, shown oversize, lie in the plane of the mirror's center of gravity at 120° intervals. They constrain the mirror against the other two translations. Note that the locations of all pads are adjustable and the adjustments have locking setscrews. This allows the mirror to be centered radially and the spring forces to be adjusted at assembly.

Since this is a flat, symmetrical mirror, rotation about the third axis (normal to the mirror) is not critical. Nevertheless, that motion also is constrained somewhat by friction at the contact areas. In some flat mirror designs, all rim contacts are omitted and friction depended upon to hold the mirror against two translations as well as the one rotation. Direct radial constraint would be appropriate for spherical or aspherical mirrors since centration is important in those cases.

If possible, the reflecting surface of a first-surface mirror should contact the fixed pads rather than the spring-loaded ones. Then wedge in the mirror substrate will not affect alignment of the reflected beam if the mirror rotates about an axis perpendicular to its face. In the case of a second-surface mirror, registering the reflecting surface is again advisable, but some light beam deviation is introduced by substrate wedge.

Frequently the springs that hold the mirror are flat blades (clips) of beryllium copper or spring steel. The force, F, delivered to the mirror by each clip is then determined by treating it as a deflected short cantilevered beam.[2,9] The following equation gives the deflection, x, required to produce a given force:

$$x = \left(1 - \upsilon^2\right)\left(4FL^3 / Ebh^3\right) \tag{35}$$

where

υ = Poisson's ratio for the spring material
F = applied force per spring
L = free length of beam
E = Young's modulus for the spring material
b = width of spring
h = thickness of spring

In this equation, h is assumed to be small compared to b.

Bonded Mounts

Techniques for mounting small mirrors by glass-to-metal bonding with adhesive such as epoxy have gained considerable popularity, especially for military and aerospace applications involving exposure to severe environmental conditions. They are frequently used for nonmilitary applications because of their ease of assembly and durability.

The simplest form of bonded interface is at the back surface of a first-surface mirror. Such a design is shown in Figure 6.54. Here, the fine ground back surface of a 2-in. (5.1-cm)-diameter, 0.33-in. (0.84-cm)-thick crown glass mirror is bonded to an elevated flat circular pad on a stainless steel bracket with epoxy adhesive approximately 0.004 in. (0.10 mm) thick. The area, Q, of the pad is 0.5 in.[2] (3.2 cm[2]). The weight, W, of the mirror is approximately 0.09 lb (0.041 kg).

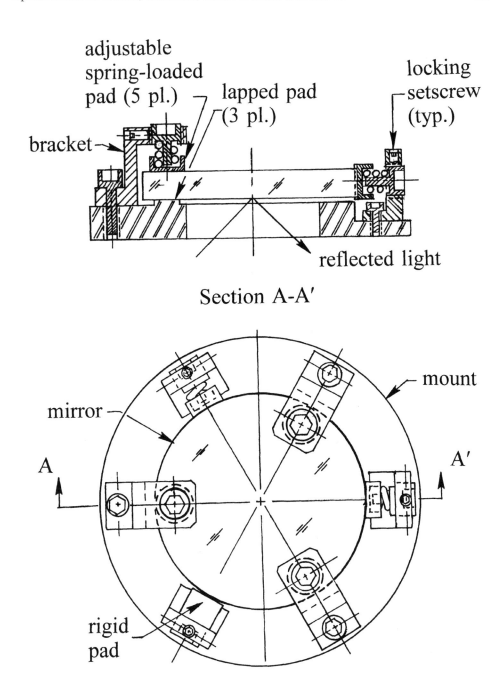

FIGURE 6.53 Schematic of a spring-loaded mounting for small, circular, flat mirrors.

The adequacy of a bond such as this can be established as follows. Under directional acceleration of "G" times gravity, the mirror exerts a tensile or shear stress upon the bond of GWS_F/Q in units of lb/in.² (or N/m²), where S_F is a safety factor. This stress can be compared to the strength of the adhesive joint, J, which, for many epoxies, is of the order of 2000 lb/in.² (1.38×10^7 N/m²). The magnitude of Q for a given S_F is given by:

$$Q = GWS_F/J \qquad\qquad (36)$$

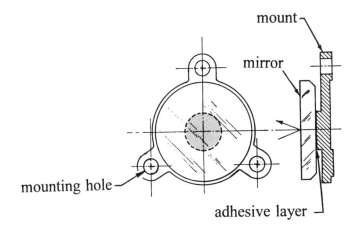

FIGURE 6.54 Typical construction of a first-surface mirror subassembly with glass-to-metal adhesive bond (shaded area) securing the back of the mirror to the mount. (From Yoder, P.R., Jr. 1985. *Geometrical Optics, SPIE Proc. Vol. 531*, Fischer, R.E., Price, W., and Smith, W.J., eds., p. 206.)

Experience indicates that S_F should not be less than 2 and perhaps should be as large as 10 to compensate for hard-to-control assembly conditions such as inadequate cleanliness of the interfacing surfaces.[2] In the example above, an acceleration as large as 1110 times gravity will allow S_F to equal 10.

Since all adhesives tend to shrink somewhat during cure and the shrinkage in a given direction is a small percentage of the corresponding maximum dimension of the bond area, it is desirable to keep the bond area as small as practical. This also may tend to speed the curing process with some adhesives. One accepted technique for minimizing bond area is to determine the total bond area required and then to divide this into three equal subareas arranged in a reasonably large triangular array on the mirror surface. This distribution is somewhat kinematic and helps to stabilize the assembly.

Since the coefficient of thermal expansion of the adhesive is significantly larger than those of the mirror or the mount, it can contract significantly as the temperature drops. It is best if excess adhesive is not allowed to extend beyond the bonding pad on the mount so as to form a fillet of adhesive "bridging" to the mirror. Experience has shown such fillets to be the cause of glass fracture at low temperatures. Careful application of a predetermined adhesive volume and/or prompt removal of excess uncured adhesive minimizes this potential problem.

Achievement of a uniform adhesive layer thickness is facilitated by such means as building localized pads or hard registration points of the proper height into the mount, by installing temporary shims between the mirror and mount or by mixing spherical beads of glass or similar material having the proper (small) diameters into the adhesive prior to application. In order to minimize shrinkage effects perpendicular to the bond area during cure, the thickness of the bond is frequently held to a small value such as 0.004 in. (0.10 mm) to 0.015 in. (0.38 mm). Exceptions to this "rule" include choice of bond thickness to athermalize the assembly (as discussed in Section 6.2 ["Elastomeric Suspension Interfaces" for lenses) or to provide a limited degree of shock resistance by virtue of the resiliency of the adhesive.

Flexure Mounts

To allow for thermal expansion coefficient variations between mirrors and mounts, flexures are frequently designed into the mounts. Figures 6.55 and 6.56 illustrate two such mounting arrangements.

In Figure 6.55, three rectangular metal pads are bonded to the edge of a rectangular aperture mirror. Three flat flexure blades attach the pads to the instrument structure (baseplate). The blades

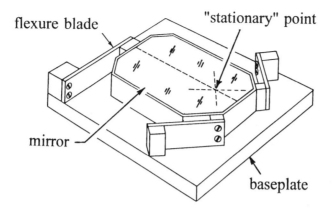

FIGURE 6.55 Concept for a flexure-mounted rectangular mirror subassembly. (From Yoder, P.R., Jr. 1985. *Geometrical Optics, SPIE Proc. Vol. 531*, Fischer, R.E., Price, W., and Smith, W.J., eds., p. 206.)

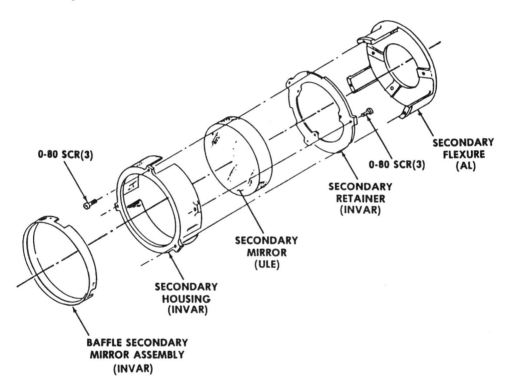

FIGURE 6.56 Exploded view of a flexure-mounted telescope secondary mirror intended for a space application. (From Hookman, R. 1989. *Precision Engineering and Optomechanics, SPIE Proc. Vol. 1167*, Vukobratovich, D., ed., p. 368.)

are stiff in the directions of their lengths and depths, but relatively flexible in the directions of their thicknesses. If the baseplate and mirror combination have different coefficients of thermal expansion, temperature changes will cause the flexures to bend, but minimal forces will be exerted upon the mirror. The bending motions are typically along arcs of radii equal to the free lengths of the flexures. If these arcs meet at a point (as shown schematically in the figure), that point will tend to remain stationary with temperature change. Ideally, this point should coincide with the center of gravity of the mirror.[2]

In Figure 6.56, a circular aperture ULE® mirror is clamped with a retainer into an Invar cell which is then attached to the ends of thin flexures machined into a circular aluminum mount.

Since the materials differ greatly with regard to CTEs, temperature changes will cause the springs to bend without unduly stressing the mirror. In this case, rotational symmetry tends to keep the center of the mirror fixed in regard to the related optical system.[2,45]

Single-Point Diamond-Turned Mirrors And Mounts

Lathe or flycutting machines with carefully oriented single-crystal diamond-tipped tools are used to fabricate highly precise mounts for conventional (i.e., nonmetallic) mirrors or metallic mirrors with integral mounting provisions.[2,46-48] The process is commonly called single-point diamond turning (SPDT).

Figure 6.57 shows an example of a stainless steel mirror mount with precisely oriented internal pads to mechanically interface with a mirror made of low expansion material such as Zerodur® as well as equally precisely oriented external pads to interface with external structure. The mirror pads are integral with flexures formed in the mount itself by electrical discharge machining techniques. These flexures compensate for thermal expansion mismatch. The critical surfaces (pads) on the mount are all machined by SPDT techniques.

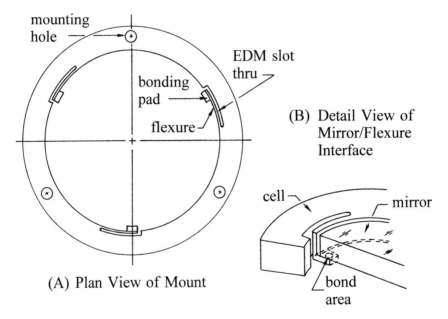

FIGURE 6.57 Concept for a rim-mounted mirror subassembly featuring integral flexure suspension means.

Figure 6.58 shows a metal mirror with integral shaft manufactured to close tolerances by SPDT techniques,[49] while Figure 6.59 shows a metal mirror with integral mount also fabricated by SPDT techniques to form an interchangeable module for a space-borne optical instrument.[50]

6.6 Mounts for Prisms

Clamped Mounts

Kinematic and Semikinematic Techniques

The position (translations) and orientation (tilts) of a prism generally must be controlled to tolerances dependent upon its location and function in the optical system. Control is accomplished through the prism's interfaces with its mechanical surround. The optical material should be placed in compression. Kinematic mounting avoids overconstraints that might distort the optical sur-

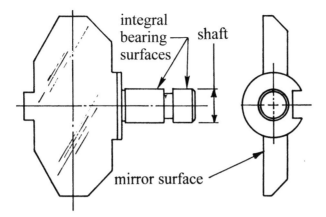

FIGURE 6.58 Metal mirror subassembly with integral gimbal axis fabricated by SPDT techniques. (Adapted from Addis, E.C. 1983. *Geometrical Optics, SPIE Proc. Vol. 389*, Taylor, W.H., ed., p. 36.)

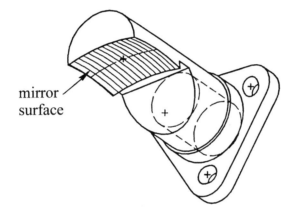

FIGURE 6.59 Diagram of a metal toroidal mirror subassembly with integral mount fabricated by SPDT technique. (Adapted from Visser, H. and Smorenborg, C. 1989. *Reflective Optics, SPIE Proc. Vol. 1113*, Korsch, D.G., ed., p. 65.)

faces.[2,10,51] Point contacts with high stresses inherent in true kinematic mounts are avoided by providing small area contacts at the interfaces.[2] Properly designed spring forces applied over these areas allow expansion and contraction with temperature changes while adequately constraining the prism against acceleration forces.[52] If contact is made on optically active surfaces, the contacting areas should be sufficiently flat and coplanar that surface deformations do not exceed the elastic deformations nominally caused by the clamping constraints.[10]

Figure 6.60 illustrates a semikinematic mounting for a cube-shaped beamsplitter prism. Here, five springs hold the prism against directly opposite pads. Although the contacts occur on refracting surfaces, they are located outside the used aperture, thereby minimizing the effects of surface distortions. This beamsplitter is used to divide a beam converging toward an image plane, each beam then forming an image on a separate detector. In order for these images to maintain their proper alignment relative to each other and to the structure of the optical instrument with temperature changes, the prism must not translate in the XY plane of the figure nor rotate about any of the three orthogonal axes. Translation in the Z direction has no effect. Once aligned, the prism must always press against the five areas indicated by the K_∞ symbols. Constraints are provided at the points labled K_i. The dashed outlines indicate how the prism will expand if the temperature increases. Registration of the prism surfaces against the locating/aligning pads does not change and the light paths to the detectors do not deviate.[52]

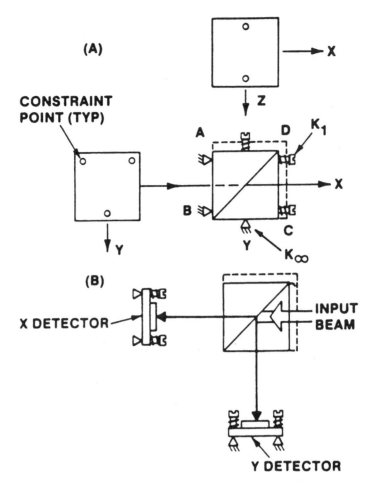

FIGURE 6.60 Schematic of a kinematic mounting for a cube-shaped, beamsplitter prism. (From Lipshutz, M.L. 1968. *Appl. Opt.*, 7, 2326.)

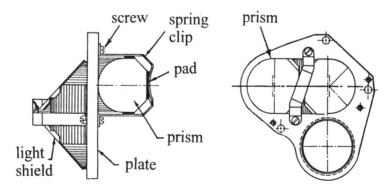

FIGURE 6.61 Schematic of a nonkinematic mounting for a Porro prism erecting subassembly. (Adapted from Yoder, P.R., Jr. 1985. *Geometrical Optics, SPIE Proc. Vol. 531*, Fischer, R.E., Price, W., and Smith, W.J., eds., p. 206.)

Nonkinematic Techniques

Spring or strap means are frequently used to hold prisms in place against the mounting interfaces in optical instruments. An example is the Porro prism erecting prism assembly shown schematically in Figure 6.61.[36] This is typical of prism mountings in binoculars or telescopes. Spring clips hold each prism against a perforated mounting shelf that is, in turn, fastened with screws and locating pins to the instrument housing. Area contact occurs over large areas on the hypotenuse faces of the prisms, while lateral constraints are provided by recessing those faces slightly into opposite sides of the shelf.

Another example of the many types of nonkinematic mounts for prisms is shown in Figure 6.62. Here, an Amici prism is held by a flat spring clip against nominally flat reference pads inside the triangular housing of a military elbow telescope. Constraint perpendicular to the plane of the figure is provided by resilient pads attached to the ground sides of the prism; these are compressed when thin plate covers are attached with screws onto both sides of the housing. Note that the spring contacts the prism on the ground bevels at the ends of the roof surfaces. The spring is loaded against the prism by a screw threaded through the housing wall. The covers and the loading screw are all sealed to protect the environment within the telescope.[2]

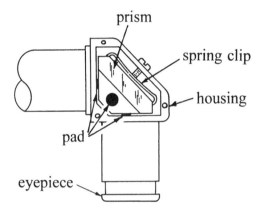

FIGURE 6.62 Schematic of a nonkinematic mounting for an Amici prism in a military elbow telescope.

Bonded Mounts

Many prisms are mounted by bonding their ground faces to mechanical pads using epoxy or similar adhesives. Contact areas large enough to render strong joints can usually be provided in designs with minimum complexity.

The critical aspects of the design are characteristics of the adhesive, thickness of the adhesive layer, cleanliness of the surfaces to be bonded, dissimilarity of coefficients of expansion of the materials, area of the bond, environmental conditions, and care with which the parts are assembled.[2] While the adhesive manufacturer's recommendations should be consulted, experimental verification of adequacy of the design, the materials to be used, the method of application, and cure conditions and duration are advisable in critical applications.

Guidelines for determining the appropriate bond area have appeared in the literature.[53] In general, the adhesive shear or tensile strength is ratioed to the product of prism weight and maximum expected acceleration divided by the bond area. If this ratio is greater than unity, some safety factor exists. This factor should be at least 2. Since adhesive layers normally shrink by a few percent of each dimension during curing, it is advisable to keep these dimensions as small as possible while providing adequate strength.

Examples of Cantilevered Techniques

Figure 6.63 illustrates a Porro prism bonded to a mechanical mounting surface in a cantilevered fashion. The prism is made of Schott SK16 glass, the mount is type 416 stainless steel, and the adhesive is 3M EC2216-B/A epoxy approximately 0.004 in. (0.1 mm) thick. The prism weight is 2.2 lb and the bond area (which covers the maximum area available on a ground face) is 5.6 in.[2]. The intended military application expected the assembly to withstand 1500 G loading. Assuming

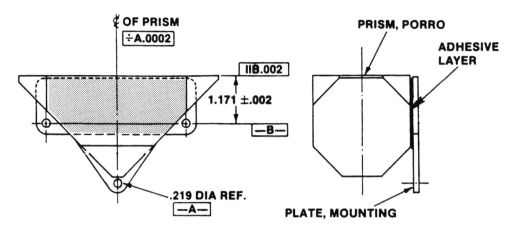

FIGURE 6.63 Schematic of a Porro prism bonded in cantilever fashion on full area (shaded) of one triangular ground face. Dimensions are inches. (Adapted from Yoder, P.R., Jr. 1985. *Geometrical Optics, SPIE Proc. Vol. 531*, Fischer, R.E., Price, W., and Smith, W.J., eds., p. 206.)

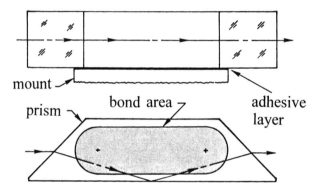

FIGURE 6.64 Schematic of a Dove derotation prism bonded in cantilever fashion on a racetrack-shaped area (shaded) of one ground face.

the adhesive strength to be 2500 lb/in.², the design safety factor would be $2500/[(2.2)(1500)/5.6]$ or 4.2. Tests of prototype hardware built to this design showed that it actually withstood 1200 G acceleration without failing. This at least partially confirmed the design.[36]

Figures 6.64 and 6.65 show two other designs for bonded assemblies with the prisms cantilevered from one surface. The former is a Dove prism in which the appropriate bond area has an elongated circle or "racetrack" shape. The latter is a Pechan prism comprised of two air-spaced elements with only one element bonded to the mount. It is best for an adhesive bond not to bridge over a discontinuity such as a cemented joint unless the elements are cemented together and the surfaces to be bonded were ground flat and coplanar after cementing. The latter example also illustrates division of a bond area into three subareas on the prism surface so as to reduce the lateral dimensions of the bond and thus to minimize the shrinkage effects. The three subareas are spaced as far apart as practical in order to stabilize the joint.

Examples of Multiple Support Techniques

Some designs for bonding prisms utilize multiple adhesive joints between the prism and structure as depicted in Figure 6.66. Here, an increased bond area and support from both sides are provided. It is necessary in such designs that the glass and metal surfaces at each interface be nearly parallel

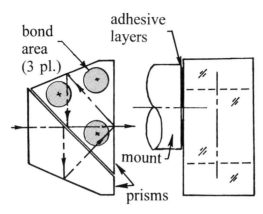

FIGURE 6.65 Schematic of a Pechan derotation prism subassembly bonded in cantilever fashion on multiple circular areas (shaded) of one ground face.

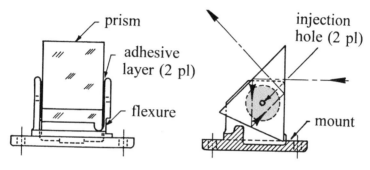

FIGURE 6.66 Schematic of Schmidt prism supported from both sides by bonded areas. (From Willey, R., private communication.)

and that the proper clearances be provided for insertion of the adhesive layers. Tolerances must be held closely enough to ensure these relationships.

Problems with differential expansion of metal and glass at extreme temperatures were avoided in this design by building a flexure into one support arm. Early models without this flexure were damaged at low temperature when the mount contracted more than the prism, causing the arms to pivot about the bottom edge of the prism, and pull away from the prism at the top of the bonds. Allowing the arm to bend slightly prevented such damage.[54]

Another design with support rendered from two sides is shown in Figure 6.67. In this case, the prism is bonded to a pad on one support arm (at left) and to the metal plug shown protruding through, but not attached to the right arm. Alignment of the prism is accomplished using mechanical references or optical fixturing during this first bonding step. After these first bonds have cured, the plug is epoxied to the right arm. With this approach, tolerances on location and tilt of the surfaces to be bonded can be relaxed, since the plug aligns itself to the prism before it is bonded to the arm.

Flexure Mounts

Some prisms (particularly large ones or ones with critical positioning requirements) are conveniently mounted by way of flexures. An example is shown in Figure 6.68.[2] Here, a large prism of unspecified shape is bonded to two cylindrical posts with multiple "necked-down" regions forming "universal joints" to compensate for nonparallelism between the surfaces to be bonded. Cruciform-shaped torsion flexures allow relative rotational motions. Temperature changes will not distort the

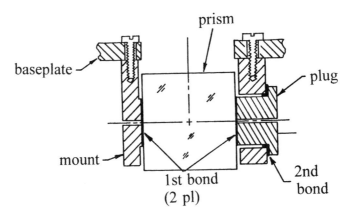

FIGURE 6.67 Schematic of another two-sided prism mount featuring two-step bonding to facilitate alignment without imposing tight tolerances.

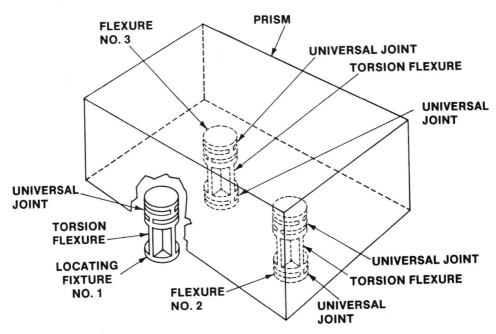

FIGURE 6.68 Schematic representation of a multiflexure mount for a large prism. (From Yoder, P.R., Jr. 1993. *Opto-Mechanical Systems Design*, 2nd ed. Marcel Dekker, New York.)

prism even if the thermal expansion coefficients of the prism, its mount, and the supporting structure are significantly different.

References

1. Smith, W. J. 1990. *Modern Optical Engineering*, 2nd ed., chap. 14. McGraw-Hill, New York.
2. Yoder, P. R., Jr. 1993. *Opto-Mechanical Systems Design*, 2nd ed., Marcel Dekker, New York.
3. Yoder, P. R., Jr. 1993. Lens mounting techniques. In *Optical Systems Engineering III, SPIE Proc. Vol. 389*, Taylor, W. H., ed., p. 2.
4. Horne, D. F. 1972. *Optical Production Technology*. Adam Hilger, Bristol, England.
5. Jacobs, D. H. 1943. *Fundamentals of Optical Engineering*. McGraw-Hill, New York.
6. Bayar, M. 1981. Lens barrel optomechanical design principles, *Opt. Eng.*, 20, 181.

7. Hopkins, R. E. 1980. Lens mounting and centering, chap. 2. In *Applied Optics and Optical Engineering, Vol VIII*. Academic Press, New York.

8. Kowalskie, B. J. 1980. A user's guide to designing and mounting lenses and mirrors, p. 98. In *Digest of Papers, OSA Workshop on Optical Fabrication and Testing*. North Falmouth.

9. Roark, R. J. 1954. *Formulas for Stress and Strain*, 3rd ed., McGraw-Hill, New York,

10. Ritchey, C. A. 1974. Aerospace mounts for down-to-earth optics, *Machine Design*, 46, 121.

11. Vukobratovich, D. 1993. Optomechanical systems design, chap. 3. In *The Infrared & Electro-Optical Systems Handbook, Vol. 4*, Dudzik, M. C., ed. ERIM, Ann Arbor and SPIE, Bellingham.

12. Delgado, R. F. and Hallinan, M. 1975. Mounting of optical elements, *Opt. Eng.*, 14, S-11.

13. Yoder, P. R., Jr. 1991. Axial stresses with toroidal lens-to-mount interfaces. In *Optomechanics and Dimensional Stability, SPIE Proc. Vol. 1533*, Paquin, R. A. and Vukobratovich, D., eds., p. 2.

14. Yoder, P. R., Jr. 1993. Parametric investigations of mounting-induced axial contact stresses in individual lens elements. In *Optomechanical Design, SPIE Proc. Vol. 1998*, Vukobratovich, D., Yoder, P. R., Jr., and Genberg, V., eds., p. 8.

15. Walker, B. H. 1993. 'Select' optical glasses, p. H-356. In *The Photonics Design and Applications Handbook*. Laurin Publishing, Pittsfield, MA.

16. *Optical Glass Catalog 3111e*, Schott Glass Technologies, Inc., Duryea, PA.

17. Yoder, P. R., Jr. 1994. Estimation of mounting-induced axial contact stresses in multi-element lens assemblies. In *Current Developments in Optical Design and Engineering, SPIE Proc. Vol. 2263*, Fischer, R. E. and Smith, W. J., eds, p. 386.

18. Hatheway, A. E. 1993. Analysis of adhesive bonds in optics. In *Optomechanical Design, SPIE Proc. Vol. 1998*, Vukobratovich, D., Yoder, P. R., Jr., and Genberg, V. L., eds., p. 2.

19. Valente, T.M. and Richard, R.M. 1991. Analysis of elastomer lens mountings. In *Optomechanics and Dimensional Stability, SPIE Proc. Vol. 1533*, Paquin, R. A. and Vukobratovich, D., eds., p. 21.

20. Yoder, P. R., Jr. 1992. Advanced considerations of the lens-to-mount interface. In *Optomechanical Design, SPIE Proc. CR43*, Yoder, P. R., Jr., ed., p. 305.

21. Betinsky and Welham. 1979. Optical design and evaluation of large aspherical-surface plastic lenses. In *Optical Systems Engineering, SPIE Proc. Vol. 193*, Yoder, P. R., Jr., ed., p. 78.

22. Yoder, P. R., Jr. 1986. Optomechanical designs of two special purpose objective lens assemblies. In *Contemporary Optical Instrument Design, Fabrication, Assembly and Testing, SPIE Proc. Vol. 656*, Beckmann, L. H. J. F., Briers, J. D., and Yoder, P. R., Jr., eds., p. 225.

23. Scott, R. M., Optical engineering, *Appl. Opt.*, 1, 387.

24. Yoder, P. R., Jr. 1972. A low-light level objective lens with integral laser channel, *Opt. Eng.*, 11, 127.

25. Hopkins, R. E., 1976. Some thoughts on lens mounting, *Opt. Eng.*, 15, 428.

26. Carnell, K. H., Kidger, M. J., Overill, M. J., Reader, A. J., Reavell, F. C., Welford, W. T., and Wynne, C. G. 1974. Some experiments on precision lens centering and mounting, *Opt. Acta*, 21, 615.

27. Fischer, R. E. 1991. Case study of elastomeric lens mounts. In *Optomechanics and Dimensional Stability, SPIE Proc. Vol. 1533*, Paquin, R. A. and Vukobratovich, D., eds., p. 27.

28. Valente T. M. and Richard, R. M. 1994. Interference fit equations for lens cell design using elastomeric lens mountings, *Opt. Eng.*, 33, 1223.

29. Yoder, P. R., Jr. 1960. Two new lightweight military binoculars, *J. Opt. Soc. Am.*, 50, 491.

30. Trsar, W. J., Benjamin, R. J., and Casper, J. F. 1981. Production engineering and implementation of a modular military binocular, *Opt. Eng.*, 20, 201.

31. 1983. *The Handbook of Plastic Optics*, 2nd ed., U.S. Precision Lens, Inc., Cincinnati, OH.

32. Ashton, A. 1979. Zoom lens systems, *Int. Semin. Adv. Opt. Prod. Technol. SPIE Proc. Vol. 163*, p. 92.

33. Parr-Burman, P. and Gardam, A. 1985. The development of a compact IR zoom telescope. In *Infrared Technology and Applications, SPIE Proc. Vol. 590*, Baker, L. R. and Masson, A., eds., p. 11.

34. Kampe, T. U. and Johnson, C. W. 1989. Optomechanical design considerations in the development of the DDLT laser diode collimator. In *Optomechanical Design of Laser Transmitters and Receivers, SPIE Proc. Vol. 1044*, Seery, B. D., ed., p. 46.

35. Quammen, M. L., Cassidy, P. J., Jordan, F. J., and Yoder, P. R., Jr. 1966. Telescope Eyepiece Assembly with Static and Dynamic Bellows-Type Seal, U.S. Patent No. 3,246,563.

36. Yoder, P. R., Jr. 1985. Non-image-forming optical components. In *Geometrical Optics, SPIE Proc. Vol. 531*, Fischer, R. E., Price, W., and Smith, W. J., eds., p. 206.

37. Manuccia, T. J., Peele, J. R., and Geosling, C. E. 1981. High temperature ultrahigh vacuum infrared window seal, *Rev. Sci. Instrum.*, 52, 1857.

38. Stoll, R., Forman, P. F., and Edleman, J. 1961. The effect of different grinding procedures on the strength of scratched and unscratched fused silica, p. 1. In *Proc. Symp. Strength of Glass and Ways to Improve It*. Union Scientifique Continentale du Verre, Charleroi, Belgium.

39. Robinson, B., Eastman, D. R., Bacevic, J., and O'Neill, B. J. 1983. Infrared window manufacturing technology. In *Infrared Technology IX, SPIE Proc. Vol. 430*, Spiro, I. J., ed., p. 302.

40. Klein, C. A., diBenedetto, B., and Pappis, J. 1986. ZnS, ZnSe and ZnS/ZnSe windows: their impact on FLIR system performance, *Opt. Eng.*, 25, 519.

41. Speare, J. and Belloli, A. 1983. Structural mechanics of a mortar launched IR dome. In *Structural Mechanics of Optical Systems, SPIE Proc. Vol. 450*, Cohen, L. M., ed., p. 182.

42. Vukobratovich, D. 1986. *Introduction to Optomechanical Design, SPIE Short Course Notes*.

43. Dunn, G. and Stachiw, J. 1966. Acrylic windows for underwater structures. In *Underwater Photo-Optics, SPIE Proc. Vol. 7*, p. D-XX-1.

44. Barnes, W. P., Jr. 1966. Some effects of aerospace thermal environments on high-acuity optical systems, *Appl. Opt.*, 5, 701.

45. Hookman, R. 1989. Design of the GOES telescope secondary mirror mounting. In *Precision Engineering and Optomechanics, SPIE Proc. Vol. 1167*, Vukobratovich, D., ed., p. 368.

46. Saito, T. T. 1978. Diamond turning of optics: the past, the present and the exciting future, *Opt. Eng.*, 17, 570.

47. Gerchman, M. 1986. Specifications and manufacturing considerations of diamond-machined optical components. In *Optical Component Specifications for Laser-Based Systems and Other Modern Optical Systems, SPIE Proc. Vol. 607*, Fischer, R. E. and Smith, W. J., eds., p. 36.

48. Sanger, G. M. 1987. The precision machining of optics. In *Applied Optics and Optical Engineering, Vol. 10*, Shannon, R. R. and Wyant, J. C., eds., chap. 6, Academic Press, New York.

49. Addis, E. C. 1983. Value engineering additives in optical sighting devices. In *Geometrical Optics, SPIE Proc. Vol. 389*, Taylor, W. H., ed., p. 36.

50. Visser, H. and Smorenborg, C. 1989. All reflective spectrometer design for Infrared Space Observatory. In *Reflective Optics, SPIE Proc. Vol. 1113*, Korsch, D. G., ed., p. 65.

51. Durie, D. S. L. 1968. Stability of optical mounts. In *Machine Des.*, 40, 184.

52. Lipshutz, M. L. 1968. Optomechanical considerations for optical beamsplitters, *Appl. Opt.*, 7, 2326.

53. Yoder, P. R., Jr. 1988. Design guidelines for bonding prisms to mounts. In *Optical Design Methods, Applications, and Large Optics, SPIE Proc. Vol. 1013*, Masson, A., Schulte in den Bäumen, J., and Zügge, H., eds., p. 112.

54. Willey, R., private communication.

7

Adjustment Mechanisms

Anees Ahmad

7.1 Introduction

This chapter discusses the design aspects of different types of adjustment mechanisms used in optical systems. Various optical elements in a sophisticated system must be precisely aligned to each other to obtain an aberration-free image. In optical systems with very tight alignment requirements, it is more cost effective to manufacture the optics and their mounts to rather loose tolerances, and then employ adjustment mechanisms to align the optics relative to each other at assembly. Another class of adjustment mechanisms is used to move one or more optical elements of a system in real time to correct the image degradation caused by environmental effects.

Certain optical systems, such as those for submicron lithography, can have 10 to 15 mirrors and lenses that must be axially positioned relative to each other and centered on a common optical axis within tolerances of a few microns. To achieve these kinds of positioning accuracies, it is impractical and extremely cost prohibitive to manufacture the optics and its mounts to micron-level machining accuracies. For such optical systems, it is more practical and economical to fabricate the optics and the mounting hardware to loose tolerances, and provide adjustment mechanisms to align the optical elements relative to each other at the time of assembly. This class of adjustment mechanisms is designed for infrequent use and generally has manual actuators. Once the optical system has been aligned, these adjustments are locked in place to retain the alignment.

Another class of adjustment mechanisms are employed to despace or tilt an optical element in real time to compensate for the degradation of the image quality due to environmental effects. These mechanisms usually have motorized actuators and position readout sensors operating in a closed loop control system. Such mechanisms are generally used to correct focus and/or magnification errors in the optical systems due to thermal effects or any other environmental degradation.

This chapter covers the three basic types of adjustment mechanisms, namely: linear, rotary, and tilt mechanisms. Each mechanism consists of a number of parts such as an interface between the moving and stationary part, an actuator, locking, and preloading components. The selection criteria for these components of the adjustment mechanisms are discussed in detail. A number of example mechanisms have also been included for the benefit of designers. Finally, some guidelines for proper design and application of adjustment mechanisms in complex optical systems are presented.

7.2 Types of Adjustment Mechanisms

The three basic types of adjustment mechanisms are linear, tilt, and rotary mechanisms. A rigid body in space has six degrees of freedom, namely, the three translations and the three rotations about x, y, and z axes. An optical element in a system may need one or more of these translation or tilt (rotation about an axis) adjustments for alignment purposes. To avoid cross-coupling effects between different adjustments, the preferable approach is to stack single axis adjustments on top of each other to achieve a multi-axis adjustment mechanism.

A typical adjustment mechanism consists of five basic components. These components are

1. An *interface* between the moving optical element and the fixed structure
2. An *actuator* to adjust the moving element relative to the fixed structure
3. A *coupling* device or method between the actuator and the moving element
4. A *preloading device* to eliminate backlash in the mechanism
5. A *locking mechanism* to retain the adjusted position

For each of these five components, a number of choices are available to a designer depending on the type of adjustment mechanism. The most commonly used components for linear, rotary, and tilt mechanisms are shown in Tables 7.1, 7.2, and 7.3, respectively. The size and shape of these components generally are dictated by the space constraints and the service requirements for that particular application. For laboratory prototypes of optical systems, a number of commercial linear rotary and tilt stages and optical mounts with built-in adjustments and actuators are available. Unfortunately, these commercial adjustment mechanisms are quite expensive and bulky to be incorporated into actual optical systems. Therefore, practical adjustment mechanisms for a particular application usually have to be custom designed to provide the desired adjustment capabilities within the given space and cost constraints. This requires synthesizing an adjustment mechanism by selecting its components from Tables 7.1, 7.2, or 7.3, and then sizing and assembling these parts to meet the desired performance specifications.

7.3 Linear Adjustment Mechanisms

General Description

Linear mechanisms are the most commonly used adjustment mechanisms in optical systems. These mechanisms are employed when an axial or centration adjustment of an optical element is required. It is clear from Table 7.1 that a designer has a wide choice in selection of the components for a particular application. The selection of a particular type of component is dictated by the perfor-

TABLE 7.1   Choice of Components for Linear Mechanisms

Interface	Actuator	Preload	Locking	Coupling
Flexure	Coarse screw	Compression spring	Set screw	Ball/cone
Kinematic	Fine screw	Extension spring	Jackscrew	Ball/flat
Ball bearing	Micrometer	Flat spring	Locknut	Ball/socket
Roller bearing	Differential micrometer	Belleville washer	V-clamp	Threads
Air bearing	DC motor/linear motor	Curved washer	Collar clamp	Flexible coupling
Dovetail slide	Stepper motor		Epoxy	Lead screw
Flat slide	Piezoelectric		Control system	

TABLE 7.2   Choice of Components for Tilt Mechanisms

Interface	Actuator	Preload	Locking	Coupling
Cross-flexure	Coarse screw	Compression spring	Set screw	Ball/cone
Kinematic	Fine screw	Extension spring	Jackscrew	Ball/flat
Spherical bearing	Micrometer	Flat spring	Locknut	Ball/socket
Journal bearing	Differential micrometer	Belleville washer	V-clamp	
	Stepper motor	Curved washer	Epoxy	
	Piezoelectric		Control system	
	Linear motor			
	DC motor			

TABLE 7.3   Choice of Components for Rotary Mechanisms

Interface	Actuator	Preload	Locking	Coupling
Cross-flexure	Coarse screw	Compression spring	Set screw	Ball/cone
Ball bearing	Fine screw	Extension spring	Locknut	Ball/flat
Spherical bearing	Micrometer	Flat spring	V-clamp	Ball/socket
Journal bearing	Differential micrometer	Belleville washer	Collar clamp	Flexible coupling
Roller bearing	DC motor/linear motor	Curved washer	Epoxy	Worm/gear
Air bearing	Stepper motor	Torsion spring	Control system	Rack/pinion
	Piezoelectric			Belt/pulley

mance requirements such as the frequency, range, and resolution of adjustment, and other design factors such as the size, cost, and the load capacity of the mechanism. In the following sections, a number of design options for different parts of a linear adjustment mechanism are discussed, and general guidelines are presented to help a designer in selecting the suitable type of components for a particular application.

Interfaces for Linear Mechanisms

In a linear adjustment mechanism, the interface between the moving optical element and the fixed structure is generally determined by such design factors as the travel range, frequency of adjustment, shock, load capacity, cost, and size. If a long travel range is required and the mechanism is going to be adjusted frequently, a bearing interface must be used between the moving element and the fixed structure. Various types of slides suitable for linear mechanisms are illustrated in Figure 7.1. Ball and roller slides have a low friction and are suitable for long travel ranges. Ball slides (Figure 7.1[c]) are less expensive than roller slides (Figure 7.1[b]), but also have lower load capacity and accuracy (straightness of travel) as compared to roller slides. A dovetail slide, shown in Figure 7.1(c), has a high stiffness and load capacity and is less expensive. The main disadvantages of a dovetail slide are stiction and high friction. This slide is generally used in prototypes for laboratory type setups, where the adjustments are made infrequently and have to be simple and economical. Table 7.4 summarizes the linearity and running friction coefficients (RFC) for various types of slides.[1]

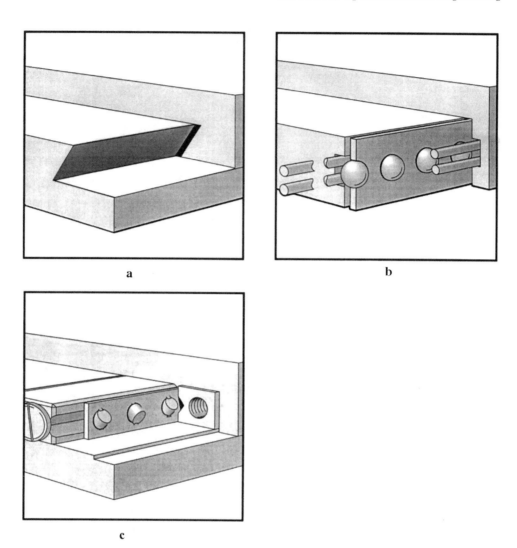

FIGURE 7.1 Types of linear slides for translation mechanisms. (a) Dove tail slide; (b) ball-bearing slide; (c) roller-bearing slide. (Courtesy of Newport Corporation, Irvine, CA.)

TABLE 7.4 Linearity, Running Friction Coefficients (RFC), and Load Capacity of Slides

Slide Type	Linearity (μm/10 mm)	RFC	Load Capacity and Stiffness
Dovetail	10	0.05–0.2	High
Ball	2	0.002	Low
Roller	1	0.003	Moderate

Hydrostatic Bearings for Linear Mechanisms

Hydrostatic bearings, which include both gas and oil bearings, are virtually free of friction and wear and have a negligible cross-axis runout. In optical systems, oil bearings are not commonly used because these are messy and present the risk of contaminating the optics. An optical system that requires an adjustment mechanism with a long travel range and high accuracy and load capacity can use a gas bearing. The pressurized gas is generally very clean and dehumidified air, but in some applications dry nitrogen or helium may be required. The main disadvantages of gas bearings are their cost and complexity. A remote and elaborate pumping and filtration system is required to supply clean and dry air. The design and fabrication of gas bearings is complex and expensive. The number and size of air jets, size of the air relief pockets, surface area of the bearing, and supply pressure of the air must all be taken into account when designing the air bearing for an application.

The simplest form of an externally pressurized bearing is a circular thrust bearing with a central jet as shown in Figure 7.2.[2] The pressurized air is forced into a recessed pocket in the middle, and it escapes along the periphery, thereby creating a very thin lubricating film of very high stiffness between the two surfaces. For incompressible flow, the load capacity for such a bearing is given by the following equation:

$$W = 0.69(p_s - p_a)\frac{\pi(b^2 - a^2)}{2\ln(b/a)} \tag{1}$$

where p_s = supply pressure
        p_a = exhaust or ambient pressure
        a = the radius of central recess
        b = the outer radius of bearing

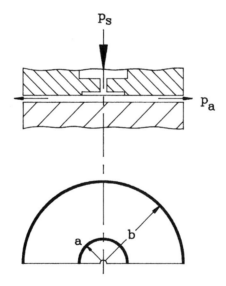

FIGURE 7.2 Schematic of a flat thrust gas bearing for linear mechanisms.

A bearing may also be designed with a ring of jets and radial grooves. It is very expensive to design and fabricate custom gas bearings; therefore, it is advisable to purchase standard bearings from commercial manufacturers, as far as feasible.

Flexures for Linear Mechanisms

Flexures are suitable for backlash-free adjustments over short travel ranges (1 to 2 mm). These have low friction and hysteresis, and do not require any type of lubrication. Flexures can be in several shapes such as a flat strip, circular, or universal. The design and fabrication of flexures are quite complex and are discussed in detail in Weinstein.[3] For each application, a flexure must be designed to have a specific stiffness which is determined by its length, width, thickness, shape and material. The materials having a high tensile modulus are more suitable for making flexures. The tensile modulus is defined as the ratio of allowable bending stress σ to the elastic modulus E. The

TABLE 7.5   Common Flexure Materials and Their Properties

Material	Yield Strength (ksi)	Elastic Modulus (Msi)	CTE (ppm/F)
Stainless steels			
302	35–40	28	9.6
440C	65	29	5.6
17-4 PH	125	28.5	6.0
Beryllium copper	85–110	19	9.9
Titanium	108	14.9	8.6
Invar 36	98	20.5	1.3
7075-T6 Al alloy	73	10.4	13.1

material with a higher σ/E ratio will have better compliance for a given length of the flexure. Some of the suitable materials for flexures along with their allowable bending stresses and elastic moduli are listed in Table 7.5.

Fabrication of flexures can be quite complex and expensive due to rigorous process control. To prevent failure due to stress concentration, the residual stresses due to machining must be minimized by selecting appropriate machining methods and proper heat treatment for stress relief. Similarly, a smooth-surface finish is desirable for a long life. When flexures are bent to move a component, a reaction force is induced in the component attached to the flexure. In applications where flexures are directly bonded to optical components, the reaction force from the flexure can produce a localized surface distortion. Therefore, the flexures must be designed for a proper stiffness to keep this distortion within acceptable limits.

The flexure design for linear (parallel) motion has been discussed in detail by Neugebauer.[4] For linear motion, the moving member is coupled to the fixed support through two parallel flat flexures as shown in Figure 7.3. These flat flexures have thin and rectangular cross sections, either solid or with cutouts as shown in Figure 7.4. These flat blade flexures are very stiff in tension and shear, but very compliant in bending. The deflection or travel due to actuator force F is given by:

$$Y = FL^3 / 2Ebt^3 \qquad (2)$$

where L, b, and t are the effective length, width, and thickness of the flexure, respectively, and E is the elastic modulus of the flexure material. The vertical shear stress due to F is negligible, and the bending stress is given by:

$$\sigma = 3FL / 2bt^2 \qquad (3)$$

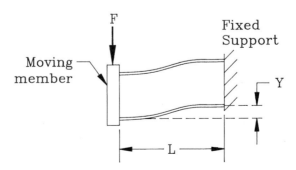

FIGURE 7.3 Parallel flat spring flexures for linear motion.

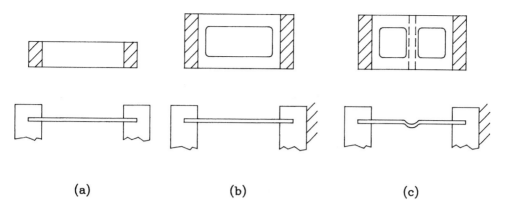

FIGURE 7.4 Types of flat spring flexure designs. (a) Solid; (b) single cut-out; (c) double cut-out.

If the weight W of the moving part is taken into account, the equations for deflection and the resulting stress become more complex. If the load W is compressive, the deflection and stress are given by:

$$\delta_c = \frac{FL}{W}\left(\frac{1}{K}\tan K - 1\right) \tag{4}$$

$$\delta_c = \frac{3FL}{2Kbt^2}\tan K + \frac{W}{2bt} \tag{5}$$

If the load W is tensile, the equations for deflection and stress are:

$$\delta_t = \frac{FL}{W}\left(1 - \frac{1}{K}\tanh K\right) \tag{6}$$

$$\delta_t = \frac{3FL}{2Kbt^2}\tanh K + \frac{W}{2bt} \tag{7}$$

The coefficient K in all these equations is defined by the following expression:

$$K = \left(\frac{3WL^2}{2Ebt^3}\right)^{0.5}$$

Actuators for Linear Mechanisms

The choice of a suitable actuator for a linear mechanism depends on travel speed, range, resolution, and frequency of adjustment, and cost, size, and weight requirements for the adjustment mechanism. For example, the motorized actuators are generally used for making frequent adjustments in real time. These include DC, linear, and stepper motors and piezoelectric devices. The main advantages of such actuators are long travel range, high resolution and velocity, and position readout capability. These actuators usually come with built-in position encoders and can be used in a closed loop control system. Therefore, the position of an optical component can be monitored, and the drifts due to environmental effects can be corrected in real time. The principle disadvantages of motorized actuators are their high cost and weight and large size.

Motorized Actuators

A number of motorized actuators are commercially available such as those shown in Figure 7.5 from Newport Corporation.[5] The actuator in Figure 7.5(a) is an ultraresolution electrostrictive type of actuator, which provides a piezo-class motion without any hysteresis and creep. It has an integrated fine pitch manual adjustment screw to center the electrostrictive fine travel where desired. The actuator shown in Figure 7.5(b) has a DC motor, which drives a precision leadscrew through low-backlash reduction gears. Both actuators provide a submicron-level positioning capability.

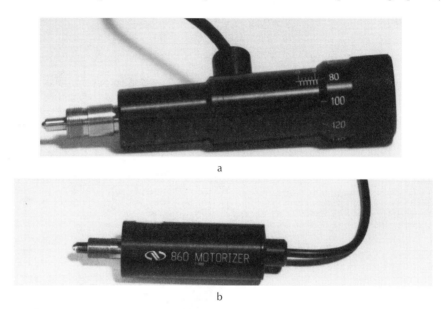

FIGURE 7.5   Types of motorized linear actuators. (a) Electrostrictive actuator. (b) DC motor. (Courtesy of Newport Corporation, Irvine, CA.)

Two other types of motors designed for high-end applications requiring micron-level positioning accuracy are the linear servomotor and piezoelectric inchworm motor. The brushless linear motors offered by Anorad Corp,[6] shown in Figure 7.6(a), consist of a fixed permanent magnet assembly and a moving coil assembly. A very small gap of the order of 0.5 mm is maintained between the coil assembly and magnets. When a current is applied to the coil, the electromagnetic force engages the moving mass with no mechanical link or friction. This friction-free force transmission eliminates jitter, stiction, backlash, and hysteresis. These types of linear motors are useful in applications requiring high acceleration (20 m/sec^2), high load capacity (up to 900 N), and high velocity (2 m/sec or more). A high accuracy planar stage combining air-bearing, linear motor, and servo-control technologies is shown in Figure 7.6(b). The model SG-2525 Servoglide x-y planar stage provides a travel range of 250 × 250 mm through direct drive motion. A closed loop servo-control system using an optical interferometer provides a long travel with dynamic yaw and rotation alignment capabilities. Table 7.6 compares the features of a brushless linear motor with those of a brush-type motor for precision linear motion applications. It is obvious that except for advantages in price and size, a brush type of motor has inferior performance in all other categories.

A piezoelectric actuator is generally used for short travel range requiring high resolution. A high voltage is needed to produce a movement of the order of a few microns. The piezo actuators have high load capacity. Typical disadvantages are hysteresis, creep, and nonlinearity of travel vs. the applied voltage. Burleigh's[7] inchworm motors, illustrated in Figure 7.7, can achieve ultraprecise

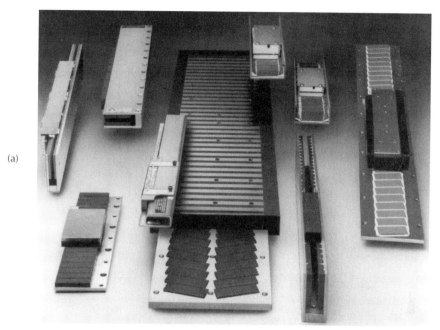

(a)

(b)

FIGURE 7.6 Brushless linear motors. (a) Single-axis motors; (b) two-axis air- bearing stage, Model SG-2525. (Courtesy of Anorad Corp., Hauppage, NY.)

positioning with a high degree of stability. These piezo-based motors provide angstrom- to nanometer-level resolution and travels of up to 200 mm. This feature of inchworm motors eliminates the need and complexity of a combination of coarse (fast speed)/fine (slow speed) positioning mechanism. The compact direct linear drive design of inchworm motors is backlash-free, nonmagnetic, and suitable for remote operation and ultrahigh vacuum applications.

A unique piezo-type actuator, depicted in Figure 7.8, has been recently introduced by New Focus, Inc.[8] under the trade name of *Picomotor*. With two jaws that grasp an 80-pitch screw, the Picomotor turns the screw much like as is done manually. A piezoelectric transducer slides the jaws in opposite directions. Slow action of the motor causes a screw rotation (right view), while a fast action due to inertia causes no rotation (left view). Traditional piezo-type actuators rely on contraction and expansion of the piezo to position or move an object. These actuators may exhibit backlash, hyteresis, and creep. The unique design feature of the Picomotor is that the piezo is only used to turn a screw, and not for holding the adjusted position. Therefore, this type of actuator is

TABLE 7.6   A Comparison of Motors Available for Precision Positioning Systems

Characteristic	Brushless Linear Motor	Brush-Type Linear Motor
Positioning accuracy	Good	Fair
Minimum step size	Good	Fair
Dynamic stiffness	Good	Good
Maximum velocity	Good	Fair
Constant velocity	Good	Fair
Settling time	Good	Fair
Friction hysteresis	Good	Fair
Lubrication/maintenance	Good	Fair
Varying load capacity	Fair	Fair
Cleanroom applicability	Good	Fair
Noise	Good	Good
Vertical applications	Fair	Fair
Durability	Good	Fair
Price <1-m travel	Fair	Fair
>1-m travel	Fair	Good

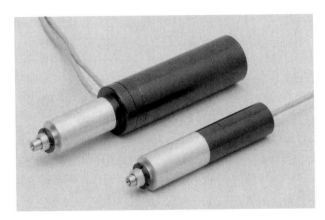

FIGURE 7.7   PZT-based *Inchworm* linear motors. (Courtesy of Burleigh Instruments, Inc., Fishers, NY.)

virtually free of backlash, creep, and hyteresis because a rigid screw is used to hold the position. Another advantage is that no applied voltage is required to hold the desired set position. Since this motor works in tandem with a fine threaded screw, its travel range is limited by the length (0.5- to 2-in. standard range) of the screw used. Some important performance specifications of Pico-motor are listed in Table 7.7.

Manual Actuators

If an application does not require frequent adjustments, it is more cost effective to use screws or micrometers shown in Figure 7.9. If a position readout is not required, the screws are more economical and compact as compared to micrometers. The screws can be coarse, fine or of differential type depending on the resolution requirements. The principle of obtaining a fine linear travel from a differential screw is illustrated in Figure 7.10. By using two threaded screws in series, each with a slightly different pitch (number of threads/inch), a high resolution adjustment can be accomplished. When the screw is turned, the resulting differential translation T_d of the nut per revolution can be calculated by the following simple equation:

$$T_d = T_c - T_f \tag{8}$$

FIGURE 7.8   A piezo-type linear actuator, *Picomotor*. (Courtesy of New Focus, Inc., Sunnyvale, CA.)

TABLE 7.7   Picomotor Characteristics

Travel range	0.5, 1, and 2 in. standard
Resolution	<0.1 μm
Load capacity	2 lb
Speed	2–3 rpm
Repeatability	<0.1 μm with feedback
Lifetime	>5 × 10⁸ pulses

where T_c = travel per revolution for coarse thread
        T_f = travel per revolution for fine thread

For example, for a differential screw with a pair of 40 and 32 threads per inch, the net travel per revolution will only be 0.00625 in., i.e., the equivalent of 160 threads per inch pitch screw. The advantages of a differential screw are obvious from this example. The threads of 40 and 32 pitch can be machined inexpensively with standard tools in any shop, while machining the threads of 160 pitch is not trivial. The differential screws have their own disadvantages, also. These screws have higher friction due to the two nuts and, therefore, require a higher turning torque. The friction and torque can be reduced significantly by using ball screws in place of conventional screw threads, but this will add substantial cost. In differential screws, it is critical that the two threads be very concentric to minimize a premature failure due to rapid wear.

Regular, fine thread and differential screws and micrometers are commercially available in a large variety of sizes, shapes, materials, travels, and resolutions. The regular and differential micrometers are used for small travels, and are quite bulky and more expensive as compared to regular screws. Therefore, in general, micrometers are used in linear translation stages for laboratory prototypes only.

FIGURE 7.9   Fine-threaded screws and micrometers. (Courtesy of Newport Corporation, Irvine, CA.)

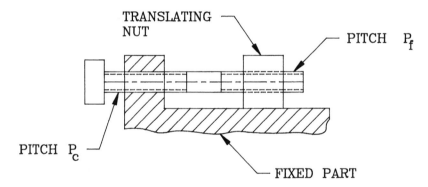

FIGURE 7.10   A differential screw with two threads of slightly different pitch can be used to produce a fine linear motion.

Coupling Methods for Linear Mechanisms

A number of options are available to couple an actuator to the moving component of an adjustment mechanism. The choice of a suitable method depends on such design factors as frequency of adjustment, shock requirements, weight, cost, and size of the mechanism. The tip of an actuator is generally rounded and polished to a high finish to minimize wear due to friction. The tip of an actuator can bear against a flat or a conical surface. The flat surface, shown in Figure 7.11(a), results in a point contact with the tip of an actuator and therefore produces high contact stresses. The ball/cone interface, shown in Figure 7.11(b), results in a line contact and produces much lower contact stresses. For infrequent adjustments of light weight components, a ball tip acting directly against a flat surface is the most simple and economical choice. The ball cone interface is more expensive to machine and is generally used for heavy components which must be adjusted frequently.

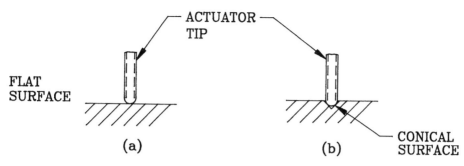

FIGURE 7.11 Two types of common interfaces for the round tip of an actuator. (a) Point contact with a flat surface; (b) line contact with a cone.

The moving component can also be directly attached to an actuator using a threaded attachment or a flexible type of coupling. The threaded coupling is not commonly used because a slight misalignment of the line of travel relative to the actuator axis induces lateral loads in the actuator and can cause rapid wear and damage to the actuator. A flexible coupling does not have this problem because slight misalignments in the mechanism are compensated by the flexibility of the coupling. The flexible couplings come in many types such as bellows, spring, Oldham, jaw, and Schmidt type. All these flexible couplings are relatively bulky and heavy, and are expensive and, therefore, are used in large mechanisms where heavy components need to be adjusted frequently.

For longer travels and heavy loads, an actuator may be coupled to the moving part through a lead screw shown in Figure 7.12. In this translation stage, the motor is coupled to the main screw while the moving platform is attached to the nut. The backlash and wobble of the nut can be minimized by preloading the nut. The friction can be reduced by using ball or roller bearings between the screw and nut. The lead screws are large in size and are expensive but they provide very linear travel over a long range. The torque T required to translate a load W is given by:

$$T = Wp/2\pi E \tag{9}$$

where p = the screw pitch or lead per revolution
        E = the lead screw efficiency

Commercial lead screws have a minimal lead error ($10~\mu m/300$ mm) and come at a reasonable cost.

Preloading Methods for Linear Mechanisms

Most mechanisms employ some form of preloading arrangement to ensure a positive movement, free of backlash, when the adjustment is made. The selection of a suitable type of preloading method depends on such design factors as the range and frequency of adjustment, load capacity, cost, and size of the adjustment mechanism. In adjustments with a long travel range, a helical compression or an extension spring may be used for preloading the moving part against the stationary part. Normally the springs are placed around the adjustment screw, or can be centered between two adjacent adjustment screws to preload both screws with one spring.

A spring is characterized by its *spring rate k*, which is the ratio of an applied force F to the resulting deflection δ, i.e.,

$$k = F/\delta \tag{10}$$

FIGURE 7.12 Commercial linear stages with lead screw coupling for a higher load capacity. (Courtesy of New England Affiliated Technologies, Lawrence, MA.)

Helical compression and extension springs are commercially available in a variety of spring rates, diameters, and free lengths. For an application, the required preload to hold the adjustment under shock loading is normally known. Also, the mechanism is designed to produce a known deflection in the spring. With this information, the required spring rate can be calculated by using the above equation, and then a suitable spring with the right free length and spring rate can be selected from a catalog. The number of coils in a spring must be sufficient to ensure that the spring wire remains within its elastic limit when the spring is at its maximum deflection. The number of coils in a compression spring also determines the minimum length, which is realized when adjacent coils come into contact. A long compression spring may buckle under stress. Therefore, long compression springs must be avoided unless they are guided by a rod or a screw through their center.[9]

If a mechanism is not locked by any other means, the natural frequency of a spring-loaded system can be calculated by:

$$f_n = \frac{1}{2\pi} \sqrt{\frac{g}{\delta_{st}}} \qquad (11)$$

where g = the gravitational constant
         δ_{st} = static deflection produced by the load

Helical springs are generally made from music wire or stainless steel wire. Music wire is very strong and hard because of the drawing process used in its production, and does not need to be further hardened after forming. Type 302 stainless steel wire springs are resistant to corrosion and are more commonly used in optical systems, although these have a lower (0.833) spring rate as compared to music wire springs.

Belleville and curved washers are used in compression mode to obtain high preloads over very small travels. A Belleville washer is a cone-shaped disk with a hole in the center as shown in Figure 7.13(a). When a load is applied, the cone flattens slightly, thus acting as a very stiff spring. These washers have very high stiffness and therefore, produce high loads as a result of relatively small

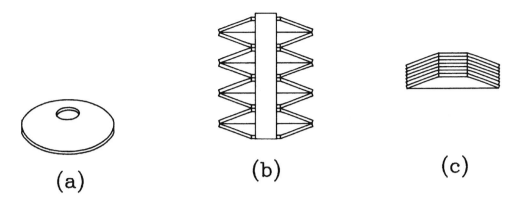

FIGURE 7.13 (a) A Belleville washer; (b) stacked in series for increased travel range; (c) stacked in parallel for higher load capacity.

compression. These washers are compact in size and are very economical and suitable for high shock applications. The travel range of mechanisms or its preload can be increased by stacking the Belleville washers in series or parallel as shown in Figure 7.13(b) and (c). When stacking these washers, a rod or a screw passing through their center is needed to retain them in the configuration shown.

Locking Methods for Linear Mechanisms

After an element has been adjusted, it must be locked in place to retain its adjusted position. The design factors affecting the choice of a locking method are the frequency of adjustment, shock, vibration, weight, size, and cost requirements. The locking can be accomplished by locking the actuator itself to prevent its accidental movement, tampering, or its drift under environmental vibrations. For micrometers or screw-type actuators, simple caps or covers can be used to prevent accidental movements. The movement due to vibrations can be prevented by using set screws to lock the rotation of actuator screws. The second option is to positively lock the moving element relative to the fixed structure, and several methods are available for doing this. If an optical element is going to be adjusted and aligned only at assembly, it can be locked in place by using epoxy bonding shown in Figure 7.14. The epoxy locking has the advantage of being a very low cost method. The main disadvantage is that the mechanism cannot be readjusted without breaking the epoxy bond, which is generally very difficult and also poses a risk of damaging the parts.

Jack screws or locknuts, shown in Figure 7.15, are economical ways of locking the mechanisms that do not require a great precision. The reason for lower accuracy is that a large force can be exerted on the adjusted element when a locknut or jack screw is tightened. This high force can cause a slight drift of the components. The advantages are that these locking components are

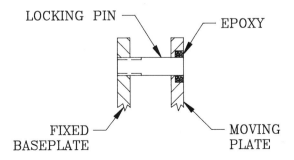

FIGURE 7.14 Typical locking scheme using epoxy bonding.

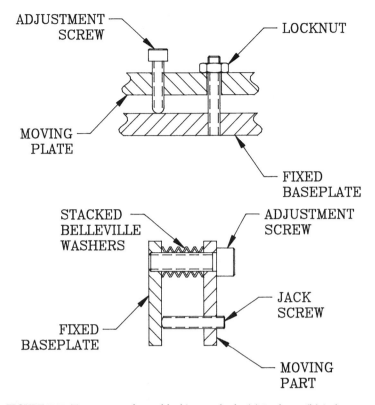

FIGURE 7.15 Two commonly used locking methods. (a) Locknut; (b) jack screw.

compact in size, and can be disassembled very easily without posing any risk of damage to the mechanism parts.

For motorized actuators, a closed loop feedback control system can be employed to retain the adjusted position of the moving element. As mentioned earlier, this method is expensive and is used only in those mechanisms which require real time adjustment.

Examples of Linear Translation Mechanisms

This section illustrates how various components of a mechanism can be assembled together to create a practical linear adjustment mechanism. These examples of a few simple linear mechanisms are meant to illustrate the principles only. As mentioned earlier, the size and shape of individual components are mostly application specific. Therefore, a designer can modify the parts of a particular mechanism, or combine various parts from different mechanisms shown in this section to design a mechanism that best meets the performance requirements. Most of the adjustment mechanisms described here are suitable for making height and axial (despace) adjustments.

In some examples here, the design of a mechanism at a single adjustment point is described. In actual practice, if a mirror mount is to be adjusted relative to a fixed structure, three similar and equally spaced adjustment points are needed around the periphery of the mirror mount. For round mirrors, windows, filters, and beamsplitters, the adjustment points must be equally spaced 120° apart, if feasible. For rectangular or square optics, the adjustment points can be located at any three corners. In such cases, if all three points are adjusted equally, a pure height or axial adjustment will result. If the adjustment is made at only one point, then a tilt motion occurs about an axis defined by a line joining the other two adjustment points. Therefore, most of the mechanisms discussed in this section are suitable for making axial as well as tilt adjustments.

Two simple linear adjustment mechanisms proposed by Tuttle[10] are shown in Figures 7.16 and 7.17. These mechanisms, though not very practical for optical instruments, illustrate the application of simple screws and sliding interfaces for linear motion. The mechanism shown in Figure 7.16 consists of two opposing bowed springs for preloading the moving part, which slides on the fixed structure. The bowed springs can be in the form of thin, rectangular sheet metal strips. The linear motion is obtained when one screw is threaded in, while the other screw is moved out simultaneously. Once the desired adjustment is achieved through an iterative process, the locknuts are tightened to retain the adjusted position. The principle disadvantages of this mechanism are high friction and poor resolution.

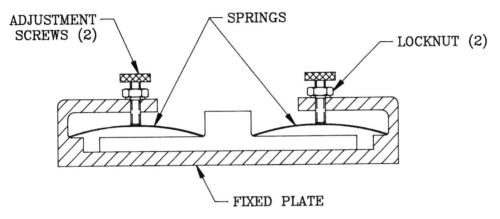

FIGURE 7.16 A simple linear mechanism with screw actuators and bowed springs for preloading.

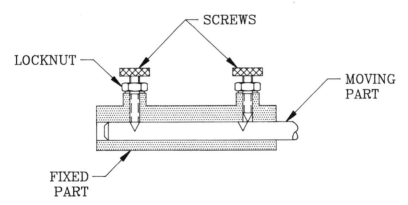

FIGURE 7.17 Two conical screws are used to slide a rod in this linear mechanism.

The linear mechanism, shown in Figure 7.17, has a sliding rod in a slightly oversized bore in the fixed structure. Two thumb screws with conical points engage the corresponding tapered holes in the rod. The center-to-center distance between the screws is slightly smaller than the corresponding spacing between the holes. When the screw on the right is threaded in while the left screw is moved out, the rod slides to the right. The opposing translation can be achieved by reversing the motion of the screws. The locking action of each screw, when threaded in, produces a side thrust on the threads. The screws must have long-enough engagement length to prevent cocking of the screws. The locknuts are provided to secure the adjusted position achieved through an iterative process. This mechanism also has friction and wear problems.

Another simple linear mechanism proposed by Elliot[11] is shown in Figure 7.18. The kinematic design of this mechanism employs two parallel cylindrical rods as guides. The moving plate has two vee and one flat contact with these rods. Simple screws or micrometers with opposing compression springs can be incorporated into this mechanism to produce linear sliding motion. Although this mechanism offers a long travel range for heavy load applications, it is too bulky to be incorporated in most of the practical optical systems.

A two-axis linear translation mechanism by Kittell[12] is illustrated in Figure 7.19. Two round-tip screws are acting on the moving part through flexures. The moving part is preloaded against the flexures by using in-line compression springs. This mechanism is suitable for low precision adjustment over short travels, and can have stiction problems.

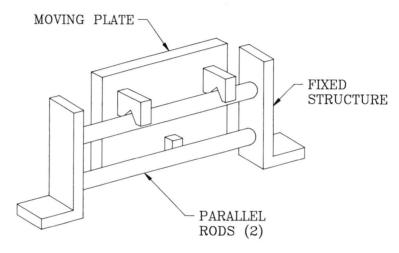

FIGURE 7.18 A linear mechanism with two parallel rods for long travels.

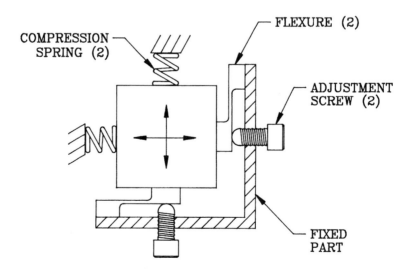

FIGURE 7.19 A two-axis linear mechanism using flexures and screw actuators.

A simple height adjustment mechanism by Ahmad,[1] depicted in Figure 7.20, has a regular screw threaded into a fixed baseplate. The moving part is spring-loaded against the screw by using a compression spring. A pair of swivel washers is provided to compensate for any nonparallelism between the moving and fixed plates. A typical mirror mount can be designed with three such adjustments on a triangular pattern. If the screws are moved equally, a pure height (or centration) adjustment is achieved. When only one of the screws is moved, a tilt results about an axis defined by the other two screws. Once the desired adjustment has been achieved, the jack screw is tightened to lock the mechanism. If an excessive force is used to tighten the jack screw, the adjusted position

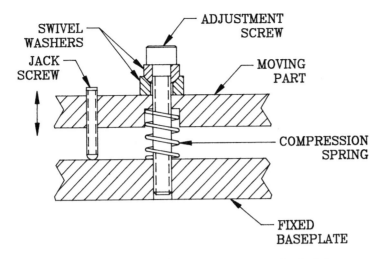

FIGURE 7.20 A height adjustment mechanism using spring-loaded screws.

may change slightly. This mechanism uses standard off-the-shelf commercial parts and is therefore, a good low cost option for linear adjustments with moderate resolution and travel.

The linear adjustment mechanism by Ahmad,[14] as shown in Figure 7.21, illustrates the principles of epoxy locking. This mechanism is essentially the same as the previous mechanism except that an extension spring, instead of a compression spring, is employed for preloading the moving part against the fixed structure. Here a fine threaded commercial screw with a ball tip is used for a finer resolution. The tip of the screw makes a line contact with a cone machined into the fixed plate to minimize the contact stresses for heavier load applications. The ends of the extension spring are retained by dowel pins going through them, thereby producing a force pulling the moving part toward the fixed plate. The locking in this mechanism is achieved by a locking pin, which is threaded into the fixed plate. The other end of the pin goes through an oversized counterbore in the moving plate, which is filled with a suitable viscous epoxy adhesive after the desired adjustment has been achieved. The clearance between the through hole and the pin must be controlled tightly so that it is only large enough to provide a free movement of the pin through the hole, but not too large so that the epoxy will run out of it during curing. Although this mechanism is more expensive than the previous one, it offers several advantages for some special applications. It is more suitable for high shock and vibration environments because the moving plate is rigidly locked against the fixed plate rather than merely relying on spring force to hold these two plates together. Moreover, the fine adjustment screws and springs can be removed for reuse after the epoxy has cured.

A simple linear adjustment mechanism,[15] which can be disassembled without destroying the adjustment, is shown in Figure 7.22. The adjustment bushing is threaded through the moving plate until the desired height is achieved. The clamp nut is then locked to retain this adjusted position. The clamp screw holding the parts together can be removed to disassemble the moving plate without losing the adjustment. For a finer resolution, this mechanism can be modified by incorporating differential screw threads as illustrated in Figure 7.23. The middle bushing has slightly different pitch threads on each section. The coarse adjustment is made by rotating the lower nut alone. When the middle bushing is rotated, a fine adjustment results due to a differential rotation in the two pairs of threads. This mechanism can also be disassembled without destroying the adjustment by removing the clamp screw. These simple mechanisms are suitable for applications requiring disassembly frequently.

Another linear mechanism employing differential threads by Kittell,[12] which is suitable for precision adjustments, is illustrated in Figure 7.24. The frictional force has been minimized by using spherical nuts in conical seats to achieve a line contact. The coarse adjustment is obtained

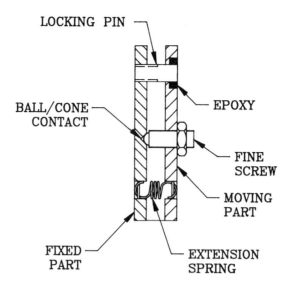

FIGURE 7.21 A linear mechanism using a fine threaded screw with epoxy locking and a ball–cone contact for a higher load capacity.

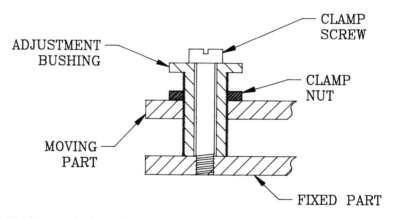

FIGURE 7.22 A linear mechanism with a threaded bushing and clamp screw suitable for applications requiring disassembly.

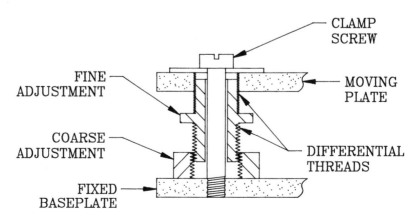

FIGURE 7.23 A linear mechanism with a bushing using differential threads for finer resolution.

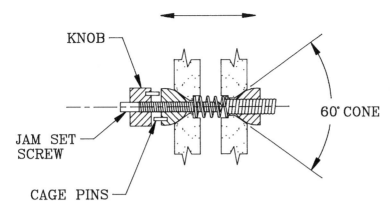

FIGURE 7.24 A high resolution linear adjustment mechanism using differential threads and spherical nuts to reduce friction.

by rotating the right side nut. When the left knob is rotated, a differential motion is achieved until the two cage pins engage. When this engagement of the pins occurs, the left nut rotates with the knob, thereby resulting in a coarse adjustment again. This clever design minimizes the time required to achieve very fine adjustments over a long travel range. A compression spring has been used in this mechanism to preload the spherical nuts into conical seats. Due to the precision required in machining the various parts of this mechanism, it is quite expensive and is recommended for high end applications only.

Ahmad[16] has described a high performance real-time focus adjustment mechanism, which is shown conceptually in Figure 7.25. An off-axis aspherical mirror is mounted in its cell through three tangent bars. The mirror cell is suspended by two pairs of flat blade rectangular flexures. A moving coil linear motor pushes at the center of the mirror cell. Since the rectangular flexures are much longer as compared to their height, they are very stiff in tension and torsion but comparatively quite compliant in bending. The flexures are arranged to be parallel to each other and orthogonal to the optical axis of the mirror. When the actuator pushes or pulls on the mirror cell, a pure axial movement of the mirror occurs along its optical axis. A closed loop control system monitors the image quality at the focal plane, and adjusts the focus in real time to compensate for environmental changes. This focus mechanism is simple in design but has a high resonance, and therefore, is suitable for high performance lithography applications requiring micron-level adjustment capabilities.

Lens Centration and Focus Mechanisms

This section describes some simple mechanisms suitable for centration and focus adjustments of single and multiple lenses. These mechanisms are used for low precision applications such as ordinary cameras and binoculars. For high precision applications, such as lithographic lens assemblies, the lens cells are normally diamond machined, and special assembly tooling, such as rotary air bearing tables, air gages, and interferometers, is employed to make the optical axis of the lenses collinear with the mechanical axis of the diamond machined cells.[17] These assembly methods are not discussed here because they do not fall into the category of adjustment mechanisms.

A simple double eccentric mount, shown in Figure 7.26, is frequently used in binocular objectives.[18] The lens is bonded into a ring with an eccentric outer diameter, which rotates inside another ring whose inner diameter is eccentric relative to the lens optical axis. Since the axis of ring rotation is offset from the axis of the lens, a centration adjustment results when the lens cell is rotated. The two eccentric rings are bonded together after making the adjustment by injecting epoxy through the radial holes. This adjustment mechanism provides a coupled low resolution two-axis centration adjustment.

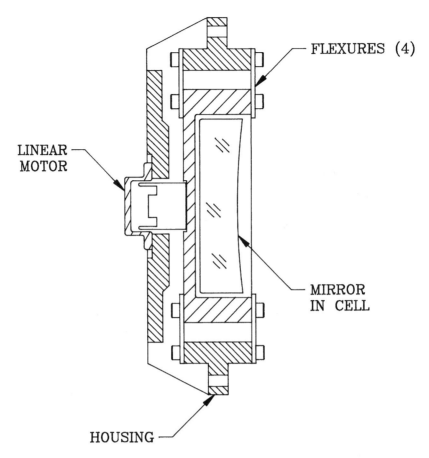

FIGURE 7.25 A real-time focus adjustment mechanism using flat blade flexures and a linear motor.

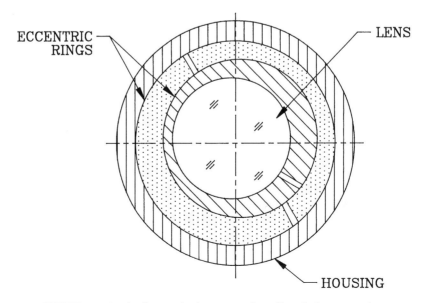

FIGURE 7.26 A pair of eccentric rings are used to adjust the lens centration.

A lens cell with four equally spaced radial set screws is shown in Figure 7.27.[18] A pair of opposing set screws is used to move the lens radially by moving one screw out while threading the other screw in. This arrangement provides an independent centration adjustment in two directions. Nylon screws can be used to avoid the risk of damage to the lens. This mechanism is suitable for lenses with thick edges, which are strong enough to withstand a compressive stress, without distortion or breakage. The desired centration is achieved through an iterative process, and its accuracy is dependent upon the pitch of screws used. Once the lens has been adjusted, it can be bonded to the cell by using dabs of epoxy, or by filling the gap between the lens OD and the cell ID with a suitable RTV elastomer. Again, this low cost lens centration mechanism is suitable for low accuracy applications.

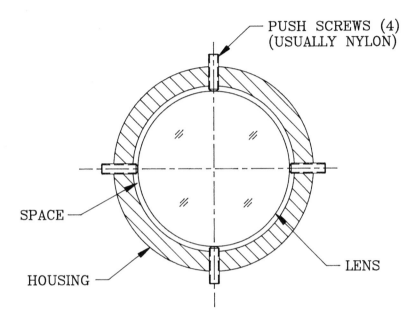

FIGURE 7.27 Lens centration using two pairs of opposing radial screws.

Another centration mechanism, suitable for higher precision applications with shock and vibration environment, has been proposed by Vukobratovich,[19] as illustrated in Figure 7.28. The lens is first assembled into its cell by bonding it in place, or by using a threaded retainer. Three equally spaced oversized holes for dowel pins are provided in the cell. The cell is then placed in a housing, which has a sufficient clearance for the lens cell OD, and has three equally spaced radial centering screws and three pressed-in dowel pins. The lens cell can be moved and centered by threading in one of the screws, which pushes the cell against the two other screws. Once the adjustment has been made, the clearance between the pins and oversized holes in the cell is filled with a suitable epoxy. The advantages of this mechanism as compared to the previous one are quite obvious. First, the lens is protected from any direct compressive loads, and the lens and cell ODs do not need to be machined precisely. The lens is not affected by the shrinkage of epoxy, also. The accuracy of the adjustment can be improved by using fine pitch screws. The negative features are a higher cost due to a number of extra parts and the permanent nature of the assembly because it cannot be disassembled easily if the lens was not adjusted correctly.

A simple focus (axial spacing) mechanism for lens assemblies has been proposed by Ginsberg[20] (Figure 7.29). The right-side lens is assembled into a cell, whose OD is threaded. This lens assembly can be translated axially with respect to the other lens group by rotating it, thereby moving it in or out of the housing. Once the correct focus is achieved, the threaded clamp ring is tightened against the housing to lock the adjustment in place. If the clearance between the internal and external threads is not controlled, excessive slop may introduce tilt errors. As no provision has been made for tilt adjustment, the common interface between the two housings must be machined square with the optical axes of both lens assemblies.

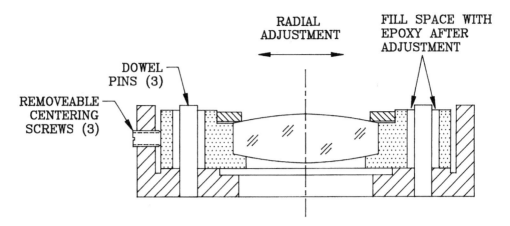

FIGURE 7.28 A lens centration scheme using epoxy and pins for precision applications.

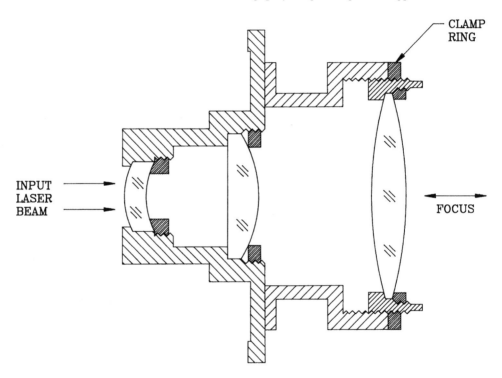

FIGURE 7.29 A lens focusing mechanism using a threaded lens cell.

Another fine-focus adjustment mechanism utilizing differential threads has been proposed by Jacobs[21] as depicted in Figure 7.30. The lens assemblies A and B are threaded into a focusing ring using two slightly different pitch threads. When the focusing ring is rotated, the relative axial spacing between the two lens assembly changes. Each lens cell must be constrained against rotation by providing a slot in the cell OD, which is guided on a fixed pin to result in a pure translation without rotation of the cell. Very fine-focus adjustments can be made by such a simple and inexpensive mechanism by selecting the right type of thread pitches, which differ only slightly. This type of focusing mechanism is commonly used in the camera objective lenses.

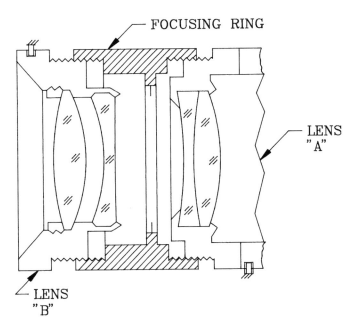

FIGURE 7.30 A fine-focus mechanism using differential threads for camera objective lenses.

7.4 Tilt Adjustment Mechanisms

General Description

Tilt adjustments of optical components such as mirrors, lenses, prisms, and diffraction gratings are frequently required to optimize the image quality in optical systems. A tilt mechanism can be designed to provide an adjustment about one axis or two mutually orthogonal axes. As mentioned earlier, tilt and linear adjustment mechanisms are quite similar in design and basically use the same kind of components. The design factors for selecting a suitable type of component for tilt mechanisms are the same as for the linear mechanisms. If an adjustment mechanism is designed with three mutually orthogonal adjustment points, it can be used to perform linear as well as tilt adjustments. When the three actuators are moved equally, a linear movement results. However, if only one of the actuators is moved, a tilt adjustment about an axis defined by the two others is achieved. The parts, which can be used as the basic building blocks for tilt adjustment mechanisms, are discussed briefly in this section. Also, some examples of tilt mechanisms are presented to demonstrate the basic design principles.

Interfaces for Tilt Mechanisms

The tilting component can be attached to a fixed structure through rotary bearings (journal, ball, roller, air), flexures (Bendix, flat blade), or a traditional kinematic interface. The trade-offs for these types of interfaces have already been discussed under linear mechanisms. A number of commercial single and two-axis tilt stages are available. These tilt stages employ a semikinematic interface between the fixed base and the moving tilt platform. The construction of a typical two-axis tilt stage is shown in Figure 7.31. The three-point interface between the tilt and fixed plates consists of hemispherical balls, which locate into a cone, a v-groove, and on a flat surface of the fixed plate. Two of the balls are rounded tips of the micrometers or fine pitch screws, and are positioned on mutually orthogonal axes with respect to the fixed ball. The interface between the two plates is preloaded by two extension springs located midway between the fixed and moving contact points.

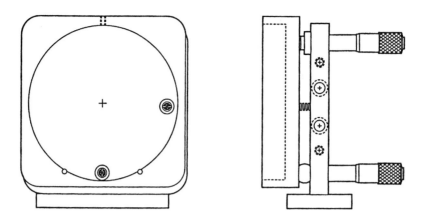

FIGURE 7.31 A typical two-axis commercial tilt stage (Courtesy of Newport Corporation, Irvine, CA.)

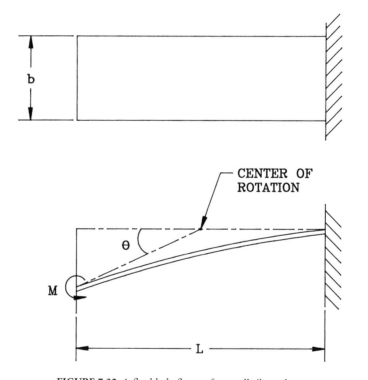

FIGURE 7.32 A flat blade flexure for small tilt angles.

Various configurations of flexures can also be used to attach the tilting part to the fixed structure. For small angles of rotation, a flexure strip shown in Figure 7.32 can be analyzed by the same equations used for beams that have end loads. If the axial load at the end is zero, the tilt angle θ is given by:

$$\theta = \frac{12ML}{Ebt^3} \tag{12}$$

where   M = the applied bending moment
          L = the length of the flexure
          E = the elastic modulus of flexure material

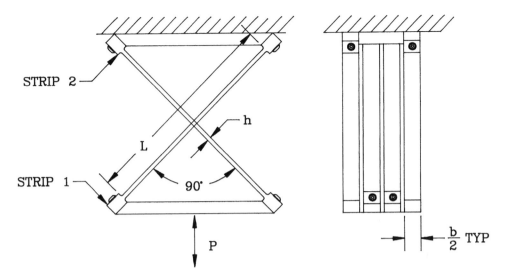

FIGURE 7.33 Schematic of a two-strip (cross) flexure for small tilts.

    b = the width of the flexure
    t = the flexure thickness

A two-strip flexure pivot, shown in Figure 7.33, can be used for small tilt angles.[3] This type of flexure is commercially available in the form of a pivot bearing. For a given tilt angle θ, the required bending moment M for a compressive load P can be calculated from the following equations:

$$M = \frac{EI\lambda}{2}\left[\frac{L\lambda}{2} + \cot\frac{L\lambda}{2}\right]\theta \tag{13}$$

If P is a tensile load, then the moment M is given by:

$$M = \frac{EI\lambda}{2}\left[\coth\frac{L\lambda}{2} - \frac{L\lambda}{2}\right]\theta \tag{14}$$

where   I = the moment of inertia of the single strip
            E = the elastic modulus of strip material

The parameter λ in these equations is defined by the following expression:

$$\lambda = \sqrt{\frac{P}{EI}}$$

A right circular flexure, shown in Figure 7.34, can be used in applications requiring a well-defined center of rotation and high stiffness.[22] For this special configuration of the flexure, the center of the cutting radius R lies on the edge of the flexure. The bending stiffness of this flexure can be estimated by:

$$\theta = \frac{9\pi MR^{0.5}}{2Ebt^{2.5}} \tag{15}$$

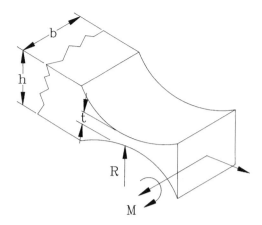

FIGURE 7.34 A typical right-circular flexure for high resonance applications.

where   M = the applied bending moment
        R = the cutting radius
        E = the elastic modulus
        b = the width of the flexure
        t = the flexure thickness

Actuators for Tilt Mechanisms

The actuators suitable for linear mechanism can also be used in tilt mechanisms. The screws and micrometer adjustments are used in manual mechanisms, and are extensively employed in commercial tilt stages and mirror mounts. Once again, the motorized actuators are large in size and expensive, and are used when real-time and frequent adjustments are required.

Coupling Methods for Tilt Mechanisms

As in linear mechanisms, the rounded tip of the actuator can bear against a flat or a conical surface in tilt mechanisms. The flat surface has high contact stresses, while a conical seat provides an increased contact area to minimize the contact stresses in those tilt mechanisms, which are adjusted frequently or require a higher load capacity.

Preloading Methods for Tilt Mechanisms

The preloading methods employed in linear mechanisms can also be used in tilt mechanisms. These include springs and washers as discussed in Section 7.3. The extension or compression springs can be centered between two adjacent actuators to achieve a uniform preloading. For smaller tilts and higher preloads, stacked Belleville or curved washers can be used in place of springs. The advantages and disadvantages of these preloading methods have already been discussed under linear adjustment mechanisms.

Locking Methods for Tilt Mechanisms

The locking methods for tilt mechanisms are similar to those employed in the linear adjustment mechanisms. These include set screws, jack screws, locknuts, and epoxy. The set screws and locknuts are used for temporary locking in coarse adjustment tilt mechanisms. The epoxy locking is economical and simple in design but is used in those mechanisms which are adjusted at initial assembly and alignment only, and do not require to be disassembled. As in linear mechanisms, a closed-loop control system can be used for motorized actuators to retain the desired tilt adjustment in real-time position control applications.

Examples of Tilt Adjustment Mechanisms

The conceptual designs of a number of tilt adjustment mechanisms are described in this section. As mentioned earlier, these examples are meant as guidelines only and a designer will most likely have to size the components of the mechanism according to the desired performance requirements. Moreover, a number of components from different mechanisms presented here can be combined to optimize the design for the application on hand. In some cases, the design of a single adjustment point is presented. As mentioned earlier, three mutually orthogonal adjustment points are needed to achieve a tilt adjustment along two mutually orthogonal axes. These three adjustment points must be spaced as far apart as possible on the optical mount to improve the angular resolution of tilt adjustment.

A very simple single-axis tilt mechanism for a flat mirror is depicted in Figure 7.35[15] for nonprecision applications. The lower edge of the mirror is rounded and sits in a v-groove in the mount. A single round tip adjustment screw threaded through the mount acts at the center of top edge of the mirror causing the mirror to pivot about its lower edge. A sheet metal spring clip is employed to preload the mirror against the adjustment screw and the v-groove. A locknut is used to retain the adjusted position and to secure the spring clip to the mount. The lateral shift of the mirror can be minimized by making the length of the v-groove approximately the same as the width of the mirror. Moreover, the in-plane movement of a flat mirror can be tolerated in most applications. The negative features of this low-cost simple adjustment mechanism are the special machining features required at the top and bottom edges of the mirror, and a low accuracy of the adjustment. Since the mirror is retained by frictional force only, it is not suitable for shock and vibration environments, where it may shift and lose its alignment.

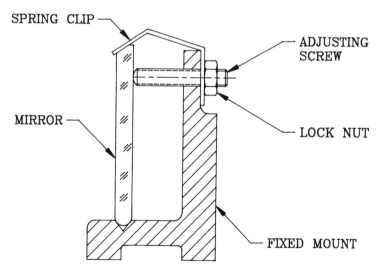

FIGURE 7.35 A low-precision single-axis tilt mechanism for a mirror.

The design of a simple two-axis tilt mechanism, similar to a commercial tilt stage described earlier, has been reported by Walsh.[23] The piano wire acts as a universal pivot between the moving and fixed plates (Figure 7.36). The two plates must be spring loaded against each other by providing two extension springs located between the adjustment screws and the piano wire. Fine threaded screws with round tips can be used to improve the resolution of tilt adjustment. One screw tip sits in a v-groove, while the other screw tip contacts the flat surface of the moving plate. It may be

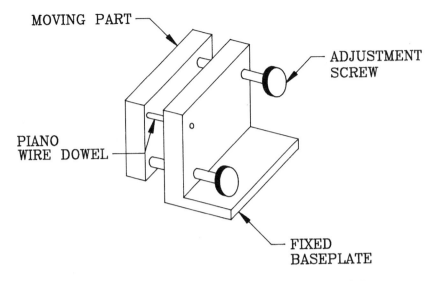

FIGURE 7.36 A simple two-axis tilt mechanism with a universal pivot.

more economical to buy a commercial tilt stage as compared to fabricating this tilt adjustment mechanism.

The design of a high precision single-axis tilt adjustment mechanism has been reported by Nemirovsky.[24] The interface between the tilt platform and the fixed base is through steel balls (Figure 7.37[A]). The top ball on the right side is spring loaded to eliminate backlash. A fine-threaded screw pushes on the lower ball, which is wedged under the middle ball, resulting in a tilt motion about the left-side ball. This mechanism has tilt resolution of the order of a fraction of an arc-second. A number of variations of this design are possible. The balls can be replaced by precision centerless ground shafting with tapered ends. The adjustment screw can have a round tip, thereby eliminating the need for a separate lower ball (Figure 7.37[B]). The screw can also be replaced by a rod with a tapered end to produce a jacking action (Figure 7.37[C]). This mechanism is useful in applications requiring single-axis tilt adjustments of a high accuracy over a limited angular travel.

James and Sternberg[25] have described the design of a simple two-axis tilt mechanism as shown in Figure 7.38. Three mutually orthogonal fine screws with ball tips provide a kinematic interface between the mirror cell and fixed structure. The mirror cell is spring loaded against the screws using a compression spring at the center. After making the tilt adjustment, the adjustment screws are locked by means of locknuts. Care must be taken to ensure that the adjustment is not lost due to excessive locking force. This is a good low cost mechanism for low accuracy applications.

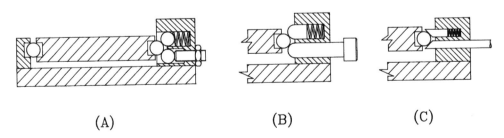

(A)   (B)   (C)

FIGURE 7.37 A high-precision mechanism for small tilt angles using: (A) two spring-loaded steel balls; (B) a round tip screw and a spring-loaded ball; (C) a tapered rod and a spring-loaded ball.

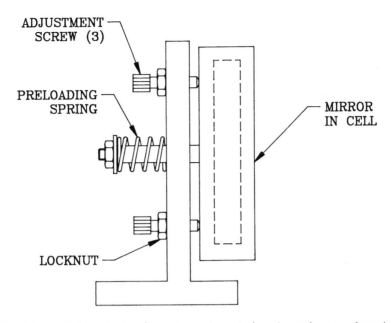

FIGURE 7.38 A low-cost tilt mechanism for a mirror using a single spring at the center for preloading.

The adjustment mechanism shown in Figure 7.39[15] can be designed to achieve very fine tilt adjustments due to a lever arrangement with a high mechanical advantage. It employs two flat blade flexures. The lower flexure is compliant in bending, while the vertical flexure is very stiff in tension. Coarse adjustment is provided by a regular nut threaded to the top end of vertical flexure. A locknut is provided to retain this coarse adjustment. Fine adjustment is made by a spring-loaded screw, which bends the horizontal flexure. By making R/r ratio equal to 10, 5 μm of vertical travel is obtained for every 50 μm of travel of the fine adjustment screw. The cost of this mechanism is comparatively higher due to the number of parts in it.

Another tilt mechanism using flat blade flexures is illustrated in Figure 7.40.[15] The two flexures are riveted or welded together to fix their lower edges. A spring-loaded adjustment screw is threaded into the fixed structure through a clearance hole in the flexures. The top edge of smaller flexure is rigidly attached to the fixed part using two screws and a pin to prevent any slippage. The top edge of the longer flexure is fixed to the tilting part in the same fashion. When the screw is threaded in to bend the flexures, a very small relative motion is produced between the free edges of flexures, thereby causing the moving part to tilt. A very fine adjustment can be achieved for a relatively large travel of the adjusting screw by properly sizing the flexures. This mechanism is rather expensive to fabricate, and is suitable for applications requiring high stiffness and accuracy over a limited travel range.

Figure 7.41 shows the design of a tilt adjustment mechanism reported by Ahmad[16] for a high resonance adjustable mirror mount. The mirror is suspended in its cell by three tangent bars with a pair of circular cross-flexures at each end. One end of each bar is secured to the invar buttons bonded directly to the mirror, while the other end is attached to the fixed structure. These tangent bars are very stiff in tension and compression, while they are relatively quite compliant in bending normal to the plane of the mirror. A small differential micrometer pushes on each invar button through a compliant flexure to eliminate lateral loads and misalignments. Each invar button is spring loaded against the tip of the micrometers. A high resolution tilt adjustment can be made by moving one differential micrometer at a time. This expensive tilt adjustment mechanism is recommended for high resonance precision applications.

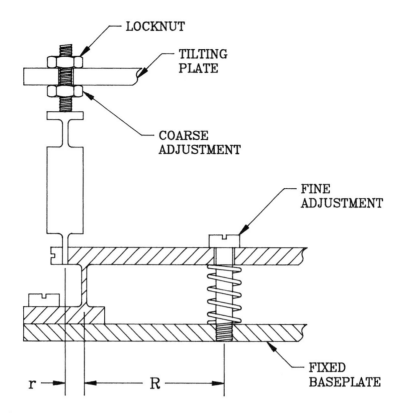

FIGURE 7.39 A tilt mechanism with coarse and fine adjustments using single blade flexures.

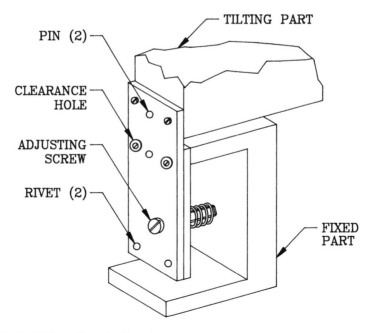

FIGURE 7.40 A tilt mechanism with a fine adjustment capability over small angles.

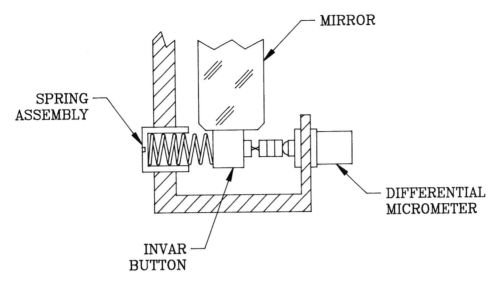

FIGURE 7.41 A tilt mechanism suitable for high resonance adjustable mirror mounts.

7.5 Rotary Adjustment Mechanisms

General Description

Rotary adjustment mechanisms are not as commonly used in optical systems as the linear and tilt adjustment mechanisms. For flat optics such as fold mirrors, windows, filters, and prisms, an in-plane rotation does not produce any change in alignment, and hence the system performance is unaffected. Similarly, for spherical optics such as lenses and mirrors, any rotation about the optical axis does not produce or correct any optical aberration. For off-axis and aspheric optics, normally linear and tilt adjustment mechanisms are employed to align the optical systems. In general, rotary mechanisms are used in scanning applications such as bar-code scanners in supermarkets and scanning telescopes and cameras for surveillance and remote monitoring, or for optical beam steering.

A number of choices are available to a designer for selecting the components of a rotary mechanism depending on the performance requirements such as frequency, range, and resolution of angular travel, shock, load capacity, cost, and size of the mechanism. A number of components that are used in linear mechanisms can also be used in rotary mechanisms, either in the same configuration or in their rotary version. These design choices along with some design guidelines are discussed in this section, followed by examples of some simple rotary mechanisms.

Interfaces for Rotary Mechanisms

The choice of a suitable interface between the rotating part and the fixed structure depends on such design criteria as the range and frequency of adjustment, shock, load capacity, cost, and size requirements. As in the case of linear mechanisms, rotary versions of the bearings discussed in Section 7.3 are commonly used to interface the rotating part to the fixed part. Spherical and journal bearings are only used for light duty cycles because of the friction and wear problems. These bearings are compact in size and are inexpensive as compared to rotary ball, roller, and air bearings, which are used in those applications where heavy loads are adjusted frequently or require continuous rotation.

A high speed rotary mechanism using a ball bearing has been described by Weinreb.[26] Figure 7.42 shows the radiometer spindle assembly for TIROS II meteorological satellite. The chopper mirror is mounted to a shaft supported by a ball bearing and is driven by a low power motor at a speed of 2750 rpm. Since this mechanism operates in space, lubrication of bearings is critical to minimize the wear for a long life and to maintain the alignment of the mirror. The ball bearing in this application is constantly lubricated in space by employing a lubricant reservoir of oil-impregnated sintered nylon.

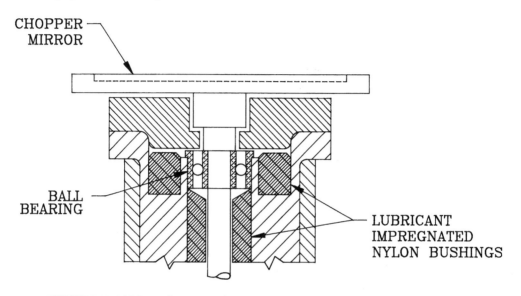

FIGURE 7.42 A high-speed rotary mechanism using ball bearings from TIROS II radiometer.

Rotary air bearings offer high stiffness and load capacity, accuracy, and cleanliness. Motorized air bearings are frictionless and use a thin film of clean dry air at a pressure of 80 to 100 psig. One such motorized air bearing by Dover Instrument Corporation is shown in Figure 7.43.[27] The bearing assembly consists of a load-carrying section, a spindle housing, and a motor housing. The air-bearing part consists of a shaft and two thrust plates. The shaft, thrust plates, and the inside surfaces of the housing are machined flat and cylindrical to tolerances of 10 micro-in. The air is fed through jeweled orifice restrictors into the clearance between shaft and the housing. The clearance between rotating elements and spindle housing is of the order of 5 to 10 μm. When compressed air passes through this clearance, a positive film pressure is created around the rotating shaft. This film of high pressure air acts like a very stiff spring to prevent any mechanical contact between the rotating parts and the housing, thereby creating a zero-friction condition. This compact bearing assembly comes with a built-in motor and encoder. These types of bearings are ideal for high speed applications requiring vibration-free rotation. The main disadvantage is that these bearings needed a supply of pressure-regulated clean air, which adds to the overall cost due to the equipment needed for air supply.

For small angular adjustments, a flexural pivot (Bendix type) offers the advantages of friction and backlash-free angular adjustment. These commercially available pivots are small in size and have low hysteresis. The application of these pivots in a mirror mount has been described by Rundle[28] in detail.

Actuators for Rotary Mechanisms

The actuators used in rotary mechanisms are very similar to those used in linear mechanisms discussed in Section 7.3. Screws are used in low cost applications for small angular adjustments. Micrometers and differential micrometers are used when a more precise adjustment with a readout

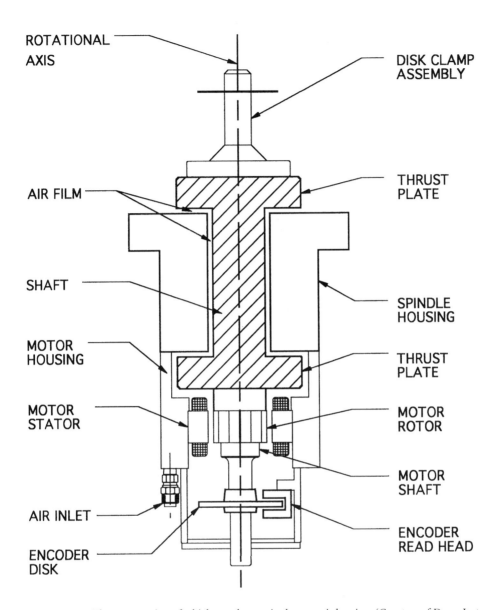

FIGURE 7.43 (A) The cross section of a high speed motorized rotary air bearing. (Courtesy of Dover Instrument Corporation, Westboro, MA.)

is required. Commercially available rotary stages use both types of micrometers extensively. A typical commercial rotary stage is illustrated in Figure 7.44. It provides full 360° rotation about a vertical axis, and typically runs on a ball bearing. The angular resolution depends on the type of actuator used. A thumb screw is used for locking the adjusted position. Once again, the motorized actuators such as DC and stepper motors are used in applications requiring large and frequent angular travels or for real-time adjustments.

Coupling Methods for Rotary Mechanisms

Some coupling methods for linear mechanisms discussed in Section 7.3 can also be used in rotary mechanisms. For small angular travels, the round tip of an actuator can bear against a flat surface, cone, or a spherical socket. The advantages and disadvantages of these arrangements have already been discussed earlier. These interfaces can only be used for very small angular adjustments, since

FIGURE 7.43 (B) Picture of a high speed motorized rotary air bearing. (Courtesy of Dover Instrument Corporation, Westboro, MA.)

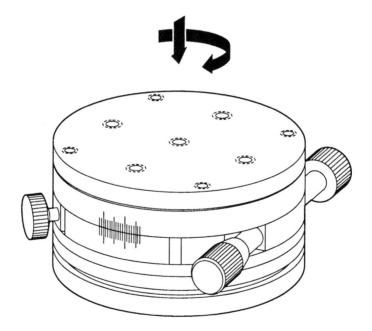

FIGURE 7.44 A typical single-axis commercial rotary stage. (Courtesy of Newport Corporation, Irvive, CA.)

the tip of the actuator slides relative to the rotating part and cannot maintain a good contact with the rotating part for larger angles.

For large, frequent, or constant angular motion, an actuator can be coupled to the rotating part through a flexible type of coupling. The choice of coupling type depends on the radial load, torque capacity, life expectancy, and maintenance.[29] Flexible-type couplings can tolerate relatively large misalignment between the axes of rotation of the actuator and the moving part. The performance of the systems using a flexible coupling depends on the inertia, backlash, friction, and linearity of the coupling used. It should be noted that couplings designed for motion control applications may not be suitable for power transmission and vice versa. The couplings illustrated in Figure 7.45 are suitable for rotary mechanisms and exhibit low inertia, zero backlash, and near-constant velocity. The principle disadvantages of couplings are their relatively large size, weight, and cost.

FIGURE 7.45 Some commonly used flexible couplings for rotary motion. (Courtesy of Renbrandt, Inc., Boston, MA.)

Sometimes linear actuators are used in rotary mechanisms because of their lower cost and compact size. In such cases the linear motion of the actuator is converted to a rotary movement through a worm and gear or a rack and pinion arrangement. In these mechanisms, the part to be rotated is attached to the rotating gear while the linear actuator moves the rack. These mechanisms are expensive because of their mechanical complexity and exhibit backlash, if not designed properly.

Preloading Methods for Rotary Mechanisms

The selection of a proper preloading method to obtain a backlash-free rotary adjustment is based on the same design factors that are discussed in Section 7.3 for the linear mechanisms. The tension or compression springs are extensively used for preloading purposes because of their low cost, and also because these can be used over a fairly large angular range. The Belleville washers are used for high preloads over small adjustment ranges. For larger angular rotations, a torsion spring can be used for preloading the rotating part. One end of the spring is attached to the rotating part, while the other end is attached to the fixed structure.

Locking Methods for Rotary Mechanisms

The design factors affecting the choice of suitable locking methods for rotary mechanisms are similar to those in the linear mechanisms as already discussed in Section 7.3. These factors include travel range and frequency, size and cost, and disassembly requirements.

The locking can be accomplished by set screws, clamps, epoxy, and locknuts. The relative advantages and disadvantages of all these locking methods have already been discussed. Epoxy locking is inexpensive, but it is more or less permanent, because the parts cannot be disassembled without a risk of damage. The other locking methods are nonpermanent and less expensive, but may introduce high stresses due to clamping force in the components being locked.

Examples of Rotary Adjustment Mechanisms

Tuttle[10] has suggested the designs of a number of simple rotary mechanisms for angular positioning as shown in Figures 7.46 to 7.49. For small angular adjustments, the lever and shaft arrangements shown in Figures 7.46 and 7.47 are simple and cost effective. Two opposing tangent screws are used in the mechanism shown in Figure 7.46. The round tip of the screw pushes on a lever that rotates on a precision journal or ball bearing through small angles, while the opposing screw is moved out. The screws are locked in place by locknuts to hold the adjusted position. This low precision, low cost mechanism is suitable for infrequent adjustments over small angular travels.

In the mechanism shown in Figure 7.47, the resolution of adjustment can be improved by using the fine pitch threads. In this mechanism, the threaded sleeve is spring-loaded against the stationary

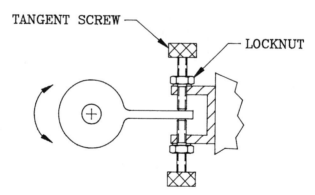

FIGURE 7.46 A simple rotary mechanism using tangent screws for small angular travels.

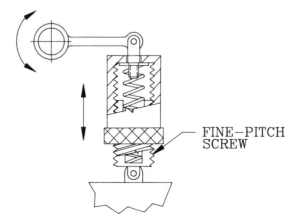

FIGURE 7.47 A rotary mechanism using a fine-pitch screw for a limited travel range.

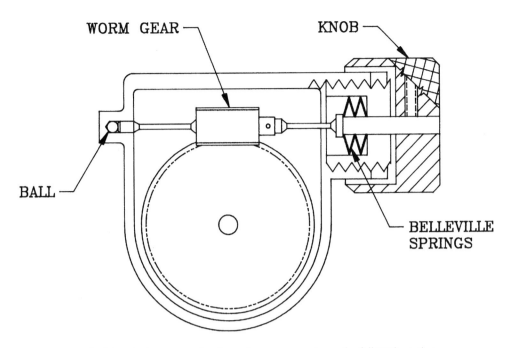

FIGURE 7.48 A rotary mechanism using a worm and gear for full 360° rotation.

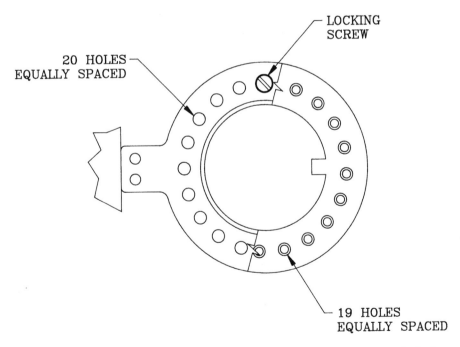

FIGURE 7.49 A very simple low-cost rotary mechanism using two concentric rings with fixed adjustment steps.

half to eliminate any backlash. When the sleeve is rotated in either direction, a change in the length of the screw produces a small angular rotation about the center of rotation. The angular resolution of this mechanism can be further improved by incorporating a differential screw. This mechanism is also suitable for applications requiring infrequent adjustments over small angles.

A worm and gear-type mechanism illustrated in Figure 7.48 provides an angular adjustment over full 360°. A worm mounted on an axially loaded shaft engages with a gear rotating on a bearing. The worm shaft with a reduced diameter is used intentionally to reduce its bending stiffness. The shaft is loaded axially with Belleville spring washers against a ball, which acts as a thrust bearing at the end of the shaft. Due to this spring load, the shaft operates in a slightly deflected shape to ensure a positive contact between the worm and gear. A knob attached to the shaft is rotated to produce the angular rotation of the gear. The angular resolution of this mechanism depends on the gear ratio selected. This mechanism is more expensive due to its mechanical complexity, but it does provide angular adjustments over full 360°, as stated earlier.

A very simple mechanism for one-time coarse angular adjustment is depicted in Figure 7.49. It consists of two concentric rings with a number of same-diameter holes on bolt circles of equal diameters. One of the rings has one hole less than the other ring. The locking is accomplished by lining up one set of holes in both rings, and then inserting a screw or pin through both rings. The number of holes n determines the angular adjustment step between any two adjacent locking positions, and is given by $360/n(n-1)$.

Figure 7.50 illustrates a rotary mechanism using a spherical bearing. The mirror cell has an integral hemispherical ball which sits in a spherical seat in the fixed structure. This arrangement not only allows the mirror to be rotated about its optical axis, but also allows it to be tilted. This mechanism is spring loaded at the center to hold the adjustment. This low precision mechanism has a high friction and is therefore suitable for infrequent angular adjustment.

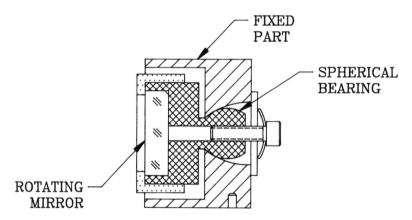

FIGURE 7.50 A spherical bearing provides a full 360° rotation in this rotary mechanism for a mirror.

7.6 Design Guidelines for Adjustment Mechanisms

An adjustable optical mount is inherently more complex, expensive, and less stable than a comparable fixed type of mount. Therefore, a careful trade-off design study must be conducted to weigh the benefits of an adjustable optical mount against the potential stability and cost drawbacks. A good summary of how and when adjustment mechanisms must be incorporated in optical systems has been presented by Ahmad[14] and Vukobratovich.[30]

The decision to use adjustable optical mounts is dictated by the sensitivity analysis performed by an optical designer. This analysis defines the assembly tolerances for an optical system to achieve its image resolution and quality requirements. For systems with high image resolution requirements, usually the alignment tolerances of the optical elements exceed the practical fabrication tolerances for the optical elements and the associated mechanical hardware. This problem is further compounded by the stack-up of machining tolerances and inspection uncertainties when an optical system has a large number of mirrors and lenses. In such applications, it is economical and time saving to manufacture the optical and mechanical parts to rather loose tolerances and provide adjustments for a few sensitive optical elements. These elements are then adjusted to compensate for the errors and uncertainties in the positions of other optical elements. Therefore, a rigorous optical sensitivity analysis must be performed to identify the optical elements that have the strongest effect on image quality of the system.

The required accuracy of adjustment is usually dictated by the type of an optical element and performance requirements of the optical system. Table 7.8 by Vukobratovich[30] gives general guidelines for the type of adjustment required and its sensitivity for different types of optical elements. Once the types of adjustments required for an optical element have been determined, the next step is to design an adjustable mount to provide the desired adjustment range and resolution. The glass optical elements are normally assembled in a metal frame, and an adjustment mechanism is provided between the frame and the fixed mounting structure. This way, a direct contact between the actuator and the glass element is avoided, and the element is protected from high contact stresses at the tip of the actuator. Providing an adjustment range, which is longer than that determined by the optical sensitivity analysis, can be very risky. Such mounts can be assembled far away from their required nominal position, and a lot of time may be wasted during alignment to obtain any image at all. In such cases, it is better to design the mount so that it can be shimmed at assembly, in case the range of adjustment is found to be inadequate. After an optical system has been aligned, the adjustable mounts must be positively locked to prevent any misalignment due to accidental tweaking or any drifts as a result of shock and vibrations. The adjustments, which are used only at initial assembly, can be designed to have lockable mounts with removable actuators. This approach not only saves on the cost of actuators, but also eliminates the risk of accidental misadjustment later on.

TABLE 7.8 Types of Adjustments Required and Their Sensitivity

Type of Optical Element	Adjustment Sensitivity	Type of Adjustment Required
Flat window	Very low	Tilt
Field lens	Low	Tilt and decenter
Flat mirror	Medium	Tilt
Prism	Medium	Tilt
Objective lens	High	Tilt, decenter, and despace
Relay lens	High	Tilt, decenter, and despace
Curved mirror	Very high	Tilt and despace

The adjustable mounts should be designed to tilt or rotate about the principal points to avoid cross-talk between axial and tilt adjustments. The cross-coupling of adjustments can be very frustrating because several iterations of tilt and axial adjustments may be needed to achieve the alignment. The adjustable mirror mounts must be designed to tilt a mirror about its vertex to avoid an unwanted image shift. The adjustment points in a tilt mechanism should be positioned in a mutually orthogonal pattern relative to the axis of the optical element. Such mounts are easy to adjust and produce predictable movements.

The adjustment mechanisms must be designed to have a large mechanical advantage such that a large axial movement or rotation of the actuator will incrementally move the optical element. This design feature can also save valuable alignment time because there is a less likelihood of accidentally overshooting the optimum position. If feasible, the mechanisms must be designed to have a coarse as well as a fine adjustment. The coarse alignment can be quickly achieved by using the coarse adjustment, while the fine part of the adjustment is only used to optimize the quality of the image.

A number of translation, rotary and tilt stages, mirror, lens, and gimbal mounts are available commercially. These mounts are economical, precise, and quite rugged, and are very suitable for prototypes and laboratory setups. Their main disadvantage is their bulky size and weight, which makes their use impractical in the systems with several optical elements that are packaged together tightly. If weight and size are not a problem in an application, it is far more economical and time saving to use commercial mounts rather than designing and fabricating the custom adjustable mounts. Locking can be added to these commercial mounts, if needed.

While the adjustment mechanisms offer several advantages, the disadvantages of the adjustable mounts must not be overlooked, and provisions must be made in their design to minimize their negative effects on the optical system. First of all, the number of adjustable optical elements in a system must be kept to an absolute minimum. This not only saves on the fabrication cost, but also maximizes the long-term stability of the system. The adjustable mounts are less rigid and less stable than the fixed type of mounts and therefore experience a drift with time. The adjustments often induce nonlinear and unpredictable effects in optical systems. The adjustable mounts are mechanically weak and are more susceptible to drifts due to shock loads, vibrations, and temperature variations.

7.7 Summary

The adjustment mechanisms play an important role in the integration and alignment of sophisticated optical systems which have tight positioning tolerances. By incorporating these mechanisms in the optomechanical design, it becomes feasible and economical to produce and assemble such optical systems to very high alignment accuracies, which in turn greatly enhances the image quality of these systems. The design guidelines for three basic types of adjustment mechanisms have been presented. The design choices for various components that make up the linear, rotary, and tilt adjustment mechanisms have been listed. The advantages and disadvantages of these choices have been presented to help a designer in choosing the most suitable parts of the adjustment mechanism for a particular application.

The designs of a number of sample linear, tilt, and rotary mechanisms have also been described in this chapter. These designs cover low cost and low resolution applications and the precision mechanisms with a fine resolution for high performance applications. Some mechanisms presented here may be employed without any modification to satisfy a particular need, while in other cases the components with desirable features can be selected from different mechanisms to design a custom mechanism with optimum features for a particular application. The size and shape of the components in a mechanism are dictated by the space constraints. Therefore, in most applications it may be necessary to select or design the components to satisfy the space requirements.

The design guidelines for how and when adjustment mechanisms must be incorporated into optical systems have also been discussed. By following these simple design guidelines, the performance and image quality of a system can often be optimized at a minimum cost. The intelligent use of adjustment mechanisms can result in lower fabrication costs and a considerable reduction in the time and effort involved in alignment of precision optical systems.

References

1. Trylinski, W. 1971. *Fine Mechanisms and Precision Instruments*. Pergamon Press, Elmsford, NY.
2. Grassam, N. S. and Powell, J. W. 1964. *Gas Lubricated Bearings*. Butterworths & Co. Ltd. (U.K.).
3. Weinstein, W. D. June 10, 1965. Flexure pivot bearings. In *Machine Design*, pp. 150– 157 (Part 1) and July 8, 1965, pp. 136–145 (Part 2). Penton Publishing, Cleveland, OH; reprinted in 1988. *SPIE Milestone Series*, Vol. 770.
4. Neugebauer, G. H. August 7, 1980. Designing springs for parallel motion. In *Machine Design*, pp. 119–120. Penton Publishing, Cleveland, OH; reprinted in 1988. *SPIE Milestone Series*, Vol. 770.
5. 1994. Newport Corporation, Irvine, CA.

6. 1994. Anorad Corporation, Hauppage, NY.

7. 1989. *Micropositioning Systems*. Burleigh Instrument, Inc., Fishers, NY.

8. 1994. New Focus, Sunnyvale, CA.

9. Moore, J. H. Davis, C. C., and Coplan, M. A. 1983. *Building Scientific Apparatus*, Addison-Wesley, Reading, MA.

10. Tuttle, S. B. February 16, 1967. How to achieve precise adjustments. In *Machine Design*, pp. 227-229. Penton Publishing, Cleveland, OH; reprinted in 1988. *SPIE Milestone Series*, Vol. 770. Bellingham, WA.

11. Elliot, A. and Dickson, J. H. 1960. *Laboratory Instruments — Their Design and Applications*. Chemical Publishing.

12. Kittel, D. 1989. *Class Notes on Precision Mechanics*. Stamford, CT.

13. Ahmad, A. 1994. Low cost adjustable mirror mount, *Opt. Eng.*, 36(6), pp. 2062–2064.

14. Ahmad, A. 1992. The Adjustment mechanisms — types and their applications in optical systems, *SPIE Crit. Rev. Ser. Vol. CR43*, pp. 254–277. Bellingham, WA.

15. Perkin-Elmer Corporation, 1975. *Mechanical Design Manual*. Norwalk, CT.

16. Ahmad, A. 1988. High Resonance Adjustable Mirror Mount, U.S. Patent 4,726,671.

17. Ahmad, A. 1990. Fabrication techniques for high resolution lens assemblies, *SPIE Proc. Vol. 1335*, pp. 194–198.

18. Yoder, P. 1991. *Optomechanical Systems Design*, IBM Class Notes, Norwalk, CT.

19. Vukobratovich, D. 1991. *Lens Centration Mechanism*, University of Arizona, Tucson.

20. Ginsberg, R. H. 1981. *Outline of tolerancing (from performance specification to toleranced drawings*, *Opt. Eng.*, 20(2), pp. 175–180.

21. Jacobs, D. 1943. *Fundamentals of Optical Engineering*, McGraw-Hill, New York.

22. Paros, J. M. and Weisbord, L. November 15, 1965. How to design flexure hinges. In *Machine Design*, pp. 151–156. Penton Publishing, Cleveland, OH; reprinted in 1988. *SPIE Milestone Series*, Vol. 770. Bellingham, WA.

23. Walsh, E. J. U.S. Patent 3,334,959.

24. Nemirovsky, R. U.S. Patent 4,880,219.

25. James, J. F. and Sternberg, R. S. 1969. *The Design of Optical Spectrometer*. Chapman Anhall, London.

26. Weinreb, M. B. 1961. *Results of Tiros II Ball Bearing Operation in Space*. NASA, Washington, D.C.

27. Charron, S. July/August, 1993. Motorized air bearings, *Motion*.

28. Rundle, W. J. 1989. Design and performance of an optical mount using cross-flexure pivots, *SPIE Proc. Vol. 1167*, pp. 306–312.

29. Brandt, E. R. July/August, 1993. Flexible couplings used in motion control, *Motion*.

30. Vukobratovich, D. 1989. *Introduction to Optomechanical Design*, SPIE Short Course Notes. University of Arizona, Tucson.

8

Structural Analysis of Optics

Victor Genberg

-8493-0133-5/97/$0.00+$.50
© 1997 by CRC Press, Inc.

255

Notations:

$$
\begin{aligned}
\text{1D} &= \text{1 dimensional} \\
\text{2D} &= \text{2 dimensional} \\
\text{3D} &= \text{3 dimensional} \\
\text{BC} &= \text{boundary conditions} \\
\text{DOF} &= \text{degrees of freedom} \\
\text{FE} &= \text{finite element} \\
\text{FEA} &= \text{finite element analysis}
\end{aligned}
$$

8.1 Introduction

Typical structural designs in most industries are governed by failure due to stress — either yield, ultimate, or fatigue. An optical structure's performance is usually determined by distortions or displacements rather than stress. Most mirrors or lenses have distortion requirements measured in wavelengths of light (about 25 micro-in.). At this level of distortion, the stresses are usually quite small. Similarly, optical systems typically have tight optical beam-pointing requirements, or tight image motion requirements, which keep stresses in metering structures low. To predict the behavior of optical structures to the level of their performance specifications, analyses must have a high degree of accuracy. Thus, analysis techniques or assumptions commonly used in other industries may not be appropriate for optics.

Closed-form equations for the analysis of plates are useful for determining the general behavior of some optics, especially for determining design rules presented in earlier chapters. When detailed mount configuration and load effects are included, closed-form techniques usually cannot provide the solutions with the desired accuracy. For this reason, the techniques used in this chapter are based on the finite element method.

8.2 Overview of Finite Element Theory

Derivation of Stiffness Matrix

The finite element (FE) method is a numerical technique for converting a system of governing differential equations over a continuous domain to a set of discrete variables defined by a matrix equation. The continuous domain (Figure 8.1) is subdivided into a system of simple elements interconnected at a finite number of points called nodes, which are located at element corners and possibly along the element boundaries. Within each element the form of the behavior is assumed as a function of the nodal variables. In structural mechanics, the displacement (u) anywhere within an element is assumed to be of the form:

$$ u = \Sigma N_j \delta_j $$

where N_j is called the shape function of node j and δ_j is the displacement of node j. Thus a continuous variable u is approximated as a function of discrete variables δ, which is the fundamental assumption in FE theory. Typically N is a simple polynomial whose order is determined by the number of nodes associated with the element.

Given the choice of shape function, the derivation of the stiffness matrix is usually found from the minimization of potential energy (Π). The strain vector (ε) is found from the appropriate strain-displacement equations involving derivatives of the displacement (u) which lead to derivatives (B) of the shape functions (N).

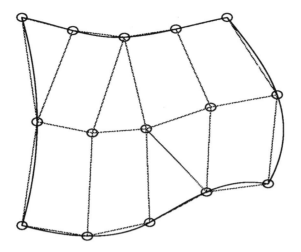

FIGURE 8.1 Finite element representation of a 2D continuum.

$$\{\varepsilon\} = [B]\{\delta\}$$

The stress vector (σ) is found from the appropriate stress–strain relations involving the material matrix (E).

$$\{\sigma\} = [E]\{\varepsilon\} = [E][B]\{\delta\}$$

The potential energy is integrated over the volume of each element:

$$\Pi = \int \{\varepsilon\}^{T} \{\sigma\} \, dV - \{\delta\}^{T} \{F\} = \int \{\delta\}^{T} [B]^{T} [E][B]\{\delta\} \, dV - \{\delta\}^{T} \{F\}$$

The nodal displacements are discrete variables and put outside of the integration. To minimize the potential energy with respect to the displacements, the partial derivative is set to zero.

$$d\Pi / d\{\delta\} = \int [B]^{T} [E][B] \, dV \{\delta\} - \{F\} = \{0\}$$

This has the form of the standard spring equation where the coefficient of displacement is the spring stiffness. Here the element stiffness matrix is

$$[k] = \int [B]^{T} [E][B] \, dV$$

integrated over the volume of the element. For any given element type, the above equation is used to find the element stiffness matrix given the material and nodal locations. The element matrices are assembled into a system level equilibrium equation of the form:

$$[K]\{\delta\} = \{F\}$$

Once valid boundary conditions (BC) are applied, the [K] matrix becomes nonsingular and is solvable by Gauss elimination or Cholesky decomposition.

The above is a very brief, simplified overview of finite element theory. For a more detailed description, some very good textbooks are available, such as Logan,[11] Segerlind,[16] Knight,[10] or Cook.[3]

Element Types

As a new user, the most difficult parts of using a FE program is in the idealization of the problem, the choice of appropriate element type, and the size of the mesh required. A description of the element types and their application to optical structures follows. The names in parentheses are the names used in the NASTRAN FE program.

Truss Element (Rod)

- Line elements which carry only axial forces, with no bending
- End conditions are perfect ball joints
- Useful for some pinned end strut mounts

Beam Element (Bar, Beam)

- Line elements which carry all forces and moments
- End conditions are perfectly welded
- Symmetric cross sections (I-beam) have shear center at neutral axis
- Asymmetric cross sections (C-channel) have shear center offset
- Short beams require transverse shear factor for accuracy
- Useful for most frame-like optical support structures

Plate/Shell Element (QUAD4, QUAD8, TRIA3, TRIA6, QUADR, TRIAR)

- 2D planar elements with membrane and bending stiffness
- Thin plate elements ignore transverse shear stiffness
- Thick plate elements include transverse shear stiffness
- Plane stress elements are typical plate/shell structures
- Plane strain elements are for 2D cross sections of long structures
- The best elements allow modeling of composite and waffle-type plates
- Accuracy is a function of mesh density, element type, element order
- Useful for thin optics (Figure 8.2) and lightweight mirrors

Solids (Hexa, Penta, Tetra)

- 3D elasticity element, usually having only translational stiffnesses
- Accounts for full 3D effects in structural behavior
- Should allow orthotropic materials
- Accuracy is a function of mesh density, element type, element order
- Useful for thick optics (Figure 8.3), bonded joints, or submodel details

Axisymmetric Solids (TRIAX6)

- 3D elasticity behavior reduced to 2D by axisymmetric conditions
- The structure and BC (and usually the load) must be axisymmetric
- Accuracy is a function of mesh density, element type, element order
- Useful for thick optics, lenses, and lens barrels (Figure 8.4)

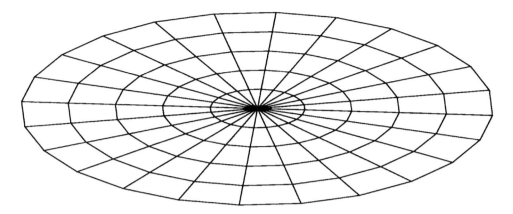

FIGURE 8.2 Shell element model of a mirror, lens, or window.

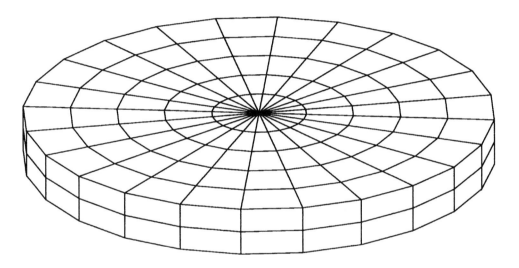

FIGURE 8.3 Solid element model of a mirror, lens, or window.

Springs (ELAS)

- Scalar spring connecting any 2 DOF
- User supplies calculated spring constant
- If connected DOF are not coincident and colinear, then hidden reactions to ground may be created
- Useful for effective joint stiffness in dynamics models.

Rigid Element (RBAR, RBE2)

- Absolutely rigid element with no elasticity
- These elements have no thermoelastic growth, so use with care
- Useful for neutral axis offsets in metering structures

Equation Element (MPC, RBE3)

- Add any linear equation to a model with a multipoint constraint (MPC)
- RBE3 can calculate average motion of several nodes
- Useful for calculating image motion in a system level model

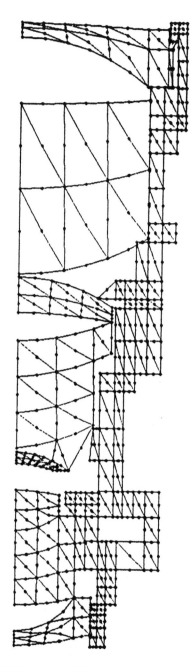

FIGURE 8.4 Axisymmetric element model of a lens barrel.

Element Accuracy

The truss, beam, spring, and rigid elements have the theoretically exact stiffness matrix. Subdividing a beam structure into more beam elements may improve visualization, but does not improve accuracy in a static analysis. In a dynamic analysis, subdivision can improve the distribution of mass with an improvement in analysis accuracy. The plate and solid elements are approximations to continuum behavior, so the element type and number do affect the solution accuracy. The first- and second-order 2D membrane elements are shown in Figure 8.5. For the following discussion, let u and ε be the displacement and strain in the x direction.

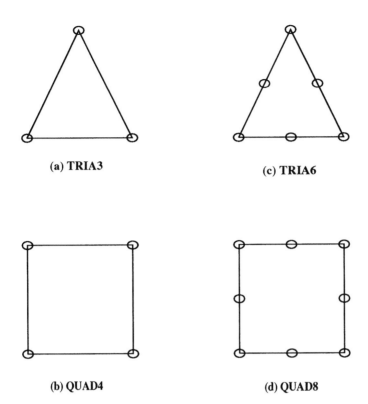

FIGURE 8.5 Shell elements.

(a) TRIA3 (c) TRIA6 (b) QUAD4 (d) QUAD8

1. 3-node triangle: (constant strain)
   $u = a + bx + cy$
   $\varepsilon = du/dx = b$
2. 4-node quadrilateral: (partial linear strain)
   $u = a + bx + cy + dxy$
   $\varepsilon = du/dx = b + dy$
3. 6-node triangle: (full linear strain)
   $u = a + bx + cy + dxy + ex^2 + fy^2$
   $\varepsilon = du/dx = b + dy + 2ex$
4. 8-node quadrilateral: (partial quadratic strain)
   $u = a + bx + cy + dxy + ex^2 + fy^2 + gxy^2 + hx^2y$
   $\varepsilon = du/dx = b + dy + 2ex + gy^2 + 2hxy$

The simple example cantilever beam shown in Figure 8.6 is modeled with the above elements in a regular pattern and with a distorted pattern in Figure 8.7(a). The mesh chosen had nearly equal numbers of nodes, and thus nearly equal size of stiffness matrix. Five load conditions were applied which have the following x direction strain (ε) patterns:

1. Membrane axial force (F_x) causes constant strain ε throughout.
2. Membrane end moment (M_z) causes ε to be linear in y, constant in x.
3. Membrane end shear (F_y) causes ε to be linear in y and linear in x.
4. Bending end moment (M_y) causes ε to be constant in x and y, linear in z.
5. Bending pressure (p_z) causes ε to be constant in y, linear in x and z.

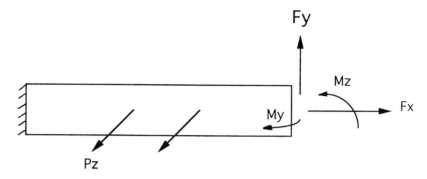

FIGURE 8.6 Thin cantilever beam with in-plane (membrane) and out-of-plane (bending) loads.

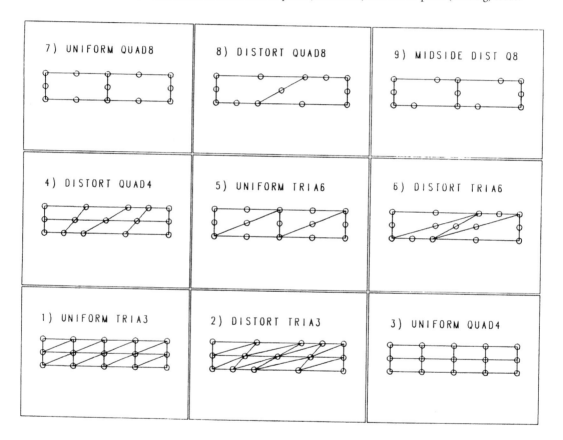

FIGURE 8.7 (a) Finite element models for the cantilever beam.

In each case, the load is picked such that the theoretical end displacement is 1.0. From the resulting displacements in Figure 8.7(b), the following conclusions can be made:

- All elements are good in constant strain.
- All elements are good in plate bending.
- The 3-noded triangle is very poor for any strain variation.
- The 4-noded quadrilateral deteriorates with distortion.
- With second-order elements the added nodes must be at midside.
- The elements with drilling DOF are much better than the originals.

Comparison of 2D & 3D Elements - Cantilever Beam Example

Membrane behavior
    1) Fx = Axial load = uniform / constant stress
    2) Mz = In-plane moment = axial strain is linear in y
    3) Fy = Shear load = axial strain is linear in y & x
Plate behavior
    4) My = out-of-plane moment = constant moment
    5) Pz = normal pressure = linear moment

Model	F.x	M.z	F.y	M.y	P.z
Conventional Plate Elements					
1) Tria3	1.00	0.30	0.32	0.98	0.98
2) Tria3-Dist	1.00	0.12	0.16	0.98	0.93
3) Quad4	1.00	0.98	0.96	0.98	0.98
4) Quad4-Dist	1.00	0.41	0.47	0.98	0.96
5) Tria6	1.00	1.00	0.96	0.98	0.97
6) Tria6-Dist	1.00	1.00	0.82	0.98	0.81
7) Quad8	1.00	1.00	1.00	0.98	0.96
8) Quad8-Dist	1.00	0.98	0.94	0.88	0.79
9) Quad8-Mid	1.00	0.59	0.59	0.47	0.49
Plate Elements with drilling DOF					
1a) TriaR	1.00	0.87	0.86	0.98	0.98
2a) TriaR-Dist	1.00	0.69	0.67	0.98	0.93
3a) QuadR	1.01	1.00	1.00	0.98	0.98
4a) QuadR-Dist	1.01	0.98	0.97	0.98	0.96

FIGURE 8.7 (b) Results for tip deflections for cantilever beam models.

Note that these results are for the MSC/NASTRAN elements. The elements in other codes may use a different formulation which gives different results. Most notably is the QUAD4 which is not the standard 4-node isoparametric formulation. The standard isoparametric 4-noded quadrilateral would not perform as well as the QUAD4 in the above test. The QUADR and TRIAR have added the drilling DOF (normal rotation or Φ_z) to the formulation, which improves the behavior under distortion. These tests are useful, but not complete. MacNeal[12] has proposed a more complete set of test cases. However, the membrane loadings in this cantilever beam are very similar to the behavior that the core struts in a lightweight mirror experience. Thus, an analyst could use this model as a prototype model in determining the best technique for representing the core structure. An analyst should run a whole series of test cases similar to this to verify the behavior of the FE code to be used in any analysis of optical structures.

Solid elements are a 3D extension of the 2D membrane behavior. Thus conclusions drawn about quadrilaterials and triangles can be extended to hexahedrons and tetrahedrons, respectively. As expected, the 4 noded-tetrahedron performs as poorly as the 3-noded triangle. Since current automeshing capabilty in 3D structures is generally limited to tetrahedron, the FE code must offer the choice of 10-noded tetrahedron if it is to be used for high accuracy optical structures. This author highly recommends the use of parametric meshing with hexahedron over any automeshing technique for highly accurate optics models.

8.3. Symmetry Techniques

Most optical structures possess some level of symmetry. Techniques which take advantage of symmetry can reduce the computer resources required for a finite element analysis. Typically, only the smallest repeating section is modeled.

Symmetry is defined as a balanced arrangement of structure about a point (spherical symmetry), a line (axisymmetry), or a plane (reflective symmetry). For a structure to be symmetric, both the structure and its boundary conditions must possess the same degree of symmetry. Applied loads may be nonsymmetric, although there are additional efficiencies when the load is also symmetric.

Axisymmetry

Most lenses and lens barrels are axisymmetric (Figure 8.4). The structure may be represented as a finite element model using axisymmetric elements as shown in Figure 8.8. If the load is also axisymmetric, as a circular line load or a uniform change in temperature, then the behavior of one cross-sectional plane represents the solution at any cross-sectional plane. Most finite element programs can solve this problem, because the theory for 2D elements can be extended to axisymmetry by the addition of a hoop stress term. The applied load (F) at a radius (R) is the net force on a full ring of structure subjected to a line load (f) in force/length.

$$F = 2\pi Rf$$

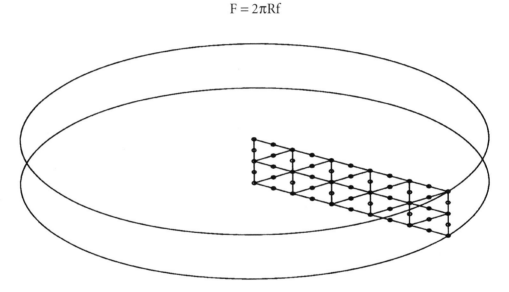

FIGURE 8.8 Axisymmetric element model of an optic.

Displacement results are limited to radial (δ_r) and axial (δ_z), while stress results include radial (σ_r), axial (σ_z), shear (τ_{rz}), and hoop (σ_θ). If the load has a variation in θ which can be represented as a Fourier series, some programs will provide a solution which is also represented as a Fourier series.

An alternative model for axisymmetric behavior is a small slice of pie as shown in Figure 8.9. Since the structure is uniform in θ, then a small slice ($<10°$) with a single element in that direction is adequate. Symmetric BC must be applied to both symmetry faces to insure the proper behavior.

$$\delta_\theta = 0, \quad \phi_r = 0, \quad \phi_z = 0$$

This model is not as efficient, since twice the number of nodes are required, but it may be possible to use some program features which might not be available in a particular FE program's axisymmetric capability list.

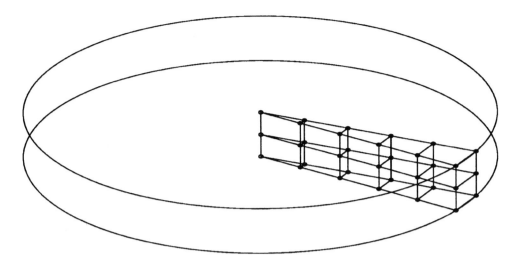

FIGURE 8.9 Solid element wedge model of an optic.

Reflective Symmetry with General Load

Reflective symmetry is very common in most man-made or natural structures. In optical structures there may be several planes of symmetry as shown in a lightweight mirror on a 3-point support in Figure 8.10.

For a single plane of symmetry, where the structure and boundary conditions are symmetric, but the applied load is not (Figure 8.11[a]), the solution is a linear combination of two solutions, a symmetric case (Figure 8.11[b]), and an antisymmetric case (Figure 8.11[c]), each with the loads cut in half. Both (b) and (c) can be solved using symmetric half-models. In Figure 8.12, the half-model (right side) is solved twice, once with symmetric loads (P_S) and symmetric BC to get a displacement vector (δ_S) in Figure 8.12(b), and once with antisymmetric loads (P_A) and antisymmetric BC to get displacement vector (δ_A) in Figure 8.12(c). The load vectors P_S and P_A can be determined from the applied loads on the right side P_R and the left side P_L by the equation:

$$P_S = 0.5\left(P_R + P_L\right)$$

$$P_A = 0.5\left(P_R - P_L\right)$$

Symmetry can be thought of as a standard reflective mirror. The symmetric BC can be found from intuition using Figure 8.11(b). If at point j on the right side, the x displacement is δ_{xj}, and for the corresponding point k on the left side the displacement is δ_{xk}, then by symmetry,

$$\delta_{xk} = -\delta_{xj}$$

If point j is a point on the symmetry plane, then points j and k are coincident (j = k). The only way the last equation can be satisfied is for

$$\delta_{xk} = -\delta_{xj} = 0$$

For symmetric loads, the displacement normal to the symmetry plane must be zero at the symmetry plane, as are rotations in the symmetry plane. If the x axis is normal to the symmetry plane,

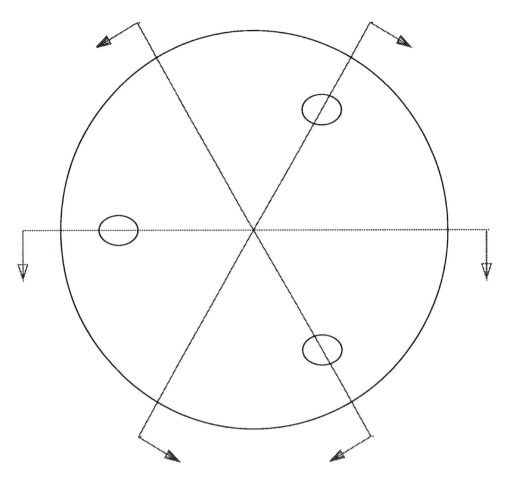

FIGURE 8.10 Planes of symmetry for an optic on a 3-point support.

$$\delta_x = \phi_y = \phi_z = 0$$

Antisymmetry is the negative of a reflective mirror. The antisymmetric BC, which can be determined by a similar argument, are the compliment of the symmetric BC. For Figure 8.11(c), the antisymmetric BC are the two displacements in the symmetric plane and the rotation normal to the plane:

$$\delta_y = \delta_z = \phi_x = 0$$

The resulting displacements δ_S and δ_A are only intermediate results. The desired results on the full structure are found by the linear combination:

$$\delta_R = \delta_S + \delta_A$$

$$\delta_L = \delta_S - \delta_A$$

The displacements on the modeled half (right side) are in a normal right-hand coordinate system. The displacements on the unmodeled half (left side) must be interpreted as being in a left-hand coordinate system.

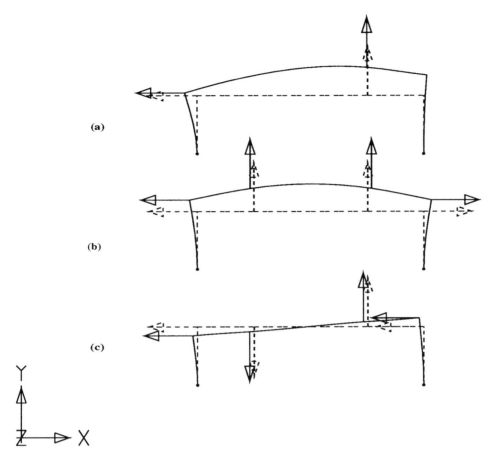

FIGURE 8.11 Symmetric structure with general loads. (a) Solution to general load condition; (b) solution with symmetric load components; and (c) solution with antisymmetric load components.

Some finite element codes allow the solution of multiple BC and linear combinations of results, making the above approach possible in a single execution of the code. The above approach can be applied twice to model two planes of symmetry with a 1/4 model and 4-load cases as shown in Figure 8.13. The double subscripts refer to the type of BC on each side of the model and the integer subscripts refer to the quadrant for which the solution applies.

$$\delta_1 = \delta_{SS} + \delta_{SA} + \delta_{AA} + \delta_{AS}$$

$$\delta_2 = \delta_{SS} - \delta_{SA} - \delta_{AA} + \delta_{AS}$$

$$\delta_3 = \delta_{SS} - \delta_{SA} + \delta_{AA} - \delta_{AS}$$

$$\delta_4 = \delta_{SS} + \delta_{SA} - \delta_{AA} - \delta_{AS}$$

The extension to three planes is also possible, requiring eight combinations of BC.

Reflective Symmetry with Symmetric Load

The application of symmetry is especially efficient for the special case when the applied load has the same symmetry as the structure. In Figure 8.14, there is no antisymmetric load ($P_A = 0$) and,

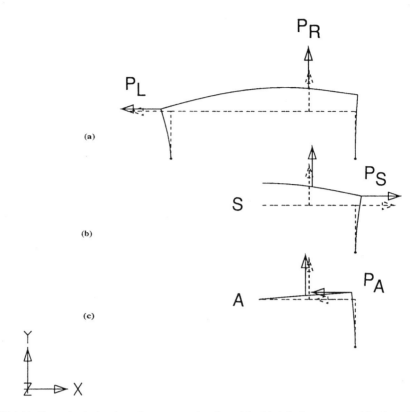

FIGURE 8.12 General solution by using symmetric submodels. (a) Solution to general load condition; (b) half model with symmetric loads and BC; and (c) half model with antisymmetric loads and BC.

thus, no antisymmetric displacement ($\delta_A = 0$). Only the symmetric case is run, and no combination is required. This is the most obvious and the most common application of symmetry.

Multiple planes of symmetry with symmetric loads are simple to use, if the finite element program allows displacements to be calculated in alternate coordinate systems. Some simple FEA programs require that all displacements be calculated in a single rectangular system, limiting symmetry to the x, y, or z plane. If displacements are calculated in a cylindrical system, then a circular mirror on a 3-point mount with a symmetric load can be analyzed with a 60° model with BC on each cut face,

$$\delta_\theta = \phi_R = \phi_z = 0$$

Note that the displacement normal to the planes is zero and the two rotations in the plane are zero.

Model Size Required

A variety of model sizes are possible in common optical structures. In the following discussion the models are pictured in Figures 8.15 to 8.19. The coordinate axes are oriented so the X axis is on one plane of symmetry and the Z axis is the optical axis normal to the plane of the optic. Symmetric ΔT includes uniform temperature change, or an axisymmetric temperature variation in the radial or axial direction.

Full Model (360°)

A full model is required for:

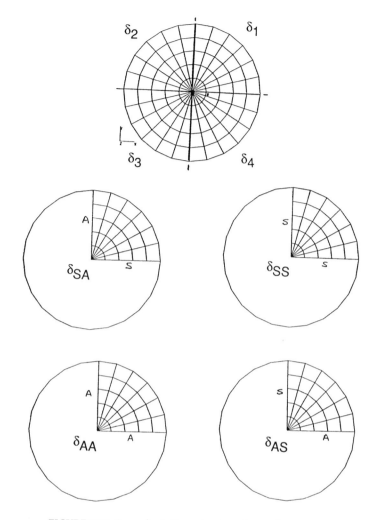

FIGURE 8.13 General solution using two planes of symmetry.

- Nonsymmetric structure (internal core or external shape)
- Nonsymmetric BC (mount location or stiffness)

Even when symmetry exists, a full model can be useful

- To find all dynamic or buckling modes in a single solution
- To get full dynamic response analysis in a single solution
- For nonsymmetric loads solved in a single subcase without combinations
- For plotting and postprocessing of the full structure

Half-Model (180°)

A half-mode applies to:

- Circular, hexagonal, and elliptic optics
- Polar, square, triangular, and hexagonal core structures
- Most mount configurations (uniform, ring, 3, 4, or 6 point)

A half-model is most efficient when:

- Symm BC: loads have 360° symmetry (i.e., Z gravity, symmetric ΔT)

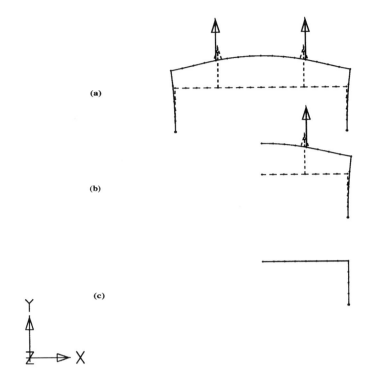

FIGURE 8.14 Symmetric structure with symmetric loads. (a) Full solution; (b) half-model with symmetric loads and BC; and (c) half-model with antisymmetric loads (null).

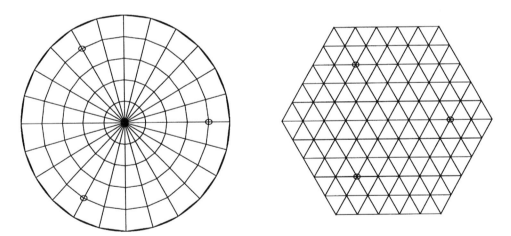

FIGURE 8.15 Full model of an optic (360°).

- Symm BC: loads have 180° symmetry (i.e., X gravity)
- Anti-BC: loads have 180° antisymmetry (i.e., Y gravity)

A half-model requires two solutions (symm BC and anti-BC) to find all dynamic modes.

Quarter Model (90°)

A quarter model applies to:

- Circular, hexagonal, and elliptic optics
- Polar, square, triangular, and hexagonal core structures

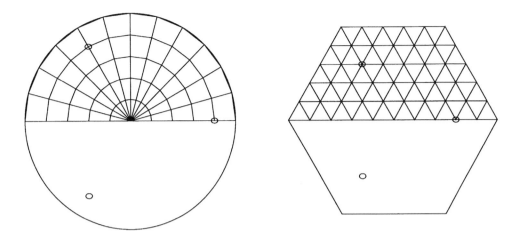

FIGURE 8.16 One-half model of an optic (180°).

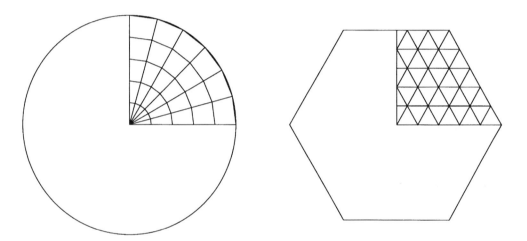

FIGURE 8.17 One-quarter model of an optic (90°).

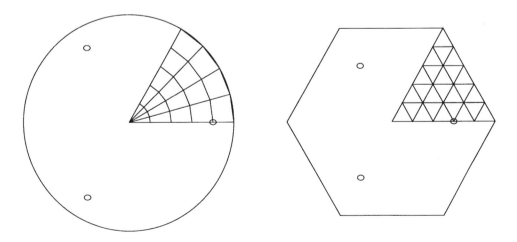

FIGURE 8.18 One-sixth model of an optic (60°).

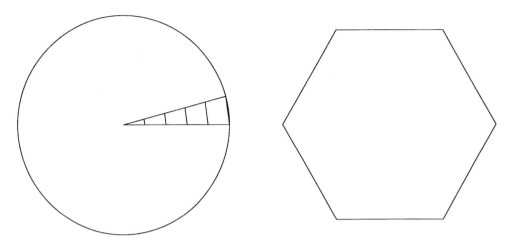

FIGURE 8.19 One-n*th* model of an optic (<10°).

• Limited mount configurations (uniform, ring, 4 point, NOT 3 or 6 point)

A quarter model is limited to the following loads:

• Symm-symm BC: loads with 360 symmetry (i.e., Z gravity, symmetric ΔT)
• Symm-anti-BC: loads with 180 symmetry (i.e., X or Y gravity)

A quarter model requires four solutions to find all dynamic modes, although some may be mode pairs.

One-Sixth Model (60°)

A 1/6 model applies to:

• Circular, hexagonal, but NOT elliptic optics
• Polar, triangular, and hexagonal, but NOT square core structures
• Most mount configurations (uniform, ring, 3 to 6 point, NOT 4 point)

A 1/6 model is limited to the following loads:

• Symm-symm BC: loads with 360 symmetry (i.e., Z gravity, symmetric ΔT)

A 1/6 model cannot find all dynamic modes, only those with 60° symmetry.

Thin Wedge Model

A 1/n model (<10°) applies to:

• Axisymmetric structure and BC (i.e., circular optic on ring support)
• Axisymmetric loads (i.e., Z gravity, symmetric ΔT)

A 1/n model can only find axisymmetric dynamic modes.

Other model sizes are possible in special cases, but the above set describes the most common applications of symmetry.

Advantages and Disadvantages of Symmetry

The obvious advantage of using symmetry in a finite element model is efficiency, but an expanded list would include:

• Faster modeling with fewer grids and elements

- Less model checking required
- Faster run times
- Less memory and disk storage required
- Smaller output files generated

Some disadvantages of symmetry include:

- Cannot get full model plots easily
- Requires multiple solutions and combinations if load is nonsymmetric
- Image side uses left-handed coordinate system for output
- Requires all combinations of BC to get all dynamic modes

In dynamics and buckling, the lowest mode is not necessarily symmetric. All BC combinations must be checked to find the lowest mode. The basic premise of symmetric is linear superposition. If the structure behaves in a nonlinear fashion (material or geometric), then symmetry may not apply. Most optical structures display some symmetry. The analysis can be much more efficient if symmetry is exploited in the modeling scheme.

8.4 Displacement and Dynamic Models for Optics

The primary concern in the structural analysis of optics is the deformation of the optical surface or the pointing of an optical surface, due to static or dynamic loads. Performance of the optical system is controlled by the motion and deformation of the optics, with stress being of secondary concern in most applications. Stress models, which require more detail than displacement models, are addressed in the next section.

Depending on the information required and the resources (manpower and computer) available, the level of detail may vary in a finite element model. The following list is ordered by increasing resolution and model size.

Single-Point Model (Solid or Lightweight Optic)

In a system level model, a "small" stiff optical element may be treated as a single node point in the structural model. For dynamics or gravity loads, the point must have the proper mass properties, including center of gravity and moments of inertia. Since line of action of forces is very important, the mount points must be modeled in their true spatial location (Figure 8.20). The optical node may be attached to the "softer" elastic mount with rigid elements. However, if thermal loads are to be analyzed, then very stiff elastic beams with the correct CTE should be used to attach to the mount. The stiff elastic beams must be stiff relative to the softer mounts, but not so stiff that they cause numerical difficulty. Suggested stiffness values are 100 to 1000 times the mount stiffness.

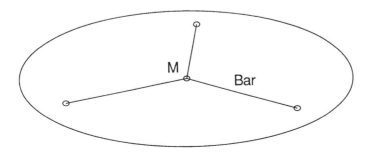

FIGURE 8.20 Rigid, lumped-mass model of an optic.

A slight improvement on this model uses three triangular plate elements with the same 4-node points used in the beam model. Now actual plate thicknesses can be modeled so fewer calculations and assumptions are required. A side benefit is that the graphical display is improved.

A single-point model can be used for optical path pointing analysis, but offers no information about the surface distortion of the individual optic.

2D Plate Model (Solid Optic)

A plate model (Figure 8.2) incorporates the bending and membrane behavior into the analysis, but does not incorporate through the thickness effects. Bending distortion due to pressure, gravity, or axial gradients, as well as radial growth due to temperature changes, can be found. However, thickness changes caused by temperature cannot be determined from a plate model. The model can account for original thickness variation in the optic by providing different property (thickness) inputs for the various plates.

For high accuracy analysis, transverse shear flexibility must be incorporated in the plate elements. In finite element documentation, this is commonly referred to as thick plate theory or Mindlin plate theory. For a solid flat circular mirror supported at the outer edge by a knife-edge support (simply supported BC), the bending and transverse shear deflection are compared in Figure 8.21. For a diameter/thickness ratio of 10, the shear deflection is 10% of the bending deflection. Although this may seem small, deflections of this magnitude are important to optical performance.

The 2D model is cheap to generate and to run. Dynamic modes are found accurately and cheaply for most conventional optics.

3D Solid Model (Solid Optic)

A finite element model composed of solid elements (Figure 8.3) can accurately predict the distortion of optics including through-the-thickness effects. These include thermal gradients and 3D variations around mounts. Transverse shear is automatically included in the solid elements. Thickness variation of the optic is accounted for by node position on the surface.

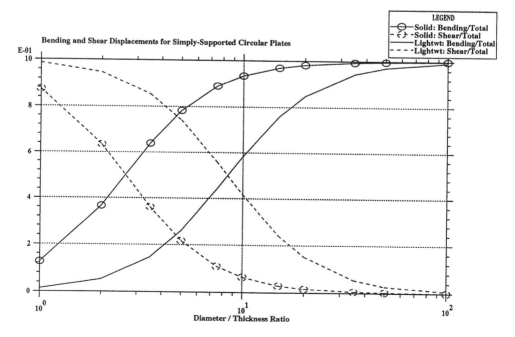

FIGURE 8.21 Bending and transverse shear components of the 1-g sag displacement of simply supported circular plates.

To get accurate bending behavior from solid models, sufficient model resolution is required in all three dimensions. One layer of 8-noded bricks through the thickness is generally too stiff, causing the deflections to be 10 to 20% too small. At the very minimum, two layers should be used through the thickness. Not all bricks are created equal. Finite element programs may use different formulations and integration rules which can modify the accuracy of a solid model. A suggested approach is to model a flat plate on a ring support using the same elements and resolution to be used on a real optic, to determine the detail required to get the accuracy desired in the analysis program used. The theoretical solution from plate theory (including transverse shear) for a circular plate of radius (R) and thickness (t) subjected to a uniform pressure (p) is

$$\delta_{total} = \delta_{bending} + \delta_{shear}$$

$$\delta_{bending} = \left[(5+\nu)pR^4\right] / \left[64(1+\nu)D\right]$$

$$\delta_{shear} = \left[3pR^2\right] / \left[8Gt\right]$$

$$D = \left[Et^3\right] / \left[12(1-\nu^2)\right]$$

Parametric meshes, as shown in Figure 8.3, provide the most accurate results in general. Most preprocessing programs can create such a mesh under user control. Some programs offer automesh capability for 3D solids. Current technology limits most automeshers to tetrahedron elements. The linear displacement 4-noded tetrahedron is notoriously stiff, causing predicted displacements to be too low. At least eight layers are required through the thickness to predict displacements with less than 10% error. These models often become too expensive to run. The quadratic displacement 10-noded tetrahedron provides much more accurate answers, requiring only two to four layers through the thickness in most cases. Automeshed optics tend to predict nonuniform and nonsymmetric response even for perfectly symmetric problems. The unexpected nonsymmetry can cause the results to be misleading. Since the geometry of most optics is very regular, parametric meshing is highly recommended.

2D Equivalent Stiffness Plate Model (Lightweight Optic)

A typical lightweight optic includes two faceplates bonded to an eggcrate core structure as shown in Figure 8.22. Common eggcrate structures may be triangular, square, or hexagonal patterns. Key dimensions in this structure are the core plate thickness (t_c) and the inscribed circle diameter (B) which define the core density ratio (α).

$$\alpha = t_c / B$$

To first order, the behavior of the core is determined by α regardless of the core pattern.[1] The other key dimensions and properties are (Figure 8.23)

t_p = faceplate thickness
H = overall height
H_c = core height = $H - 2t_p$
E = Young's modulus
ν = Poisson's ratio
ρ = mass density

A lightweight mirror can be represented as a single layer of plate elements with equivalent properties. The membrane (in-plane) behavior is found from the cross-sectional area:

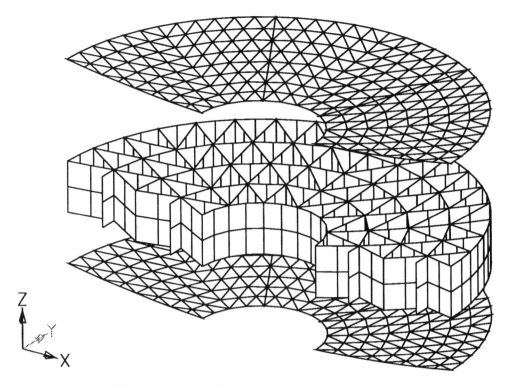

FIGURE 8.22 FE model of a lightweight mirror with triangular core.

$$t_m = 2t_p + \alpha H_c$$

The cross-sectional moment of inertia per unit length is

$$I_b = \left[H^3 - (1-\alpha) H_c^3 \right] / 12$$

The ratio of I_b to the moment of inertia of a solid plate of thickness t_m is

$$R_b = 12 I_b / t_m^3$$

This ratio is always greater than 1, and for most practical designs is on the order of 100 to 1500. Modern designs tend to have higher ratios with more structural efficiency. Transverse shear distortion is even more important in lightweight mirrors than in solids. The transverse shear ratio (R_s) is

$$S = \left[H^2 - (1-\alpha) H_c^2 \right] / \alpha$$

$$R_s = (2/3) 12 I_b / \left[S t_m \right]$$

A common approximation is to consider the core to be fully effective in carrying shear, with the faceplates carrying none. Thus the shear ratio is

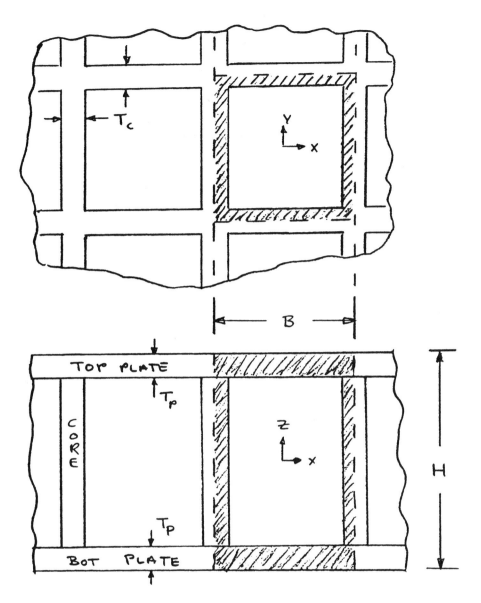

FIGURE 8.23 Lightweight mirror with square cells.

$$Rs = \left[Ht_c\right]/\left[Bt_m\right]$$

If first-order stresses are to be recovered from these effective plates, the stress recovery points must define the extreme fibers of the mirror:

$$c_1 = H/2 \quad c_2 = -H/2$$

The mass density of the effective plate must be modified, also. The mass is normally calculated from the membrane thickness (t_m) and the mass density (ρ). In this approximation, the membrane thickness is calculated from the cross-section of a cut, which represents core struts in y direction. To account for the core struts in the x direction the effective mass density must be increased.

$$\rho_e = \left[2t_p + 2\alpha H_c\right]\rho \Big/ t_m$$

To use this approach, the finite element program must allow separate inputs for membrane, bending, and shear properties. A solid, homogeneous plate with a single thickness used for both membrane and bending cannot accurately model a lightweight mirror using this technique.

The advantage of this modeling scheme is the obvious speed and simplicity. The model is easily generated without regard for the internal core geometry. Only the external features such as diameter and mount location need to be considered. The disadvantage of this technique is a slight loss of accuracy, especially in local shear effects around mounts. A plate model cannot predict through the thickness effects. If 3D effects are significant, then a 3D model should be used. As the ratio of diameter/height grows, the optic acts more plate-like and the accuracy of the plate model improves.

3D Equivalent Stiffness Solid Model (Lightweight Optic)

For some optics, a plate model is not accurate enough. Another simplified modeling scheme is available which includes 3D effects. In this scheme, the faceplates are modeled as solid plates of thickness t_p located in their true position. The core is then represented as solid elements of reduced properties (Figure 8.24). Effective isotropic properties are

$$E_e = \alpha E$$

$$G_e = \alpha G$$

$$\nu_e = \nu$$

$$\rho_e = 2\alpha\rho$$

A single layer of solid elements is typically too stiff, so at least two layers through the thickness would improve accuracy.

The isotropic effective properties underestimate the axial stiffness. The use of orthotropic materials could improve that behavior. The orthotropic properties are:

$$E_{xe} = E_{ye} = \alpha E$$

$$E_{ze} = 2\alpha E$$

$$G_{xy} = 0$$

$$G_{xz} = G_{yz} = \alpha G$$

$$\nu_{xy} = 0$$

$$\nu_{zx} = \nu_{zy} = \nu$$

If the program accepts material constants, then the above terms can be used directly. In NASTRAN, the material matrix [C] must be input on MAT9 entries.

$$\{\sigma\} = [C]\{\varepsilon\}$$

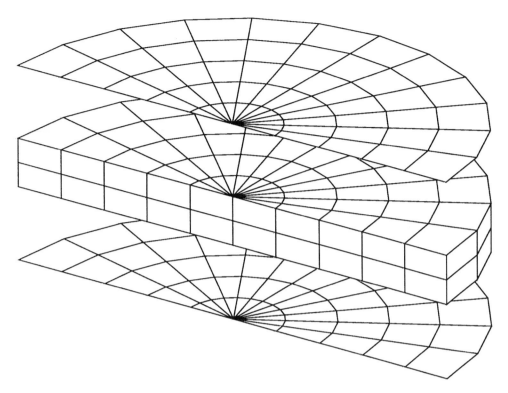

FIGURE 8.24 3D equivalent stiffness model of a lightweight mirror with the core modeled as reduced stiffness solids.

Since C is a symmetric matrix, it is only necessary to define the nonzero terms in the lower triangular portion:

$$K = \alpha E / \left(1 - v^2\right), \quad C_{11} = \left(1 - 0.5v^2\right)K,$$

$$C_{21} = 0.5v^2 K, \quad C_{22} = \left(1 - 0.5v^2\right)K,$$

$$C_{31} = vK, \quad C_{32} = vK, \quad C_{33} = 2K,$$

$$C_{44} = 0, \quad C_{55} = \alpha G, \quad C_{66} = \alpha G$$

This modeling scheme preserves the efficiency advantage of not modeling the detailed core structure. Again the penalty is a slight loss in accuracy, especially in shear effects around point loads and mounts.

In either the 2D or 3D equivalent models, design trades on core geometry, faceplate thickness, and even mirror height are easy to perform. Once a particular design has been chosen from the trade study, then a full 3D model (next section) should be created to verify the predictions.

3D Plate Model (Lightweight Optic)

To obtain high accuracy in the prediction of distortion of lightweight optics, a 3D plate model is required. In this model (Figure 8.25), the faceplates and each individual core strut are modeled as solid, homogeneous plate elements. Since the detailed core geometry is modeled, this approach is quite time consuming. Preprocessing programs can speed up the model generation depending on

(a)

(b)

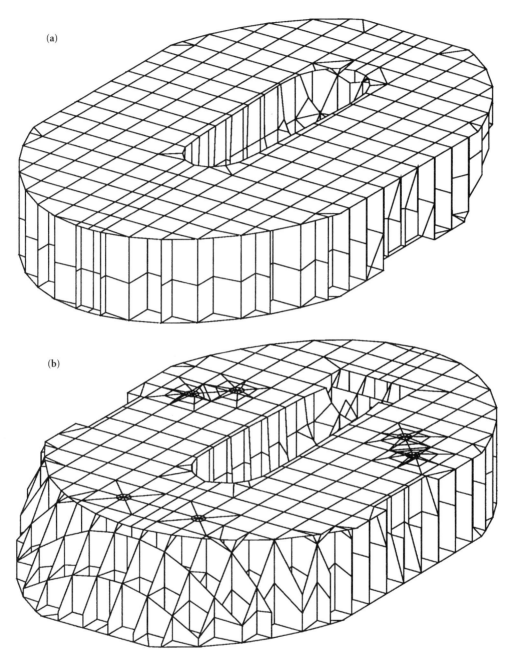

FIGURE 8.25 (a) 3D shell model of a lightweight scan mirror — top view; (b) 3D shell model of a lightweight scan mirror — bottom view.

particular capabilities. Extra detail is required around mounts to accurately model mount geometry, adding to the model generation time.

To accurately predict mirror stiffness, the neutral axis of the plate elements must coincide with the midplane of the faceplates. Thus, either the faceplate nodes lie in the midplane of the faceplate, or offsets are required for the elements. A single layer of plates to represent the core is not accurate enough for most applications. Two or more layers should be used in the core. Meshing on curved surfaces is sometimes difficult and prone to inaccuracies. A lightweight mirror could be modeled in a flat geometry quite efficiently, then curvature could be added to the final model by moving grids in the axial direction, either in a preprocessor or a separate Fortran code. Each node's flat axial position (z_o) must be modified to a curved position (z_n) by using the radius of curvature (R_c) and the angular position (θ) to the node from the center of curvature:

$$z_n = z_o + (1 - \cos\theta) R_c$$

In some analyses, a high degree of accuracy is required on the net mass of the optic. Depending on modeling practices, a 3D plate model may over- or underpredict the true mirror mass. If nodes are located at the faceplate midplane, then the core elements are too tall, overpredicting mass. In many core structures, the joints have extra material (posts or fillets) due to fabrication techniques, so a model will underpredict the weight. The user must adjust the mass density of the core or add nonstructural mass (positive or negative) to the core plates to adjust the mass.

The lightweight mirror depicted in Figure 8.25 had such complex geometry that only a full 3D plate model could predict accurate results. The analytical prediction shows excellent correlation to the experimental results for a 1-g load on edge with a 3-point back mount (Figure 8.26).

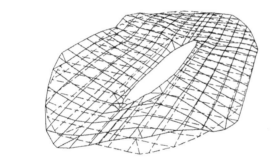

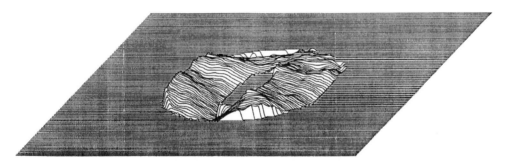

FIGURE 8.26 Comparison of analysis and test surface figures for the mounted scan on edge.

Figure 8.27 shows a more uniform lightweight mirror with extra detail at the core-to-faceplate intersections. In this analysis, the quilting effect due to adhesive joint (dark thick lines) shrinkage was predicted (Figure 8.28). Again, only a full 3D plate model could predict this type of 3D behavior; the equivalent models could not.

Comparison of Models (Lightweight Optic)

A lightweight mirror sitting on a 4-point mount was modeled using 1/4 symmetry to compare the modeling techniques. A full 3D model is shown in Figure 8.29(a) as all plate elements. The 3D equivalent stiffness model is shown in Figure 8.29(b) using a polar mesh for convenience. The faceplates are modeled as plates with the core modeled as equivalent solids. The 2D equivalent

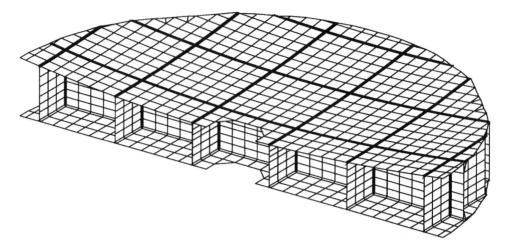

FIGURE 8.27 Lightweight mirror with adhesive bonds between core and faceplates.

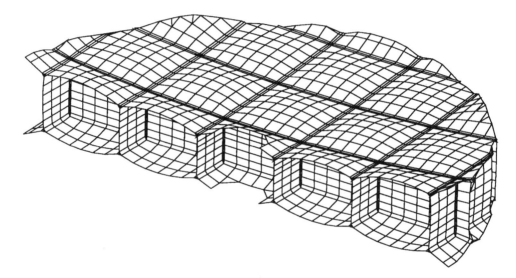

FIGURE 8.28 Deformed shape (scaled) for adhesive joint expansion.

stiffness plate is shown in Figure 8.29(c) as a single layer of plate elements, again with a polar mesh. Deformed shapes are shown in Figure 8.30 where the displacement contours are very similar, except at the mount points where shear has its largest effect. If the rigid body motion is removed (Figure 8.31[a]), the contours are very similar, since the local shear displacement has been subtracted. After the power is removed (Figure 8.31[b]), the contours are nearly identical. A surface fit of the deformed surface shows that they perform optically the same to within 3%. The natural frequencies are very similar, also. For early trade studies, this level of accuracy is usually sufficient. For final designs with accurate stresses, a 3D model is usually required.

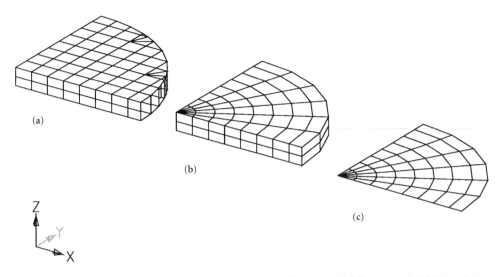

FIGURE 8.29 Three models of a lightweight mirror. (a) 3D shell model with full core detail; (b) 3D equivalent stiffness model with core as reduced solids; and (c) 2D equivalent stiffness model with effective stiffened plates

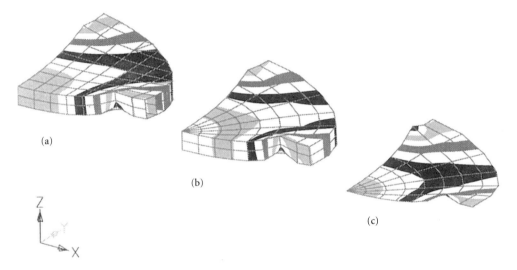

FIGURE 8.30 Comparison of deformed shapes with z displacement contours.

8.5 Stress Models for Optics

Even though displacements govern most optical designs, stresses must be checked. Most optics consist of brittle materials, such as glass or ceramic, which have different failure modes than ductile materials, such as steel or aluminum. To get the accurate stress values from a finite element model, high model resolution is required in the areas of rapid stress gradients. Interpretation of stress output can be sometimes confusing in graphical postprocessing programs to casual users.

Ductile Failure (Most Metals)

A metal mirror will suffer stress failure in a ductile manner in most applications. There are multiple levels of stress failure depending upon the design requirement.

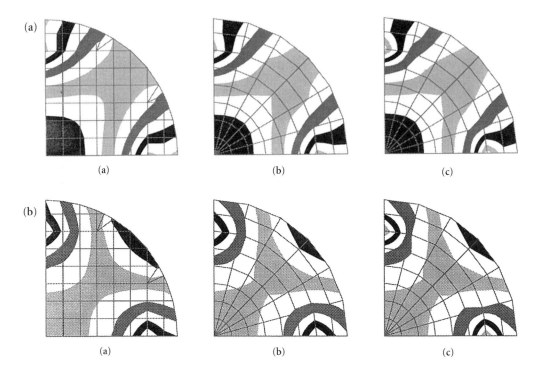

FIGURE 8.31 (a) Displacement contours after best-fit plane removed; and (b) displacement contours after power removed.

(a) • Fatigue: if the part fails due to repeated cycles of stress

  $\sigma > S_e$ (where S_e is the endurance stress limit)

 • Microyield: if the part suffers 0.000001 (1 ppm) permanent strain

  $\sigma > S_m$ (where S_m is the microyield stress limit)

 • Yield: if the part suffers 0.002 (0.2%) permanent strain

  $\sigma > S_y$ (where S_y is the yield stress limit)

 • Ultimate: if the part fractures

  $\sigma > S_u$ (where S_u is the ultimate stress limit)

These material properties are obtained from uniaxial tension samples. In a typical structure, the stress state is multiaxial. To compare the multiaxial stress to a uniaxial property the equivalent stress most commonly used is the Von Mises stress (σ_{vm}):

$$\sigma_{vm}^2 = \left[\left(\sigma_1 - \sigma_2\right)^2 + \left(\sigma_2 - \sigma_3\right)^2 + \left(\sigma_3 - \sigma_1\right)^2 \right] \Big/ 2$$

where the principal stresses ($\sigma_1, \sigma_2, \sigma_3$) may be obtained from the directional stresses ($\sigma_x, \sigma_y, \sigma_z, \tau_{xy}, \tau_{yz}, \tau_{zx}$) by use of a Mohr's circle diagram. A Mohr's circle diagram is shown in Figure 8.32 for a two-dimensional state of stress. For this case the center (C) and radius (R) of Mohr's circle are given by:

$$C = \left(\sigma_x + \sigma_y\right) \Big/ 2$$

$$R^2 = \left[\left(\sigma_x - \sigma_y\right) \Big/ 2 \right]^2 \pm \tau_{xy}^2$$

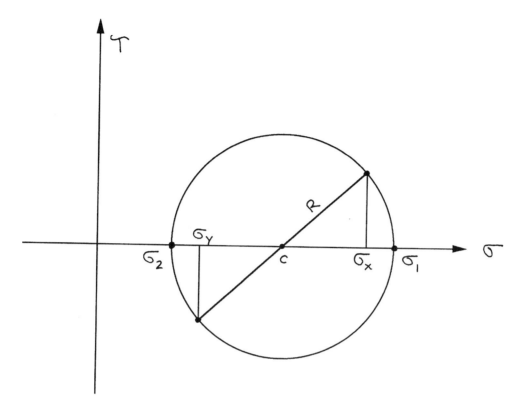

FIGURE 8.32 Mohr's circle for a 2D stress state.

The maximum principal stress (σ_1) is the most positive value, and the minimum principal stress (σ_2) is the most negative value.

$$\sigma_1 = C + R$$

$$\sigma_2 = C - R$$

For example, failure due to yield occurs when:

$$\sigma_{vm} = S_y$$

Thus, when postprocessing a metal part, the user should be plotting Von Mises stress. Note that the Von Mises stress is always positive, even when the directional stresses are negative. Thus, plots of Von Mises cannot distinguish between tension and compression. Although the Von Mises stress is often the largest stress, it is possible for a directional stress or principal stress to have up to a 15% larger magnitude.

In some FE codes, Von Mises stress may not be an option for 3D solid elements. The equivalent stress provided is called octahedral shear stress (τ_{oct}) which can be related to σ_{vm} as:

$$\tau_{oct} = \sigma_{vm}\left[\sqrt{2/3}\right] = 0.577\sigma_{vm}$$

When comparing t_{oct} to failure, use the shear failure stress, which is

$$S_{sy} = 0.577S_y$$

Brittle Failure (Most Glass and Ceramics)

Brittle failure is not as well understood as the ductile failure. Juvinall[9] suggests the use of the modified Mohr theory for the failure of brittle materials. The ultimate stress in compression is usually several times larger than the ultimate stress in tension. A simplifying assumption is that only the tensile failure need be considered. For a multiaxial stress state, the largest tensile stress is the maximum principal stress (σ_1) found from Mohr's circle calculations. Since brittle materials exhibit no yield, failure occurs from fracture when:

$$\sigma_1 = S_u$$

In finite element postprocessing of brittle optics, the analyst should be plotting the contours of maximum principal stress, not Von Mises stress. In many applications the applied load may have a positive or negative value. If the load is reversed in direction, the Mohr's circle is reflected about the origin. Thus, a plot of σ_1 would check failure for a positive force; a plot of σ_2 would check failure for a negative force of the same magnitude.

Fracture Mechanics Approach

If a crack with sharp corners exists in a part, then linear elasticity predicts the stress to be infinite at the crack tip. Any linear finite element code will verify that the stress is infinite. In a series of analyses with successively smaller elements, the program will predict successively higher stresses, while chasing infinity. The results will not converge to a reasonable solution. The fracture mechanics approach can be used to predict when an existing crack will grow, thus causing failure.

According to the theory of fracture mechanics, an initial crack will propagate if the stress intensity factor (K_I) is greater than the material's fracture toughness (K_{IC}). The value of fracture toughness, which is temperature dependent, has units of pressure times square root of length. The stress intensity factor K_I (as opposed to the stress concentration factor K_t) is a function of the initial crack size (a) and the surrounding stress field (σ):

$$K_I = C\sigma\sqrt{a}$$

For a small crack in a large, thin plate, $C = \sqrt{\pi} = 1.8$. If this crack size is not small relative to the plate dimensions, or occurs at the edge of the plate, C increases. For thick plates or solids, the relationship is not as simple.

If cracks are visible in an optic, then the actual geometry of the crack and the part, along with the state of gross stress predicted by the finite element model, should be used to predict K_I. If cracks are not visible, then cracks smaller than the visible threshold should be assumed. For a polished surface with no visible cracks, existing crack size could be as large as 0.001 to 0.005 in. depending on the inspection technique. In this case, a reasonable prediction of K_I is

$$K_I = (1.9)\sigma\sqrt{a}$$

In the above model, the crack detail is not modeled. The relatively coarse mesh is used only to predict the "gross" stress (σ).

If a more accurate analysis is required, the stress intensity can be predicted directly by a more detailed model which "zooms" in on the crack. This local model may be part of a larger system model, or it may be a stand-alone model which gets its BC from a system level model. There are several ways of predicting K_I from a detailed model, but only the two most common will be mentioned here. If the finite element program has a crack tip element, then this element is embedded in a model of standard elements of the same type (i.e., 2D plane stress, 2D plane strain, or 3D solid). The alternative method, available in all FE codes, is to model the crack area with the standard elements. The model is run once with the initial crack area (A_1 = a t) and run again with a slightly larger crack [A_2 = (a + Δa) t] . The difference in strain energy (U) is used to predict the strain energy release rate (G) and the stress intensity factor.

$$G = \Delta U / \Delta A = \left(U_2 - U_1\right)/\left(A_2 - A_1\right) = \left(U_2 - U_1\right)/\left(\Delta a\, t\right)$$

$$K_I = \sqrt{EG}$$

The strain energy release rate is very general and typically more accurate than the crack tip element.

Model Detail Around Stress Concentrations

Stress levels change very rapidly around stress concentration effects. When trying to predict stress in a high gradient area like a fillet, the model detail must be fine enough to describe the fillet geometry accurately. More elements are required when using first-order elements (corner nodes only) than when using higher order elements which have one or more nodes along an edge. These higher order isoparametric elements can be used to more accurately describe the geometry as well as to more accurately predict a rapidly varying stress. The analyst must exercise care to verify the accuracy of the stress predictions. Some recommended steps are listed below.

1. Prototype model: Find a theoretical solution to a problem which is similar to the actual problem. Run studies of element type and size to find the required model detail to get within a desired accuracy. Use the prototype results to model the actual problem.
2. Convergence study: When analyzing the actual problem, run an additional analysis with more detail until the change in stress from run to run is within a desired bound.

The conventional approach of adding model resolution by making more elements of smaller size is called "h" convergence. If the size and number of elements are held constant, but their order is increased, then the method is called "p" convergence. Some FE codes offer "p" elements which automatically increase the order and cycle through the solution to reach a desired accuracy of stress. This automated technique offers a higher quality answer for a moderate increase of computer resources.

A more economical approach to the problem is to combine classical stress concentration factors (K_t) with the FE results. In this approach, small details such as fillets are ignored in the model detail. The FE model is used to predict nominal stresses (σ_n) which are then multiplied by a classical K_t to estimate the peak stress (σ_p).

$$\sigma_p = K_t \sigma_n$$

Discrete mount points on lenses and mirrors represent zones of high stress gradients. Extra detail is required in the mount area to accurately describe the stress state. The model shown in Figure 8.22 has sufficient detail for accurate deflection analysis or dynamic analysis. However, to obtain accurate stress results on the same lightweight mirror, the model shown in Figure 8.33 is required. The mirror is flipped over so that the detail around the mount is visible.

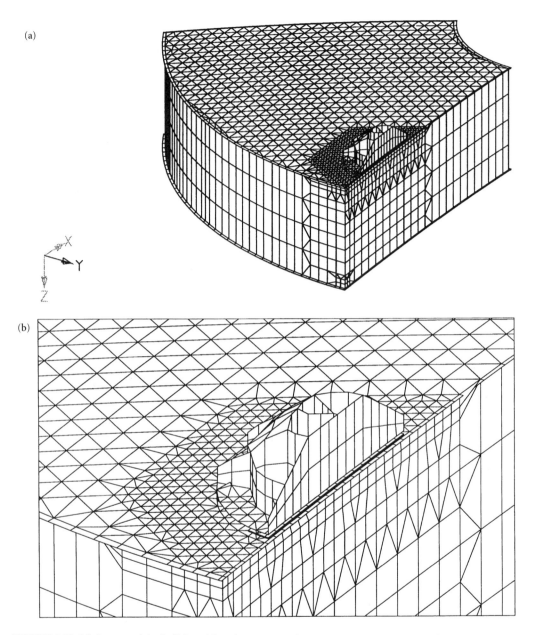

FIGURE 8.33 (a) Stress model of a lightweight mirror on 3-point support — 1/6 segment; and (b) zoom in on detail around mount.

Stress Plots

Most graphical postprocessing programs are not written by the same people who write the FE analysis programs, even if they work for the same company. For that reason, some care must be exercised when interpreting stress plots. For the lightweight mirror segment in Figure 8.34(a), various stress plotting techniques show a wide range of peak stress results all from postprocessing the same FE results file with the same postprocessing program. Thus, the stress plotting technique can add significant error beyond the FE approximation error. The vertical end loads on this model are chosen to give a peak top plate stress of 100 psi at the symmetry plane over the knife edge mount. In Figure 8.34(b), node point values are obtained from averaging centroid stresses of all elements connected to a node. This technique does not consider element orientation or stress coordinate system when averaging the stress. This is the easiest to program, but provides the most inaccurate stress values. In this example, the small core stresses are averaged with the high plate stress to produce a peak stress of 70 psi (30% error) at the wrong location. The results in Figure 8.34(c) are more accurate if the user selects only the top surface elements for averaging. Without the core averaged in, the peak plate stress is 85 psi (15% error). If the FE program provides corner stresses, the graphics program can use them to provide more accurate plots. The common technique of averaging corner stresses at a node improves the stress plot, only if the elements lie on a smooth surface with no breaks or joints and the stresses are measured in the same coordinate system. The best use of corner stresses is to contour each element independently, providing disjoint contours from element to element. The analyst can then see the magnitude and location of stress discontinuities.

The most accurate stress technique averages stress only over continuous surfaces. Whenever a break or joint is encountered the stress is not averaged. Also, element coordinate systems must be accounted for. At a common node, the directional stress from adjacent elements must be converted to a common coordinate system before averaging. The averaged directional stresses are then used in a Mohr's circle calculation to find new values of principal stress or Von Mises stress. An example of the proper technique is the MSC/NASTRAN GPSTRESS module which produced the stress results in Figure 8.34(d) of 100 psi (0% error).

Averaging principal stress or Von Mises stress from element to element is wrong and can result in large errors. For example, suppose two adjacent elements had a state of uniaxial stress where element 1 had $\sigma_x = +100$ and element 2 had $\sigma_x = -100$. In both elements the Von Mises stress is +100, and thus the average Von Mises is +100. If the directional stresses are averaged first, the average $\sigma_x = 0$ and thus the recalculated Von Mises stress is 0, also.

Smooth contour plots are the most appealing, but should only be used for data which is presented as nodal values. When plotting element centroid values the most accurate depiction is a solid, single fill color plot per element. The averaging of centroid values to get the smooth contours always misses the peak response values that are the goal of the analysis.

An analyst should run experiments with his software to determine the accuracy of the FE results. Additional tests are required to determine how the graphics program alters or interprets those results for plotting. This author's rule of thumb is "The prettier the stress plot, the less accurate the result."

8.6 Adhesive Bond Analysis

Many optics are attached to their mounts with a thin, somewhat compliant, adhesive layer. Even a 3-point mount may not be perfectly kinematic with an adhesive bond. The bond area required to handle the service loads can be large enough to require an analysis of the bond layer effects on the performance of the optic. Bond layers cause distortion of optics due to:

- Bond layer relative growth due to a mismatch of CTE
- Bond layer shrinkage during curing
- Bond layer growth due to moisture absorption

Typical bond layers are

- Very thin (<0.1 in.)
- Very compliant with low modulus (E < 1000 psi)
- Rubber-like and nearly incompressible ($v > 0.49$)

Each of the above features causes some difficulty in a FE analysis. The sudden change in element size required to describe very thin layers can cause geometrical modeling problems. Due to computer limitations, the optic and mount cannot be modeled with such a fine resolution. Typically,

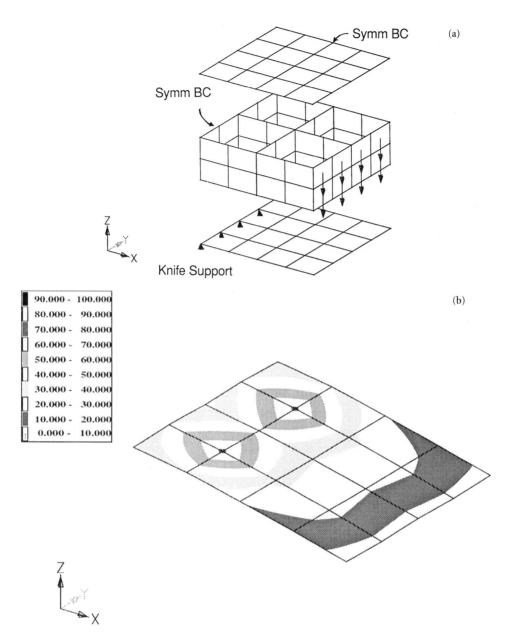

FIGURE 8.34 (a) Ligthweight mirror segment with a bending load; (b) stress contours averaging centroid stress of all connected elements;

the thin adhesive elements will have high aspect ratios due to modeling constraints. The low modulus adhesive causes a large stiffness change relative to the much stiffer optic and mount material. Finally, the high Poisson ratio can cause numerical problems because typical element formulations have a term in the denominator of the stiffness matrix of $(1 - 2v)$. As v approaches 0.5, this denominator term goes to zero, causing a divide by zero. Because of these problems, special modeling techniques have been developed for the bonded joints.

If the adhesive is a stiff material with a Young's modulus close to the optic's modulus, then the special techniques used in the following sections do *not* apply. More conventional modeling rules will apply for such cases.

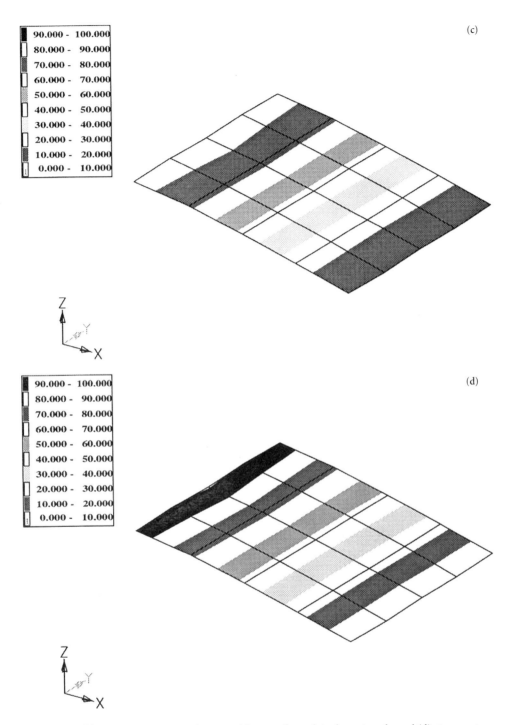

FIGURE 8.34 (c) stress contours averaging centroid stress of top plate elements only; and (d) stress contours averaging corner stress of top plate elements only.

Material Relationships

The following material definitions are used in this section:

E = Young's modulus (measured from a uniaxial tensile test)

v = Poisson's ratio (the radial contraction during uniaxial tension)

G = shear modulus (measured from a constant volume test)
B = bulk modulus (measured from volume change in constant shape)
M = thin layer modulus (the limiting modulus for very thin layers)
α = coefficient of thermal expansion

For isotropic materials, the shear modulus can be obtained from:

$$G = E/[2(1+v)] = gE$$

where

$$g = 1/[2(1+v)]$$

The bulk modulus can also be obtained from the Young's modulus:

$$B = E/[3(1-2v)] = bE$$

where

$$b = 1/[3(1-2v)]$$

Using the full 3D elasticity stress–strain equations, the stress through the thickness is

$$\sigma_z = E/[(1+v)(1-2v)][(1-v)\varepsilon_z + v(\varepsilon_x + \varepsilon_y)] - E\alpha T/(1-2v)$$

For a very soft, thin layer of adhesive between two much stiffer structures, it can be assumed that the stiff structures prevent any in-plane strain in the adhesive.

$$\varepsilon_x = \varepsilon_y = 0$$

As shown in Figure 8.35, the above is true everywhere except within a thin edge zone with width approximately two times the bond thickness. For most bond joints this is negligible compared to the surface area. Using the approximation and neglecting thermal effects, the stress–strain equations reduce to:

$$\sigma_z = (1-v)E/[(1+v)(1-2v)]\varepsilon_z = M\varepsilon_z = mE\varepsilon_z$$

where

$$m = (1-v)/[(1+v)(1-2v)]$$

and

$$M = mE$$

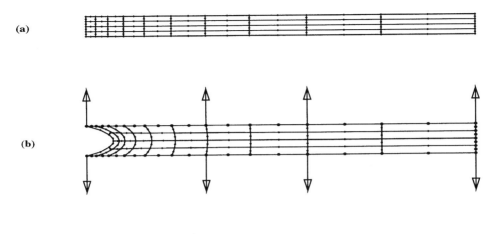

(a)

(b)

FIGURE 8.35 Adhesive bond distortions for high Poisson's ratio materials. (a) Axisymmetric element model; (b) deformed shape under tension.

For free thermal growth under a uniform temperature change, using the same assumptions of no in-plane strain and $\sigma_z = 0$,

$$\varepsilon_z = \left[(1+v)/(1-v)\right]\alpha \, \Delta T = a\alpha \, \Delta T$$

where

$$a = (1+v)/(1-v)$$

Curves of all four coefficients (g, b, m, a) vs. v are given in Figure 8.36. The change in properties over the range of conventional materials ($0.15 < v < 0.3$) is relatively small. However, high Poisson materials ($0.45 < v < 0.5$) have very high values of m, causing the special thin layer effects. The steep slope of the m curve shows the sensitivity to minor changes in v, also seen in the following values.

$$v = 0.49 \qquad m = 16.7$$

$$v = 0.499 \qquad m = 167$$

$$v = 0.4999 \quad m = 1667$$

Each additional 9 adds another power of 10 to the thin layer modulus. The limiting value on coefficient of thermal expansion is 3, which states that all of the volumetric growth is normal to the bond plane as an apparent linear expansion.

The above development assumes that the bond layer is very thin. The obvious question is "How thin is thin?" A series of finite element numerical experiments were run to generate curves (Figures 8.37) of behavior vs. a nondimensional diameter/thickness (D/T) ratio. In these curves, the apparent modulus (E') for the particular thickness is compared to the true Young's modulus (E) and

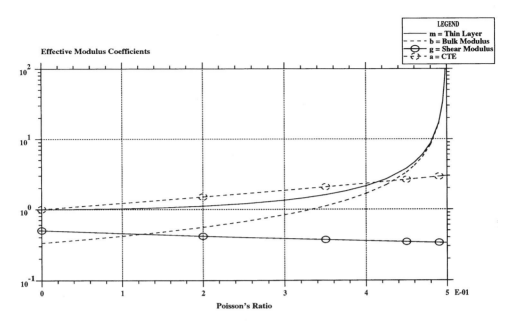

FIGURE 8.36 Effective joint properties as a function of Poisson's ratio.

the thin layer modulus (M), and the apparent coefficient of thermal expansion (α^*) is compared to the the actual coefficient (α). For high Poisson materials ($\nu > 0.45$), E^* can be significantly different than E for D/T ratios as small as 10. However, the limiting value of M is not obtained until D/T gets above 100 to 1000. For noncircular sections, the diameter (D) in the curves can be approximated with an effective diameter (D_e)

$$D_e = 4(\text{Bond Area})/(\text{Bond Circumference})$$

For models of adhesive joints, E^* and α^* should be used for the material properties normal to the bond area. The shear behavior and other in-plane behavior are unchanged by the thin layer, so G and the other E values should be the original values. This results in an orthotropic material with a diagonal material matrix:

$$E_x = E_y = E, \ \ E_z = E^*$$

$$G_{xy} = G_{yz} = G_{zx} = G$$

$$\alpha_x = \alpha_y = \alpha, \ \ \alpha_z = \alpha^*$$

Adhesive Bond Joint Models

Several possibilities exist for models of bond joints depending on the purpose of the analysis.

Option 1: Detailed 3D Solid Model

When investigating the stress state in and around a bond joint, a detailed model with solid elements is required. To get proper edge effects, at least four layers of elements are required through the bond thickness with smaller elements near the free edge. With this level of detail, the elements can accurately predict the Poisson stiffening. Use the original material properties, not the apparent properties. Elements with bubble functions provide better results than standard isoparametrics for

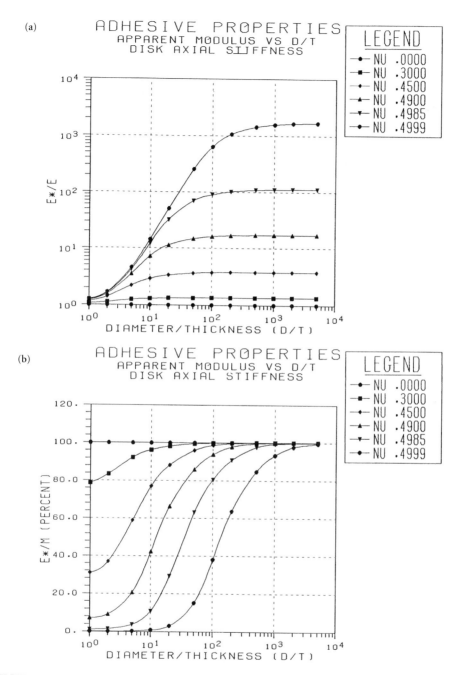

FIGURE 8.37 (a) Apparent modulus ratio (E*/E) vs. diameter/thickness; (b) apparent modulus ratio (E*/M) vs. diameter/thickness.

high Poisson values. If a Poisson ratio of 0.5 is desired, then special incompressible elements must be used, since regular elements will be singular at that value. With this high local detail, this option is usually reserved for a "break-out" submodel or a multilevel superelement model.

Option 2: Coarse 3D Solid Model

In this model, only one layer of elements, usually with a high aspect ratio, are modeled through the thickness. This type of model is used to get average or net effects over a bond, rather than a distribution of behavior. This model is too coarse to predict edge effects in the bond. A typical

(c)

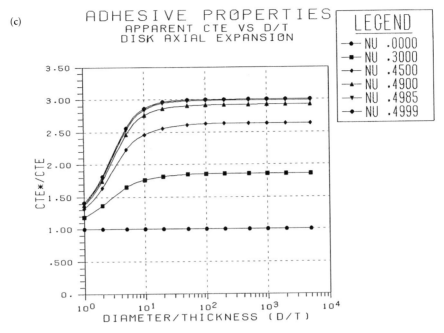

FIGURE 8.37 (c) apparent coefficient of thermal expansion ratio (CTE*/CTE) vs. diameter/thickness.

application is the prediction of bond shrinkage effects on optical surface distortions. Effective orthotropic bond properties are required to get accurate behavior.

In-plane modulus $= E$
Out-of-plane modulus $= E^*$
Shear modulus $= G$
In-plane coefficient $= \alpha$
Out-of-plane coefficient $= \alpha^*$

Option 3: Single Beam Normal to the Bond Plane

This approach would be used in a very coarse model trying to predict the first-order deflections or dynamic response. The bond surface geometry is used to predict the cross-sectional beam properties. Since bond layers are typically very thin compared to their surface area, the resulting beams are very short. For short beams, transverse shear effects can dominate the bending behavior. Figure 8.38 shows how the displacement due to bending and the displacement due to shear vary with beam length for a cantilever beam. The two contributions are equal when the length of the beam is 0.7 of the height of the beam. For bond layers, the beam length/height ratio is much less than 0.1, so the shear represents the full effect. The material properties of the beam should use E^* for axial effects, G for shear effects, and α^* for coefficient of thermal elasticity. Note that the isotropic relationship does not hold for the effective properties, so G is not equal to $E^*/2(1 + v)$. If the default shear factor is zero, a nonzero value must be input to the program. This model will predict the thermoelastic displacement of the optic normal to the bond, but will not predict the distortion of the optic due to in-plane bond effects.

Option 4: Equivalent Spring Model

The bond layer is represented as a set of springs which produce the equivalent stiffness of the bond layer. This scheme is used in a system level dynamics model which is trying to keep the number of nodes to a minimum, yet includes all of the soft elements in the system which contribute to the lower modes, especially those which involve the image motion. This model is very similar to the

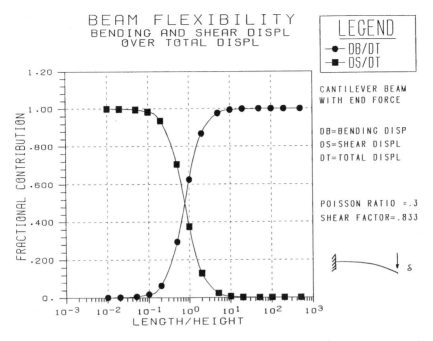

FIGURE 8.38 Relative bending and shear deflections for a cantilever beam with a square cross section.

beam model in Option 3 except that the springs offer no thermoelastic effects. The effective spring constants are

$$K_x = K_y = GA/t$$

$$K_z = E^* A/t$$

$$K_{\theta x} = K_{\theta y} = E \times I/t$$

$$K_{\theta z} = GJ/t$$

where A, I, and J are the cross-sectional properties of the bond, and t is the thickness of the bond. Note that whenever using the spring elements, the node points should be coincident in space and their displacement coordinate system must be aligned. Otherwise, hidden springs to the ground may be created which will cause errors in the results. Exceptions to this rule are possible, but should only be used by a trained analyst.

Bond Joint Failure Analysis

According to linear elasticity, the state of stress in the adhesive at the end of the joint bondline is infinite. Detailed stress models of the joint will chase infinity as the model is refined. To predict failure, fracture mechanics theory is required. As in Section 8.5, an initial crack is assumed. The stress intensity factor is calculated and compared to the fracture toughness of the adhesive. Buettner[2] compares the use of crack-tip elements to predict the stress intensity with conventional elements and strain energy release rate to predict K_I in a bond joint. Also in Buettner,[2] the multimesh extrapolation is shown to be a useful technique in predicting the error and improving the calculation of stress intensity. In Devries,[4] it is shown experimentally that high Poisson materials will fracture first at the edge for a small diameter-to-thickness ratio (D/t < 10), but will fracture at the center for high diameter-to-thickness ratio (D/t > 40). The reason fracture initiates at the center is that the stress intensity is higher at the center than the at the edge for the high Poisson material.

8.7 Mounts and Metering Structures

The analysis of optical systems requires careful modeling of the mounts and metering structures to get an accurate description of optical surface distortion and image motion. This section deals with modeling techniques important to precision optical structures, rather than more common stress-limited structures.

Determinate Structures

A structure is "statically determinate" if the force distribution can be determined solely by the equations of static equilibrium. An optical mount which is statically determinate is also called "kinematic", "exact", or "strain-free", because of the properties associated with these systems.

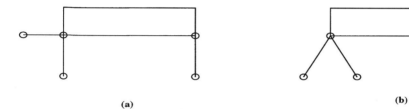

(a)   (b)

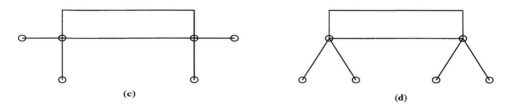

(c)   (d)

FIGURE 8.39 Mount configurations in a 2D plane using pinned jointed members: (a) and (b) are statically determinate; (c) and (d) are statically indeterminate.

Figure 8.39 (a) and (b) shows examples of two statically determinate, stable mounts in a 2D space. The elements have pinned ends and carry only axial force, with no moments or shear. In each case, the three unknown element forces can be determined from the three equations of equilibrium:

$$\Sigma F_x = 0$$

$$\Sigma F_y = 0$$

$$\Sigma M_z = 0$$

When additional, or "redundant", members are added to a determinate system (c and d), more unknowns are added; but no new equations are added, making the system "indeterminate". The system can be solved only by adding the equations of elasticity to the system. Not all 3-force systems are valid statically determinate, stable mounts. Figure 8.40 gives four examples of systems which are unstable mechanisms.

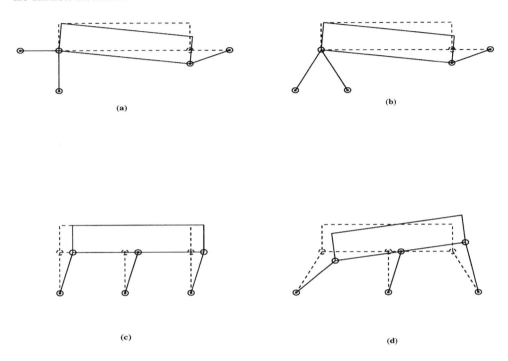

(a)

(b)

(c)

(d)

FIGURE 8.40 Unstable mount configurations in a 2D plane.

In 3D space, six rigid body motions are possible, requiring six constraints. There are also six equations of equilibrium available to determine those six element forces. Figure 8.41 shows three sets of constraints which are stable and statically determinate. Figure 8.41(a) is the common ball, slot, and flat mount. Figures 8.41(b) and (c) could be a 3-slot mount, or a 3-bipod mount. Figure 8.41(a) has no symmetry, 8.41(b) has a single plane of symmetry, and 8.41(c) has three planes of polar symmetry. As noted in Section 8.3, a significant improvement in solution efficiency is possible if symmetry is used.

The real significance of a determinate mount is not in the ease of solution of the mount forces, but in the uncoupling of the optic's internal behavior from its mount behavior. As noted below, this uncoupling effect is important to precision structures. No matter how, or for what reason the mount support moves, the optic moves only in a rigid body sense, with no distortion of the optic itself. Thus the error created is only a pointing error and not an image quality error.

As shown in Figure 8.42(a), the support motion for a determinate mount causes only a rigid body "strain free" motion of the optic, as opposed to the redundant mount in 8.42(b). This mount displacement could be due to mechanical or thermal loads, either static or dynamic. Initial imperfections, fabrication errors, and tolerance buildup also cause only the rigid body motion without a surface distortion. These pointing errors are more easily corrected in the system than the image quality errors. In a statically indeterminate design (Figure 8.42[d]), uniform temperature changes can cause strain and distortion due to a difference in the coefficients of thermal expansion. The determinate designs, on the other hand, allow for a strain-free thermal growth as shown in Figure 8.42(c). This is especially important in optical systems fabricated at room temperature, but used

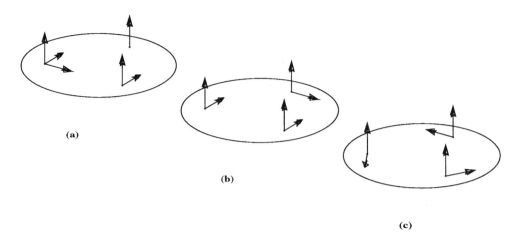

(a)

(b)

(c)

FIGURE 8.41 Statically determinate mount schemes in 3D space. (a) 3-2-1 which is common for a ball, groove, slider mount; (b) three grooves or bipods in a rectangular configuration; and (c) three grooves or bipods in a polar configuration.

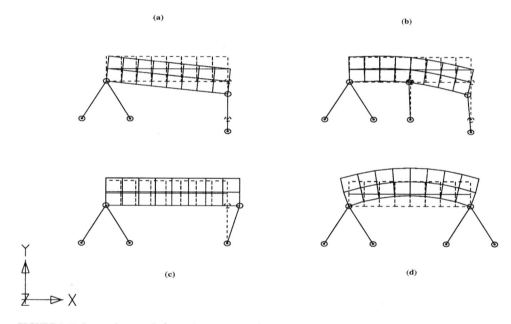

FIGURE 8.42 Determinate vs. indeterminate mount schemes. (a) Determinate mount with strain-free mount motion; (b) determinate mount with strain-free thermal growth; (c) indeterminate mount with distortion due to mount motion; and (d) indeterminate mount with distortion due to thermal growth.

at high or low temperatures. Note that the thermal gradients within the optic will still cause distortion, but it will be uncoupled from any distortion in the mount.

Real mount systems usually try to be as determinate as possible, but are not perfectly kinematic. In the analysis model, the mount can be made "exactly" determinate. This feature is useful in debugging and checking of the finite element models. In a kinematic system, the analyst should always check the mount forces or reactions for any load condition. If the load condition is a uniform temperature change, any nonzero reaction is a modeling error and causes unreal distortions in the optic for this and other load conditions. If the load condition is mechanical, the sum of the applied forces and moments on the optic should exactly equal the sum of reaction forces and moments at the optic mounts. These can be easily checked by using the six equilibrium equations.

From a reliability point of view, the determinate mount may have some drawbacks. Since there is no redundancy in the system, the failure of one element causes failure of the full system. In an optical system, a precision operation is usually more important than stress, so the statically determinate scheme is often used.

Models of Determinate Mounts

A common mount scheme for large mirrors is the use of six struts in a bipod pair arrangement (Figure 8.43). To make each strut an axial force-only member, the ends are ball-in-socket joints. This scheme is exactly represented as the truss members (ROD). For most applications, this is a good first-order model which can be used in design trade studies. In a real mount, the ball joint often has friction which causes some extra forces and moments to be introduced in the optic as a second-order effect. Up until slip, the friction could be modeled as rotational springs. If slip or slop is to be included, then a nonlinear analysis with gap elements is required. Also, once the moments are added, the strut must be modeled as a bending element (BEAM).

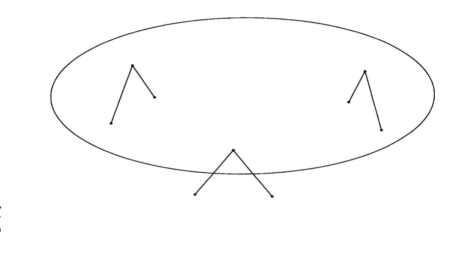

FIGURE 8.43 Three-bipod mount determinate scheme.

A variation on strut arrangement which eliminates the nonlinearities associated with slip and slop in the ball joint is a flexure mount. In this design, each ball is replaced by a necked-down section in the strut which transmits only a small, but highly predictable moment to the mirror. Again, to a first order, a truss element model is possible, but a beam model is required to include the small transmitted moments. This remains a linear analysis, static or dynamic, to very high load levels. Since the loads may be in compression, a buckling analysis is required to verify that the system will not buckle due to the necked down regions.

Many other kinematic mount schemes exist, including the blade flexures, finger flexures, pin-in-hole, ball-in-groove, and flat-to-flat point mounts. To a first order, each can be represented as a determinate mount, but may require the incorporation of second-order effects when a very high precision is required.

Zero G Test Supports

Since many optical systems are used in space, it is necessary to simulate a zero-gravity situation when testing an optic here on earth. These test systems try to support a mirror by applying a distributed load over the back surface. Since these are only approximations, it is necessary to analyze the effect these supports have on mirror distortion. The following are first-order models which predict the primary behavior of the optic. If higher accuracy is required, then more detail is required in the mount model.

An inflatable, fluid-filled rubber bag is used to support the mirror on its back surface. This can be represented as a uniform pressure over the loaded surface. A first FE analysis run is required to calculate the FE model weight and to calculate the net support force caused by the support. By comparing the net vertical force to the model weight, a scale factor on pressure is determined which will exactly balance the mirror weight. Since an FE model requires a nonsingular stiffness matrix, a set of kinematic BC are required to create a valid model. A check of the reaction forces in the second run will determine how accurately the scale factor was chosen. In some real applications, a small, nonzero 3-point support force is desired so that the exact mirror position is determined, rather than floating in an unknown vertical position or tilt. Again, this residual force can be obtained by a proper choice of the scale factor.

An alternative approach is to use a multipoint constraint (MPC) equation. From Figure 8.44, an equation can be written that states that the volume change ΔV in the bag is zero.

$$\Delta V = 0 = \Sigma A_j \delta_j$$

where δ_j is the normal displacement of node j and A_j is the nodal area associated with node j. In an FE model, this area is easily determined by applying a unit pressure to the loaded surface and obtaining the area from the calculated load vector. Since the net volume change is zero, this equation constrains the mirror from having a net vertical displacement. Note that this is a single constraint on the vertical motion which does not prevent a rotation about the horizontal axes. In most models, the rotation would be eliminated by symmetry of BC. This technique removes the requirement to balance the loads as in the reverse pressure method above.

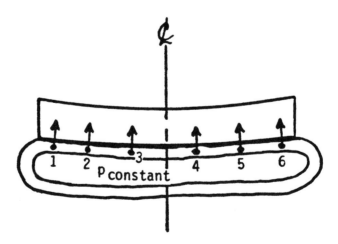

FIGURE 8.44 Optic supported on a zero-g simulation airbag.

Note that neither technique accounts for the in-plane elastic stretch of the bag or its frictional drag on the mirror surface. In actual tests, precautions are taken to minimize this effect. If the mirror surface is highly curved and a liquid-filled bag is used, then the change of hydraulic head from center to edge should be incorporated in the model.

A whiffle tree (Figure 8.45) may be used to simulate a zero-gravity state. In this scheme a system of pin-jointed levers is used to support the mirror in a uniform manner. This scheme is tuneable by varying the lever arm lengths. Note that this is exactly determined by a set of lever equations (MPC) or the MSC/NASTRAN RBE3 element (called a whiffle tree element). In a 2D representation, each node point has two equations (sum vertical forces = 0 and sum moments = 0) to determine the two forces at the lever ends. In 3D, an additional moment equation is used to find the third vertical force on the corners of the triangular lever arm. Note that this model ignores the frictional effects in the joints.

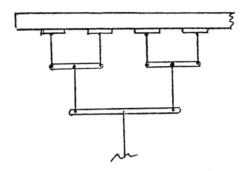

FIGURE 8.45 Optic supported on a zero-g simulation whiffle tree.

Metering Structures

The structure which holds all of the optical elements together to form a system is called a metering structure, because the relative position and motion of the optics are critical to the system level performance. With the wide variety of metering structures possible, only some simple modeling guidelines will be offered in this section.

The most important rule is that the model must contain all effects which cause the distortion of optics or alters their position. This means that any nonkinematic effects at the mounts which pass forces or moments to an optic should be included if the magnitude is significant. This may require the submodel of a mount to determine the magnitude of the unwanted forces and a subsequent optics analysis to determine the magnitude of the effect on surface quality. The effects of friction in ball joints or the moments passed through the flexures are examples.

In a metering structure, the line of action of a member force is important. If the neutral axes of members do not intersect, the resulting moments may cause significant rotations of members. These moments and rotations can cause pointing errors which may not be predicted by a model in which the neutral axes incorrectly intersect. An example is a force from one member which does not pass through the shear center of an attached c-channel. Another example is an offset lap joint which creates bending under axial tension, which might be modeled as an in-line butt joint which has no bending (Figure 8.46). Ring-stiffened cylindrical shells behave differently whether the ring is inside or outside of the shell. The model must include the offset between the ring and shell to get the proper behavior. To get all effects modeled correctly, the analyst must understand and use the concepts of neutral axis and shear center for the cross sections involved. The finite element program must allow for the proper independent offset of neutral axis and shear center. Rigid bodies are often required when the lines of action do not intersect at a point.

If the analysis is to include thermal loads, then the rigid bodies must be used with care since they have no thermoelastic growth. Figure 8.47 shows a ring and stringer-stiffened shell offset with rigid elements. If all the material has the same coefficient of thermal expansion, then a uniform temperature change causes stress-free uniform growth. In the example, the rigid offsets have no growth, so the radial growth is not uniform, causing the distortion shown in the contour plot (Figure 8.48). If the structure was a flat stiffened plate, rather than a shell, the thermal growth would be uniform. Thus, some modeling conditions are affected more than others. A good check of the rigid body effect is to convert all materials to the same coefficient of thermal expansion. Apply a uniform temperature change, then plot the stress in the structure. Any nonzero stress indicates a modeling problem. One fix is to replace the rigid elements with very stiff elastic elements

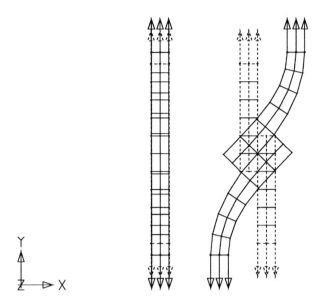

FIGURE 8.46 Comparison of bending caused line-of-action forces for in-line butt joints vs. offset lap joints.

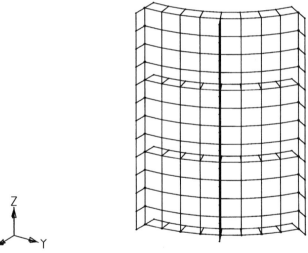

FIGURE 8.47 Curved shell with internal rings and external stringers modeled as beams with rigid offsets from the shell midsurface.

which have thermoelastic growth. Very stiff in this case means a stiffness of three to five orders of magnitude larger than the surrounding stiffness, but not so large as to create numerical problems. Whenever using stiff elements or rigid elements, the analyst should carefully check all the warning messages and error checks (EPSILON) put out by the analysis program.

When using gravity loads or dynamic loads, the center of gravity (cg) is important. A small error in the cg of a large optic can cause a significant moment error in a metering structure resulting in pointing errors. The correct mass moments of inertia are required for all large lumped masses to get a proper dynamic response.

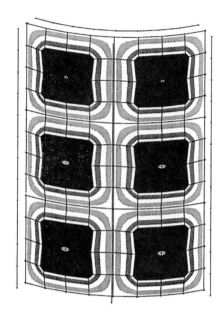

FIGURE 8.48 Radial displacement contours caused by a uniform temperature increase when using rigid offsets in a shell.

8.8 Optical Surface Evaluation

Typical finite element results include the deformed shapes of optical surfaces, usually in contour plot form. Although these data are useful, the distortion of the surface is often masked by a large rigid body motion which may be aligned out of the system. A useful postprocessing feature is to fit selected polynomials to the deformed surface to decompose it into meaningful components.

Polynomials

Let a function be defined as a summation of a polynomial series:

$$D(z, \theta) = \Sigma\Sigma \left[A_{nm} P_{nm}(z) \cos(m\theta) + B_{nm} P_{nm} \sin(m\theta) \right]$$

ue63/

where

D = displacement normal to surface = axial (radial)
z = radial (axial) position
θ = circumferential position
n = radial (axial) wave number
m = circumferential wave number
P = polynomial function of (n,m,z)
A = cosine coefficient
B = sine coefficient

Zernike polynomials represent the common aberrations over conventional optics, such as power, astigmatism, coma, trefoil, etc. (Figure 8.49). For Zernike polynomials (z = radial position):

$$P_{nm}(z) = \Sigma C_p(n,j) z^{n-2j}$$

$$C_p(n,j) = \left[(-1)^j (n-j)! \right] / \left\{ j! \left[(n+m)/2 - j \right]! \left[(n-m)/2 - j \right]! \right\}$$

where the series is summed for m < n and for alternate values of n.

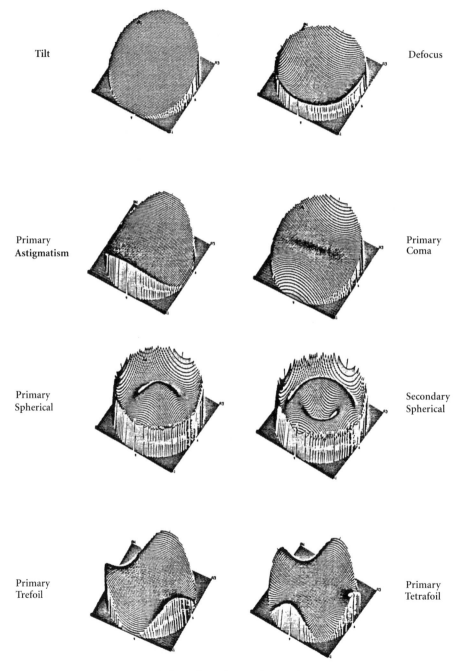

FIGURE 8.49 Typical Zernike polynomial terms for conventional optics.

Legendre-Fourier polynomials are similar to Zernike polynomials, but fit cylindrical optics (Figure 8.50). For Legendre-Fourier polynomials (z = axial position):

$$P_{nm}(z) = \Sigma C_p(n, j) z^{n-2j}$$

$$C_p(n, j) = \left[(-1)^j (2n - 2j)!\right] \Big/ \left[2^n j! (n - j)! (n - 2j)!\right]$$

where the series is summed for all values of n and m.

Since both series are orthogonal and complete, their representation is exact. If the series is truncated, then it becomes an approximation in general. The series terms may be represented as the coefficients of cosine and sine (A,B) as above, or as magnitude and phase (M,Φ) where:

$$M_{nm} = SQRT\left[A_{nm}^2 + B_{nm}^2\right]$$

$$\Phi_{nm} = (1/m) ATAN2\left[A_{nm}/B_{nm}\right]$$

ATAN2 is the standard FORTRAN arc tangent function with two arguments.

Surface Fitting

Given a deformed shape of an optical surface from a FE solution, the error between the FE solution and a polynomial approximation is defined by Genberg[5] as:

$$E = \Sigma W_i (\delta_i - D_i)^2$$

where

$$
\begin{array}{rl}
i & = \text{node number} \\
\delta_i & = \text{FE displacement of } ith \text{ node} \\
D_i & = \text{polynomial displacement of the } ith \text{ node} \\
W_i & = \text{area weighting of the } ith \text{ node}
\end{array}
$$

In a typical FE model the mesh varies throughout the model so each node point does not represent the same amount of surface area. W_i is the area weighting factor which can be determined from the load vector calculated from a unit pressure over the surface. If the series for D is written symbolically as:

$$D_i = \Sigma c_j f_{ji}$$

then

$$E = \Sigma W_i \left(\delta_i - \Sigma c_j f_{ji}\right)^2$$

To find the best-fit polynomials, minimize the error E with respect to the coefficients c_j.

$$dE/dc_j = 2\Sigma W_i \left(\delta_i - \Sigma c_j f_{ji}\right) f_{ji} = 0$$

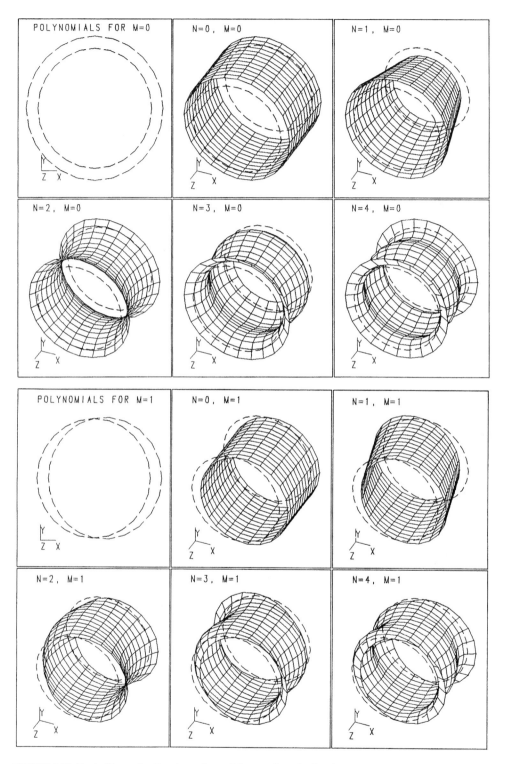

FIGURE 8.50 Typical Legendre-Fourier polynomial terms for cylindrical optics.

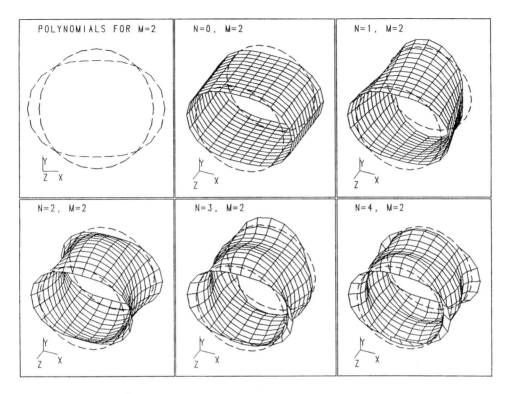

FIGURE 8.50 (continued)

Collecting the terms and writing in a matrix notation:

$$[H]\{c\} = \{p\}$$

where

$$H_{jk} = 2\Sigma W_i f_{ji} f_{ki}$$

$$p_j = 2\Sigma W_i \delta_i f_{ji}$$

This is a linear system with a square, symmetric coefficient matrix solvable by Gauss elimination. The best-fit coefficients (c) can be represented as the original series coefficients (A,B) or as the magnitude and phase (M,Φ) for each polynomial term. These polynomials are orthogonal over a full circular geometry, so their coefficients are constant regardless of the number of terms used in the series. If the geometry is irregular or obstructions exist, the coefficients may vary with the number of terms used in the series. It is useful to calculate the error term, both root-mean-square (RMS) error and peak-to-valley (P-V) error, after each term in the series is subtracted from the original deformed surface. This error indicates the amount of surface distortion not accounted for by the previous polynomials.

Interpretation

Both polynomial series considered above include the rigid body motion. In the Zernike series, for example, the first two terms are rigid body and the third is power:

Bias: n = 0, m = 0
Tilt: n = 1, m = 1
Power: n = 2, m = 0

Any optical system which has pointing and focusing capability can adjust out the bias, tilt, and power terms. Thus these terms do not affect image quality in such a system. All of the higher order terms represent aberrations in the surface which do affect the image quality. The measure of surface quality is the error calculated after the first three terms have been removed, not the error in the original FE deformed shape. Figure 8.51 shows a mirror on a 3-point, delta frame mount system oriented at 45° with gravity. Contours on the original deformed shape are dominated by bias, tilt, and power. The contours after these three terms are removed show the actual surface quality which affects imaging. Figure 8.51(f) gives the magnitude and phase angle of each of the terms, as well as the RMS and P-V error after the term is removed from the surface. In this example, the original P-V error is 90 waves, but after the bias, tilt, and power are removed, the P-V error is 8 waves. Obviously, the original FE surface displacements are not an accurate measure of surface quality, until postprocessed by a surface-fitting program.

The decomposition of a surface into a series is useful for understanding the important factors in the optics behavior. In the above example, all of the trefoil effect can be attributed to the 3-point mount and all of the coma is due to the in-plane gravity. If the mirror had a lightweight square core, then its effect would show up as tetrafoil. In many cases, this can be useful in improving the performance of a system.

To be a useful surface-fitting program the following features should be incorporated:

- Submodels using symmetric BC
- Multiple, nested coordinate systems defining the node location and output
- User coordinate system to define the polynomial centers and orientation
- Apertures and obstructions to limit the amount of surface fitted
- Units conversion to arbitrary output (i.e., wavelengths)
- Linear scaling and combining of loadcases (displacement vectors)
- Residual surface output format suitable for plotting

8.9 Modeling Tricks for Optical Structures

In this section, three modeling tools useful in optical structures are discussed.

Image Motion Calculation

For the small displacements and rotations common in optical structures, the image motion calculation is a linear equation. This is true for rather complex assemblies of flat mirrors, curved mirrors, and lenses. For the simple system shown in Figure 8.52, the image motion shown (δ_i) is affected by the rotations (Φ) of the two mirrors (a,b) and their respective path lengths (L):

$$\delta_i = \left(L_2 + L_3\right)\Phi_a - 2\left(L_3\right)\Phi_b$$

For more complex systems, all of the mirrors' motions (translations and rotations) enter into the image motion equation. For powered elements, the effective lengths are modified. An optical analysis program may be needed to determine the appropriate coefficients of each optic's unit motions.

Absolute motion of the image is not important to the system performance, but the relative motion of the image to its receptor is. In a copier, the relative motion is measured as smear. Given the receptor's motion (δ_r), the smear (δ_s) is

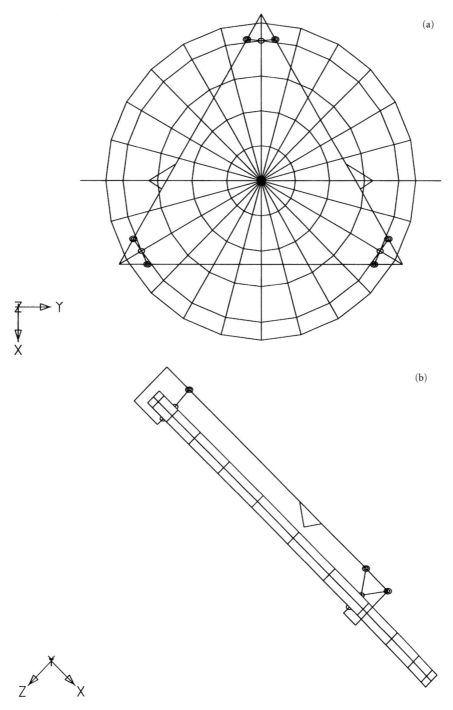

FIGURE 8.51 (a) Flat turning mirror on a delta frame mount — front view; (b) flat turning mirror on a delta frame mount — side view.

$$\delta_s = \delta_i - \delta_r$$

If the optical system is modeled, then the FE program will determine the mirror and receptor motions. If the program has the capability to include the user written equations (MPC), then δ_i or δ_s can be directly calculated and output by the FE program. For the simple system shown, this

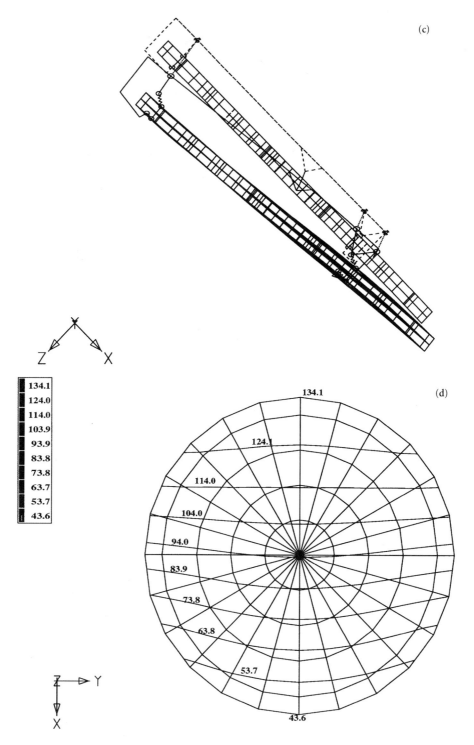

FIGURE 8.51 (c) flat turning mirror on a delta frame mount — deformed side view; (d) flat turning mount — contours of normal displacements.

may not represent a big savings in data postprocessing outside of the FE code. However, for systems with many elements, and considering all six degrees of freedom of each optic, this can represent

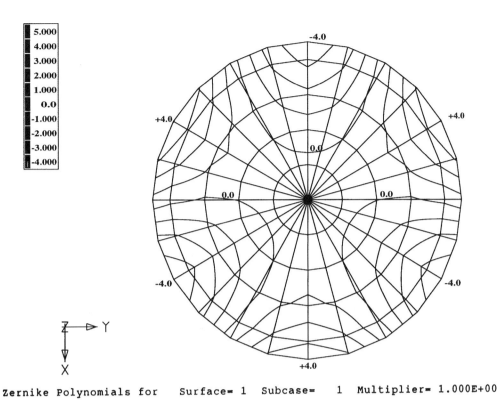

Zernike Polynomials for   Surface= 1 Subcase=   1 Multiplier= 1.000E+00

| Order | | | Surface | Functional | Phase | Residual | Residual |
k	n	m	Aberration	(Units)	(Deg)	RMS	P-V
			Input Surf			24.8842	90.4800
1	0	0	Bias	91.58552	0.0	24.8842	90.4800
2	1	1	Tilt	98.43091	179.9	2.2940	9.7651
3	2	0	Power	-5.43184	0.0	1.5996	8.0006
4	2	2	Pri Astigmatism	0.07498	89.8	1.5995	8.0074
5	3	1	Pri Coma	0.02916	-180.0	1.5995	8.0147
6	3	3	Pri Trefoil	8.68164	0.0	0.2345	1.0534
7	4	0	Pri Spherical	0.53040	0.0	0.1599	0.9565
8	4	2	Sec Astigmatism	0.01820	0.0	0.1599	0.9564
9	4	4	Pri Tetrafoil	0.02178	0.0	0.1599	0.9445
10	5	1	Sec Coma	0.00148	-0.1	0.1599	0.9451
11	5	3	Sec Trefoil	0.70869	60.0	0.1102	0.5993
12	5	5	Pri Pentafoil	0.00523	-0.1	0.1102	0.6029
13	6	0	Sec Spherical	-0.05257	0.0	0.1096	0.6029
14	6	2	Ter Astigmatism	0.00095	-89.8	0.1096	0.6033
15	6	4	Sec Tetrafoil	0.00390	0.0	0.1096	0.6033
16	6	6	Pri Hexafoil	0.74367	30.0	0.0331	0.1716
17	7	1	Ter Coma	0.00113	0.0	0.0331	0.1717
18	7	3	Ter Trefoil	0.06975	60.0	0.0311	0.1999
19	7	5	Sec Pentafoil	0.00493	36.0	0.0311	0.1968
20	8	0	Ter Spherical	0.00171	0.0	0.0311	0.1968
21	8	2	Qua Astigmatism	0.00117	-89.9	0.0311	0.1971
22	8	4	Ter Tetrafoil	0.00232	44.9	0.0311	0.1960
23	8	6	Sec Hexafoil	0.14599	0.0	0.0228	0.0908

FIGURE 8.51 (e) contours after best-fit plane and power removed; and (f) surface fit results with the magnitude and orientation of the Zernike polynomials and the resulting surface RMS and peak-valley after the term is removed.

a very useful capability. The amount of data processed increases significantly in the dynamic analysis where the smear is calculated at each time step. Nowak[14] shows a laser printer system with 13 independent optical elements analyzed for the frequency response as well as transient response.

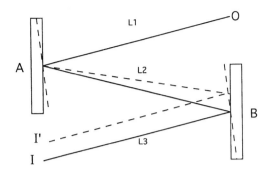

FIGURE 8.52 Image motion due to mirror rotations.

Poor Man's Spot Diagrams

Wolverton[23] describes a procedure to provide a simulated spot diagram directly from the FE output, without any additional postprocessing. The procedure involves the following steps:

- Define a spot center at twice the focal length
- Create a node at the spot center for each node on the optical surface
- Add rigid beams from the optical surface to the spot center
- Add a circle of grounded beams to represent the blur circle
- Plot the deformed nodes at the spot center to get a spot diagram

Note that this spot diagram is a standard FE output without further postprocessing. The individual rigid elements (or ray bars) are not connected to each other. They represent massless cantilevers off the optical surface with no effect on the surface response. The spot diagram is located at twice the focal length (2L) since an incoming ray sees twice the node rotation (Φ) (angle of incidence + angle of reflection):

$$d_i = L\left(2\Phi\right) = \left(2L\right)\Phi$$

Optical Pathlength Calculation

When a planar optical wave passes through a planar window of a constant index (n), it stays planar as it exits. If the window has a variation in optical index, then the pathlength may be different for different rays, causing the exiting wave to be nonplanar. The index may vary with temperature and stress level.

The variation from planarity is called the optical pathlength difference (OPD). To find the OPD, two finite element runs are necessary. First, a heat transfer analysis is conducted using the techniques discussed in the next chapter. The result is a new temperature (T) at every node. In the second analysis, the OPD is output directly using the following analogy to thermoelastic expansion:

$$OPD = L\left[n + \left(dn/dT\right)\Delta T\right] - L\left(n\right)\Delta T = L\left(dn/dT\right)\Delta T$$

which has the same form as the thermal expansion:

$$\delta = L\,\alpha\Delta T$$

In the second model:

- Set the material properties to E = 1, G = 0, ν = 0.
- Set the coefficient of thermal expansion (CTE) to dn/dT.
- Constrain all in-plane displacements to zero.
- Constrain the first (incident) surface normal displacements to zero.
- Apply the new temperature vector as a thermal load.
- The normal displacement of the second (exit) surface is the OPD.

This second surface displacement vector may be postprocessed for Zernike coefficients if desired. An example of a window in a vacuum chamber with an incident laser beam is shown in Figure 8.53. The resulting OPD map correlated closely to the experimental results.[6]

The above calculated OPD is one contributor to the total OPD. Other effects such as the true thermoelastic expansion may be added, but are typically much smaller than the index change effect. This calculation is exact within FE theory for the planar windows, but may be a useful approximation for lenses, also. The suggestion is that the element boundaries should be parallel to the path of the rays to improve the quality of the approximation.

Plastic optics may absorb moisture and change in index. The moisture absorption can be analyzed by analogy to heat transfer with the appropriate property changes. The OPD due to index change can be calculated using the method above where the temperature is replaced with moisture concentration. A similar procedure can be used to predict the OPD due to stress effects (birefringence).

8.10 Ray Tracing

A finite element program presents its analysis results at the node points. A ray tracing program which bounces random rays off of the deformed optical surfaces requires an accurate displacement and slope information at ray–surface intersection points which are, in general, not at the node points as shown in Figure 8.54. High accuracy is required on the slope data because it usually has a large effect on the final ray position at the image plane. Other postprocessing programs for optical evaluation may require the deformed surface data on a regularly spaced square pattern which typically do not line up with the finite element node points (Figure 8.55). This section presents a highly accurate technique for obtaining deformed surface data at points on a finite element model which are not at node locations. The technique, which is more accurate than spline fitting, extracts the maximum information from a finite element model because it uses the same theory to post-process the data as was used in the finite element program to solve the problem. The technique applies to a variety of mirror surface geometries including flat, spherical, or cylindrical. Additional details are available in Genberg.[7]

In a ray trace program evaluating a deformed surface, the plate or shell bending behavior is required for accurate slope information at intermediate ray–surface intersection points. The technique described applies to the low order plate and shell elements (QUAD4 and TRIA3). This restriction is not usually a major limit on the model. If the optic is modeled of solid elements, then a "thin" coating of plate elements is applied to the reflective surface. Since most solid elements do not have rotational stiffness, this coating of thin plates provides the necessary node point rotation data required for the cubic interpolation.

Nonrectangular quadrilaterals are converted to two triangles. The rectangles and triangles are used because their Jacobian transformation matrix [J] is a constant throughout the element and not a function of spatial location as in a general quadrilateral. The transformations can be calculated once and then stored for the search routine. The element transformation matrices from spatial coordinates $\{x\}$ to parametric coordinates $\{\xi\}$ are calculated for the resulting rectangles and triangles. For element m,

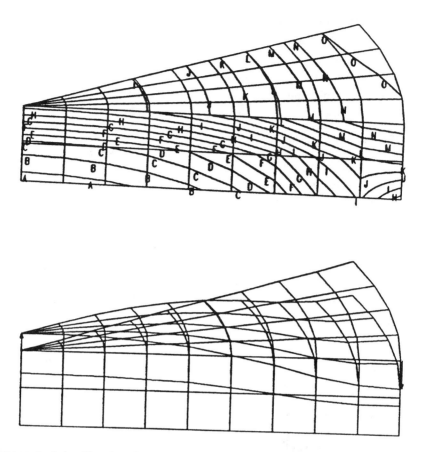

FIGURE 8.53 Optical pathlength analysis due to temperature-dependent index of refraction due to laser beam heating in a test chamber window. (a) Temperature contours due to laser beam heating; (b) wavefront change due to index change.

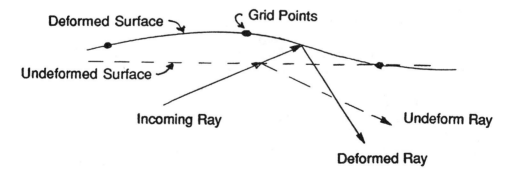

FIGURE 8.54 Ray tracing from deformed analysis results.

$$\left[J_m \right] = \left[dx/d\xi \, \ldots \right]$$

The origin of element m in parametric coordinates $\{\xi_0\}$ can be found from the geometric center $\{x_0\}$

$$\left\{ \xi_0 \right\} = \left[J_m \right]^{-1} \left\{ x_0 \right\}$$

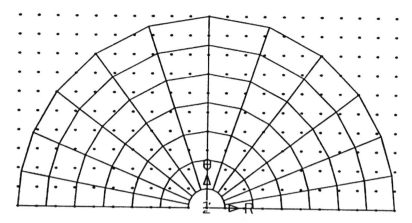

FIGURE 8.55 Uniform grid (x-y) pattern of results interpolated from a polar FE model of an optic for optical postprocessing.

For every ray point $\{x_p\}$, a search over all elements is conducted. The point is converted to element m's parametric coordinates $\{\xi_p\}$

$$\{\xi_p\} = [J_m]^{-1}\{x_p\} + \{\xi_0\}$$

then tested to see if contained within element m.

$$-1 < \xi_p < +1$$

When the proper element is found containing the point, the interpolation takes place using the element's shape functions (N) and the element's nodal results, such as displacements (δ). In-plane motion is found from membrane behavior

$$u(x_p) = \Sigma N_j(\xi_p) \times \delta_j$$

The out-of-plane displacement $w(x_p)$ and slopes w_x and w_y are found from plate bending behavior as given in Yang.[24] The equations have the general form of

$$w(x_p) = \Sigma\left[f_j W_j + g_j W_{xj} + h_j W_{yj}\right]$$

where W_j, W_{xj}, and W_{yj} are the nodal displacements and rotations. The cubic shape functions (f, g, and h) depend on the plate formulation chosen. It is suggested that fully compatible elements be used in the interpolation so that the surface slope is continuous in both directions at all element boundaries. This continuity condition provides for smoother behavior for rays that bounce from nearby points across the element boundaries.

The two-mirror system shown in Figure 8.56 was used to test the ray-trace algorithm which used interpolation over the shell surface. The incoming rays are collected at the focal plane by very low angle-of-incidence rays grazing off mirrors which are nearly cylindrical. The interpolation was verified by forcing a known functional displacement over both mirror surfaces. The interpolated rays were then compared to the rays from the perfect functional surface, again with an excellent agreement. Additional details are presented in Genberg.[7]

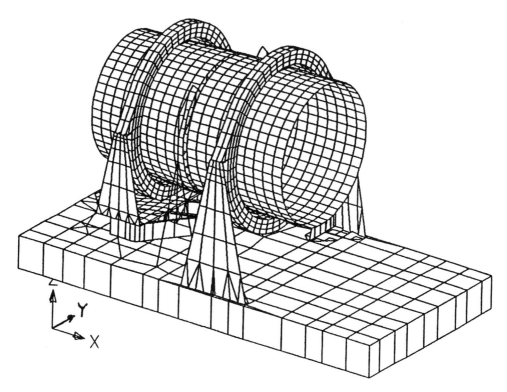

FIGURE 8.56 Cylindrical optics which were ray-traced from deformed FE results.

8.11 Model Checkout

Included below is a checklist for the model creation and checkout. It is suggested that a novice user refer to this list before and during the analysis process.

1. Is FE necessary for this problem?

    • Can a closed form solution be found?
    • Use both if possible, since each verifies the other.

2. Why is the analysis required?

    • What are the analysis goals?
    • Conceptual design vs. detail design verification?
    • Statics vs. dynamics, deflection vs. stress?
    • Accuracy required vs. time and resources available.

3. Check FE program documentation.

    • Does the FE program have the required capabilities?
    • Read about the solution method and element types.
    • Check the current error list for program bugs.

4. Idealize the problem.

    • What is the important behavior — beam vs. shell vs. solid?
    • Consider constraints/loads/element types/material.
    • Consider symmetry — structure, BC, loads.

5. *Always* run a *prototype* model!!

- Small model with important features of the true problem.
- Prototype problem should have a theoretical solution.
- Compare with theory to determine the accuracy vs. mesh density.
- Test program input/output/restarts/plots/alters — all features that may be new to the analyst.

6. Model true problem.

- Keep the model simple; don't overkill with too much detail.
- Add detail later as required.
- A small model is easy to debug and understand.
- A big model has more errors which are hard to find.

7. Know your model generator (preprocessor).

- Are circles really circles or are they parametric cubics?
- How symmetric are the generated models?
- What is the accuracy of generated node points?
- Know about equivalencing/element normals/coordinate systems.

8. Run graphical checks on model.

- Plot the model using hidden lines and shrunken elements.
- Check free boundaries for unwanted cuts in the model.
- Check element normals for reversal for pressure loads/stress.
- Check element geometry for warp/skew/aspect ratio.
- Display of loads/BC/constraints.

9. Use checkout runs to validate the model.

- 1-g static loads in all directions.
  — Check max displacements/reactions.
  — Check mass properties of the model, compare to known values.
  — Look for symmetry of the response where appropriate.
  — Perform sanity checks, compare to hand solutions.
  — Is the response realistic and sensible?
  — Check epsilons to see if small.
  — Compare the sum of loads to the sum of reactions.
  — Plot deformed shape/stress plots for discontinuities and peaks.
- Uniform thermal soak with all materials having a uniform CTE.
  — Check stress caused by the offsets, rigid bodies.
  — Deformation should be a stress-free growth.
- Rigid body error checks.
  — Remove constraints to the ground.
  — Give one node a unit translation/rotation, to see if stress free.
- Natural frequency analysis.
  — Check for near mechanisms (Freq = 0).
  — Check for reasonableness.
- Compare to any test data of similar structures.

10. Run production analyses.

- Run statics before dynamics.
- Run linear before nonlinear.
- Make all sanity checks/comparisons as above.

11. Understand your postprocessor.

- Stress averaging vs. extrapolation and fitting?
- Over what set is averaging done?
- Does it use nodal values/centroid values?
- Does it know which stress component is which?
- Does it know the element coordinate systems?
- Does it label the output correctly?
- Can it interpret the displacements in the local systems?
- How does it treat the midside nodes?

12. Interpret the answers.

- Look at the analysis results file *before* creating the plots.
  — Look for the warning/error messages.
  — Check epsilon/maximum displacement/sum loads/sum reactions.
- Look at the stress gradients and strain energy density.
  — Are model refinement and reanalysis required?
- Are results linear or is a nonlinear analysis required?
  — Are displacements large?
  — Is stress above the yield?
  — Is buckling possible due to high compression?
- Is redesign required based on the analysis results?
  — Use design sensitivity and optimization.

13. Document the model assumptions and analysis results.

- Keep a notebook — sketches, calculations, section properties.
- Keep listing of the input data file with lots of comments.
- Make many plots of the model and results with labels.
- Keep the input file or data base for important analyses.
- Document the labor, cpu time, and calendar time for future estimates.
- Report the assumptions, model description, results, and conclusions.

The most common sources of errors in FE models are listed below with some recommended checks to locate these errors:

- Bad geometry — find by plots and the mass properties.
- Bad elements — use shrink plots, free boundary plots, normal checks.
- Bad beam orientation — check v vector, section properties, stress points.
- Bad MPC/rigid bodies/offsets — compare sum of the loads to reactions, run rigid body error check, and thermal soak with a uniform CTE.
- Bad BC — same as above, also check for nonsymmetry in the results.
- Bad properties — check for the wrong units, wrong exponents, mixed units.

8.12 Optimum Design

High performance mirrors such as those used in the orbiting telescopes or large, ground-based observatories require a light-to-moderate weight, low stress, and small deflections under static and dynamic loads. The design approach in the past has been through parametric studies to achieve the "best" design within the trade space studied. In this section, the automated optimum design techniques based on nonlinear programming will be discussed as applied to optical structures, in general, and the lightweight mirrors, in particular.

Nonlinear programming techniques were first applied to structural design by Schmit.[15] Early work was limited to the problems where the designer could write the analysis equations as a subroutine and embed them in a general purpose optimization program such as DOT.[20] This limitation prevented the technique from becoming a popular design tool for complex structures. When the theory became available for design sensitivity[19] of general purpose structures through finite elements, the optimization gained favor quickly.

Design Problem Statement

Any design problem can be stated as a general nonlinear programming problem.

Minimize $F(X)$
Subject to: $g_j(X) \leq 0$
and $XL_i < X_i < XU_i$
where $F =$ objective function
            $g =$ inequality constraints on behavior
            $X =$ vector of design variables
            $XL, XU =$ lower and upper bounds on variables

If equality constraints are present, they may be treated as two inequality constraints.

$$h_j \Rightarrow 0 \geq g_j \leq 0 \text{ and } g_{j+1} \geq 0$$

Note that the functions F and g are nonlinear functions of X. In a finite element code, the constraints on displacement and stress are found numerically (not analytically). A constraint on displacement written as:

$$\delta \leq \delta_U$$

where δ_U is an upper limit on displacement, can be converted to the general form as:

$$g = (\delta - \delta_U)/\delta_U \leq 0$$

Design Sensitivity

Nonlinear programming methods are iterative in nature, moving from one design to a better design. An efficient optimization code requires the first derivatives of the responses to determine a proper move direction in the design space. Finite difference operations are too time consuming for most applications. The efficient alternative is the use of implicit derivatives for the design sensitivity of constraints with respect to the variables.[19] In a static analysis, the system equation

$$[K]\{\delta\} = \{F\}$$

is varied by the implicit derivatives

$$[K]\{d\delta/dX\} + [dK/dX]\{\delta\} = \{dF/dX\}$$

To find the response derivative, an additional "load case" is applied to the system equation, where the right-hand load terms are easily calculated.

$$[K]\{d\delta/dX\} = [dF/dX] - [dK/dX]\{\delta\}$$

Note that the additional load vectors are just another column of multiplication, vs. the alternative of new decompositions of the stiffness matrix as required by a finite difference approach to the response derivatives. Finite element programs which provide these sensitivities internal to the code are efficient in a general design optimization program.

Design Variables

Almost all FEA programs which offer the design optimization offer the *sizing* variables which include beam cross-sectional properties and plate thicknesses. These variables effect the "property" cards (pbar or pshell), but not the node locations. A more general capability would include the *shape* variables which change the node point locations. In a continuum structure such as a mirror, the individual node points should not normally be independent variables, but rather overall shape parameters are the variables. Shape optimization can be approached with a variety of techniques, but two methods are prevalent:

1. Basis vector technique
2. Automesh technique

In the basis vector method, a valid mesh of the nominal structure is created. This mesh is perturbed in various directions which represent the candidate designs. The node and element numbering are unchanged in each candidate vector. The optimizer then finds the scale factors for the linear combination of all candidates which yields the "best" design. This is highly efficient, but is limited in the amount of variation possible before a remesh is required.

The automesh technique allows a greater amount of variation in the design because an automatic remesh is redone at every design step. However, an automatic mesh requires a good error evaluation technique which tests the accuracy of the automesh and modifies the mesh for sufficient accuracy. This extra iteration loop, combined with the automeshing algorithm, can be quite time consuming when buried inside a shape optimization loop. Another bothersome feature of automeshing is that symmetric response is not maintained for the symmetric structures such as optics. Any level of asymmetric response for the symmetric checkout loads usually signals a modeling error.

Design Constraints

Optical systems must survive and operate in a variety of environments. For example, during transportation and handling, the stresses must be less than the allowable stress, during launch the natural frequency must be greater than a minimum value, and during operation the surface deformations must be less than an allowable value. A design approach which optimizes for the static stress by providing a soft mount will often violate the dynamic response with low natural frequencies. To obtain a truly optimum mirror, both the static and dynamic constraints must be considered simultaneously. If the finite element code is to be useful, it must have the combined analysis capability. In fact, a very desirable feature is to include the frequency response, transient responses, and buckling as simultaneous analysis and constraint options along with the static and natural frequency constraints.

Since the optical surface performance is often difficult to relate to the raw finite element displacements, some user function capability is required. For a mirror which has a large tilt, but whose surface remains perfectly smooth, the results will show large finite element displacements. However, if the optical system has a pointing capability, the smooth surface will perform satisfactorily (see Section 8.8). What is needed is the ability to find relative motions by writing the responses as equations, or by letting the user include the subroutines, such as surface fitting subroutines, to calculate the response functions. This would allow the constraints to be placed on the RMS error after the rigid body motion and power have been removed.[17]

Algorithms

Many iterative algorithms have been created for the solution of general nonlinear programming problems. In the DOT optimizer,[20] the method of modified feasible directions and the method of sequential linear programming are chosen for their efficiency and robustness. The key issue when combined with a finite element program is the efficiency, especially as related to the number of full FE analyses required per design optimization. In order to reduce the number of full FE analyses, the best procedure is to create an approximate problem which is a first-order Taylor series expansion of the design responses:

$$R(X) = R(X_0) + dR/dX(X - X_0)$$

where R is any response quantity, X_0 is the current design, and dR/dX is found from the design sensitivity. This approximate problem is optimized to get a new design. A full FE analysis is then run on the new design, along with design sensitivity, to create a second approximate problem. At each cycle, the constraints are checked, sorted, and the inactive ones temporarily dropped. Using this approximation technique, the typical designs require five to ten full FE analyses to reach an optimum design.[13]

Lightweight Mirror Design Issues

The previous fabrication and assembly techniques limited the lightweight mirrors to regular, uniform spacing, with a uniform wall thickness. Therefore, the mirror cores were restricted to square, triangular, or hexagonal cells of constant size (B) and a constant wall thickness (t_c). Recent advances in waterjet cutting have allowed a very general core structure to be a possibility.[22] Now the core can be created with an irregular geometry (spacing, shape, and thickness) over the whole mirror which provides an extensive new design freedom.

In the past, the mirrors were polished to a high figure by polishing laps rubbing on the surface. The pressure forced the center of the cell to deflect relative to the cell edge, causing a nonuniform pressure with associated nonuniform material removal. The core print-through effect on the finished surface was labeled quilting. The cell spacing (B) was determined by the polishing quilting displacement (q) which is a function of the cell geometry and faceplate thickness.

$$q = \text{function}\left(B^4/t_p^3\right)$$

New procedures using ion figuring[22] can place a finished surface of very high quality on a mirror without the use of surface pressure. This allows a greater freedom in the core geometry with larger cell sizes.

In a solid mirror, the only structural design variable is the thickness. Conventional lightweight mirrors can be described structurally by a few parameters as defined in Section 8.4:

> H = overall height
> t_p = faceplate thickness
> t_c = cell wall thickness
> B = effective cell spacing

In most applications, the mirror diameter and curvature are specified by the optical requirements. Since the usual goal is the lightest weight mirror which satisfies all the performance criteria, the design problem could be stated:

> Find the design = X = (t_p,t_c,B,H) which will minimize W = weight

subject to:

q < quilting limit (polishing)
σ < stress limit (handling, transportation, launch)
δ_{pv} < peak-to-valley displacement limit (test, use)
δ_{rms} < rms displacement limit (test, use)
f_n > natural frequency limit (transportation, launch)

Two general approaches to the design optimization of lightweight mirrors are possible with today's capabilities in finite element analysis. Depending on the mirror complexity and the program's capability, either the sizing or shape design may be used.

Sizing Optimization

Sizing optimization is limited to changes in the effective plate thickness (PSHELL entries). Thus any 3D model can use as many independent design variables as desired to change the core thickness or faceplate thickness. The mirror height (H) and cell size (B) cannot be changed since that involves changes in the node position.

If a mirror is regular enough such that a 2D equivalent stiffness plate model can be used accurately, the equations in Section 8.4 show that the mirror height (H) and cell spacing (B) can be treated as sizing variables.

For highly irregular geometry of the core, a 2D equivalent stiffness "plate" model is very difficult to create and has questionable accuracy. For these irregular mirrors a 3D shell model is required for the design/analysis.

Shape Optimization

A more general and more accurate capability for design optimization is the combination of sizing and shape optimization. With this capability, a full 3D shell model of the mirror is used. The design variables could include the faceplate and individual core wall thicknesses as sizing variables, with cell strut intersections, mount locations, and overall height as the shape variables.

Genberg and Cormany[8] present a comparison of a conventional mirror design with a shape-optimized lightweight mirror. The elliptic mirror (27 in. × 14 in.) shown was to be mounted at 3 points on the back surface with the gravity acting normal to the face. Due to the space requirements, the mirror thickness was limited to 2 in. The design problem can be summarized as

> Objective function:
>   Min wt = minimize total weight on mirror
>
> Design constraints:
> δ_{pv} < 4 m-in. = max P-V under 1 g

As a reference, four design solutions are presented.

1. A regular square core mirror with a hexagonal outline
2. A solid elliptic mirror
3. An unconventional lightweight mirror resulting from parametric studies
4. An unconventional mirror using the optimization techniques

The square core mirror was presented as a design option by an unknown source, so the amount of design effort is unknown. The solid mirror represented the cheapest solution from both a fabrication and a design effort measure. The "parametric" mirror was the "best" design available after a large amount of the design effort from experienced engineers supported by several finite element analyses. The optimized mirror was the result of two trial runs with the new GENESIS program[21] combining the sizing and shape optimization. A plot of each of the finite element models appear in Figure 8.57. Symmetric half-models were used for the efficiency. Comparing the unconventional designs, the optimizer moved the mount locations (shape variables) and changed many of the core strut thicknesses (size variables). It was the combination of these changes that was successful and could not be found from the parametric studies. A summary of the resulting designs was

Design	Displacement (m-in.)	Weight (lb)
Square core	8.8	31.6
Solid	8.0	53.6
Parametric	6.0	38.1
Optimized	3.8	30.1

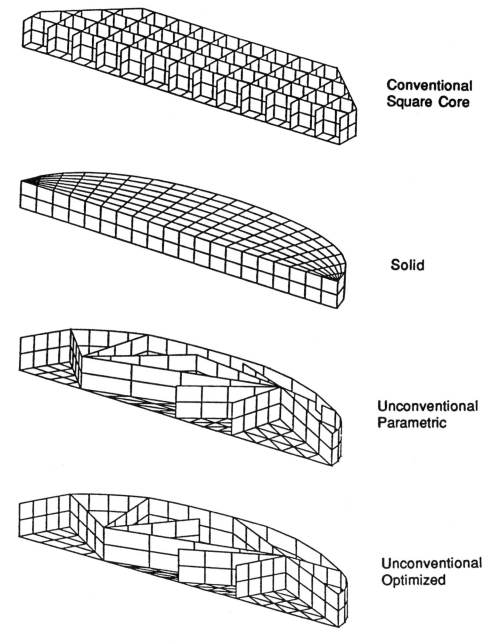

Conventional Square Core

Solid

Unconventional Parametric

Unconventional Optimized

FIGURE 8.57 Elliptic turning mirror design.

The significant result is that the optimized design was the only design to meet the design requirement, but it did so with the lightest weight. From a design cost viewpoint, the parametric design required about three times as much labor and cpu time as the optimized design. The conclusion to be drawn is that the optimization techniques can produce better designs with less effort.

Another conclusion that can be drawn from the above study is that the new design freedom available from new fabrication (waterjet cutting) and processing (ion figuring) techniques has provided more design variables than an experienced design engineer can handle. Only the automated optimization techniques can utilize the many new variables successfully.

Optimization Summary

Since these lightweight mirrors must survive a variety of handling, transportation, launch, and in-use load conditions, all effects must be considered in the design process. A general design capability embedded in a finite element program must include the following tools as a minimum:

1. Sizing and shape variables
2. Static analysis with multiple load and boundary conditions and constraints on the displacements and stress
3. Natural frequency analysis with the constraints on frequency

Additional tools which are highly desirable include:

4. Frequency response with constraints on the displacement and stress
5. Transient response with constraints on the displacement and stress
6. Buckling analysis with constraints on the critical load
7. User-defined equations for the response functions
8. User defines subroutines/programs for the response functions

The above analyses must be available as simultaneous solutions, so that the design is not optimized for the static loads alone, and then separately for the natural frequency constraints. The design algorithm must work on all design constraints simultaneously.

For lightweight mirrors, the basis vector approach to the design variables is efficient and sufficiently general for most mirror designs. The use of automesh is not a viable tool unless there is also an error estimator to revise the mesh for sufficient accuracy. This automesh capability allows a wider design variation within a given run, but is more time consuming than the basis vector approach.

8.13 Summary

An optical structure's performance is more often limited by structural distortions than by stress. Thus, the typical assumptions and modeling techniques used for stress-limited structures may not be appropriate for determining the critical behavior of an optical structure. Even small bending moments caused by neutral axis offsets can seriously degrade optical performance. Several modeling techniques applicable to precision optical structures were discussed in this chapter including symmetry, adhesives, and surface evaluation.

References

1. Barnes, W. P. 1969. Optimal design of cored mirror structures, *Appl. Opt.*, 8, 1191.
2. Buettner, A. and Quesnel, D. 1992. Evaluation of fracture specimen using multimesh extrapolation. In *Proc. 1992 MSC World Users' Conf.* MacNeal-Schwendler, Los Angeles.
3. Cook, R. D., Malkus, D. S., and Plesha, M. E. 1989. *Concepts and Applications of Finite Element Analysis*, 3rd ed., Wiley, New York.

4. Devries, K. L., Gramoll, K. C., and Anderson, G. P. 1986. Analysis of adhesive tensile test methods, *Polym. Eng. Sci.*, 26.

5. Genberg, V. L. 1983. Optical surface evaluation. In *Proc. Structural Mechanics of Optics*, Vol. 450-08. SPIE, Bellingham.

6. Genberg, V. L. 1987. Optical pathlength calculations via finite elements. In *Proc. Structural Mechanics of Optics II*, Vol. 748-14. SPIE, Bellingham.

7. Genberg, V. L. 1993. Ray tracing from finite element results. In *Proc. Optomechanical Design*, Vol. 1998-08, SPIE, Bellingham.

8. Genberg, V. L. and Cormany, N. Y. 1993. Optimum design of lightweight mirrors. In *Proc. Optomechanical Design*, Vol. 1998-07. SPIE, Bellingham.

9. Juvinall, R. C. and Marshek, K. M. 1991. *Fundamentals of Machine Component Design*, 2nd ed. Wiley, New York.

10. Knight, C. E. 1993. *The Finite Element Method in Mechanical Design*. PWS-Kent, Boston.

11. Logan, D. L. 1991. *A First Course in the Finite Element Method*, 2nd ed. PWS-Kent, Boston.

12. MacNeal, R.H. and Harder, R. L. 1985. A proposed standard set of problems to test finite element accuracy, *Finite Element Anal. Design*.

13. Moore, G. J. 1992. *MSC/NASTRAN Design Sensitivity and Optimization User's Guide*. MacNeal-Schwendler, Los Angeles.

14. Nowak, W. 1988. The analysis of structural dynamic effects on image motion in laser printers using MSC/NASTRAN. In *Proc. 1988 MSC World Users' Conf.* MacNeal-Schwendler, Los Angeles.

15. Schmit, L. A. 1960. Structural design by systematic synthesis, p. 105. In *Proc. of 2nd Conf. on Electronic Computation*. ASCE, New York.

16. Segerlind, L. J. 1984. *Applied Finite Element Analysis*, 2nd ed. Wiley, New York.

17. Thomas, H. L. and Genberg, V. 1994. *Integrated Structural/Optical Optimization of Mirrors*, AIAA paper No. 94-4356CP.

18. Timoshenko, S. and Woinowsky-Krieger, S. 1959. *Theory of Plates and Shells*, p. 74. McGraw-Hill, New York.

19. Vanderplaats, G. N. 1984. *Numerical Optimization Techniques for Engineering Design*, McGraw-Hill, New York.

20. Vanderplaats, G. N. 1991. *DOT Users Manual*. VMA Engineering, Goleta.

21. Vanderplaats, G. N. 1993. *GENESIS Users Manual*. VMA Engineering, Goleta.

22. Wilson, T. and Genberg, V. L. 1993. Enabling advanced mirror blank design through modern optical fabrication technology. In *Proc. Fabrication and Testing of Optics and Large Optics*, Vol. 1994-27. SPIE, Bellingham.

23. Wolverton, T. and Brooks, J. 1987. Structural and optical analysis of a Landsat telescope mirror. In *Proc. 1987 MSC World Users' Conf.* MacNeal-Schwendler, Los Angeles.

24. Yang, T. Y. 1986. *Finite Element Structural Analysis*. Prentice-Hall, Englewood Cliffs, NJ.

9

Thermal and Thermoelastic Analysis of Optics

Victor Genberg

Notations:

1D = 1 dimensional
2D = 2 dimensional
3D = 3 dimensional
BC = boundary conditions
CFD = computational fluid dynamics
CTE = coefficient of thermal expansion
DOF = degrees of freedom
FD = finite difference
FE = finite element
FEA = finite element analysis

-493-0133-5/97/$0.00+$.50
1997 by CRC Press, Inc.

9.1 Introduction

The goal of most thermal analyses of optics is to provide temperature profiles for subsequent thermoelastic analyses which in turn provide surface distortions for optical performance predictions. The flow of data for a typical design is shown in Figure 9.1. There have been many problems involving the interaction and data flow between the steps in Figure 9.1. This chapter will address some of those issues. Other applications of thermal analysis include the design of thermal control systems and the analysis of temperature-dependent properties such as the index of refraction. Analogies for adhesive curing and hygroscopicity are discussed in the final section.

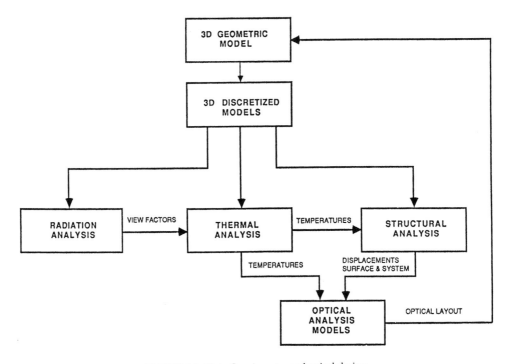

FIGURE 9.1 Data flow in optomechanical design.

9.2 Heat Transfer Analysis

Modes of Heat Transfer

In the analysis of precision optical systems, conduction, convection, and radiation must all be considered. Since small temperature gradients may be important in the resulting optical performance, the analyst should check for all three modes of heat transfer.

The Fourier heat conduction equation in 1D is

$$q = -k\left(dT/dx\right)$$

where   q   = heat flux
        k   = thermal conductivity
      dT/dx = spatial thermal gradient

For an optical structure, conduction must be considered as the most important factor in the determination of gradients in the optic and the support structure. The conductivity of most materials is fairly constant over a broad range of temperatures. The heat balance equation for conduction in a homogeneous and isotropic material is

$$k \, \Delta T + Q = \rho c \left(dT/dt \right)$$

where Q = volume heat generation
          ρc = heat capacity/unit volume
        dT/dt = change of temperature with time

Other terms may be added to the energy balance equation. For convection,

$$q = -hA \left(T - T_a \right)$$

where h = convection film coefficient
          A = surface area
          T_a = ambient temperature

Convection is a surface effect requiring a fluid such as air. The convection coefficient is difficult to predict and is temperature dependent. For some applications, the net heat flow due to convection is small so an approximate value of h is sufficient.

Radiation is another surface effect:

$$q_{ij} = \sigma F_{ij} \varepsilon \left(T_i^4 - T_j^4 \right)$$

where q_{ij} = heat flow from surface i to surface j
          σ = Stefan-Bolzman constant
          ε = emissivity
          F_{ij} = radiation view factor from surface i to surface j
          T_i = temperature of surface i in absolute units

Note that the radiation term is to the fourth power, requiring nonlinear solution algorithms. Radiation is especially important in spaceborne applications since there is no air present for convection, and the view to deep space is at absolute zero. Unfortunately, the calculation of view factors is very expensive in computer resources and will be addressed in a later section.

Conduction

For most real problems, a numerical solution of the conduction equation is required. Most heat transfer books teach the use of a finite difference approach to heat conduction as in Krieth.[5] However, finite difference lacks the advantage of finite elements in analyzing complex geometries. Almost every finite element text addresses both structural and thermal derivations of finite elements as in Hubner[4] or Segerlind.[6]

The finite difference (FD) approach is very straightforward for rectangular geometries with a regular mesh pattern. The FD approach can also be used on optics with a polar mesh pattern effectively. However, when additional detail around mounts is required, or nonregular geometry is required as in most real real optical systems, then the FD approach has difficulty.

A simple example will show the advantage of using finite elements in real geometries requiring irregular geometries. In the simple problem of Figure 9.2, the resulting thermal contours should be uniformly spaced horizontal lines, because the heat flow is one dimensional. Using an irregular mesh required by most real problems, the finite difference mesh provides erroneous answers, whereas the finite element mesh provides the correct results. The conventional finite difference

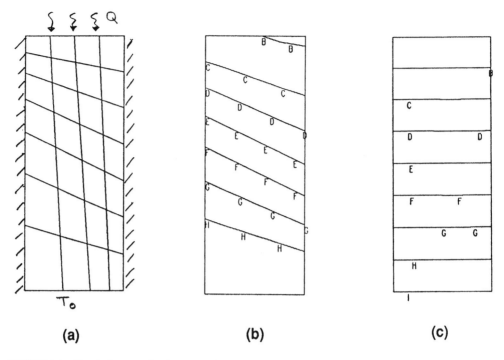

FIGURE 9.2 Simple heat conduction example. (a) Skewed mesh pattern; (b) finite difference results; and (c) finite element results.

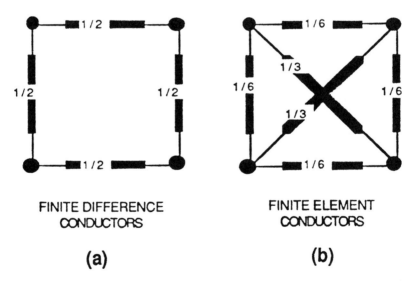

FIGURE 9.3 Conduction links for a square mesh. (a) Finite difference conductors; (b) finite element conductors.

conduction links are compared to the finite element conduction links in Figure 9.3. The presence of the diagonal links in the FE method allows for the accurate solution of distorted meshes.

Many thermal analysts use programs such as SINDA to solve heat transfer problems. SINDA is often referred to as a finite difference program, however, it is more accurately described as a matrix iterative solver. The calculation of the input coefficient matrix can be done using either finite difference or finite elements. It is possible to use an FE program such as NASTRAN to calculate the conduction and capacitance matrices, output the matrices from NASTRAN, reformat the matrices as SINDA input, then solve the system in SINDA. The resulting solution will be a finite element solution from SINDA. This approach has the advantage of the highly automated, graphics-oriented FE model generation programs, yet allows the analyst the use of traditional solution tools in the traditional thermal solvers.

Another problem associated with the traditional matrix solvers is that there is no graphical display of thermal contours since there is no associated geometry. If the FE approach described above is used, then the temperature contours can be displayed on the FE model. An additional step may be required to convert the solver output to a format which can be read into the FE postprocessor. These approaches are illustrated in flowchart form in Figure 9.4.

The finite element derivation for thermal analysis parallels the structural derivation leading to an element conductivity matrix:

$$[k] = \int [B]^T [E][B] \, dV$$

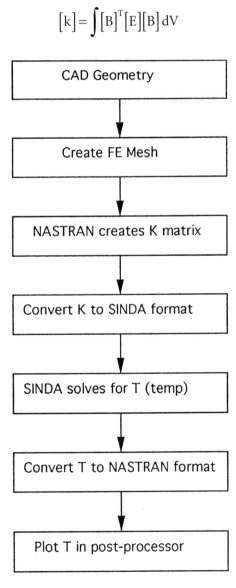

FIGURE 9.4 Thermal analysis flowchart.

where the [E] matrix is a diagonal matrix of material thermal conductivity and the [B] matrix is matrix of derivatives of assumed shape functions as described in Section 8.2 ("Derivation of Stiffness Matrix"). Assembly into a system level equation leads to:

$$[K]\{T\} = \{Q\}$$

where [K] is a symmetric system conduction matrix, {T} is the vector of nodal temperatures, and {Q} is the vector of applied nodal heat loads. Once boundary conditions (fixed temperatures) are applied, the system can be solved by many standard techniques.

Convection

Convection is a common heat transfer mechanism whenever fluids are present. Optics in such common machines as the office copier or a laser printer are affected by convection heat transfer, especially when fans are used to cool the machine. Although convection is not a consideration for orbiting telescopes in space, it must be included in the analysis of ground-based tests of such systems.

The most accurate calculation of convection requires the use of computational fluid dynamics (CFD) codes, which may be finite difference or finite element based. In these techniques the fluid is modeled in 2D or 3D space. Boundary conditions of pressure and temperature are applied, causing fluid motion and the resulting heat flow. These techniques can be very computationally intensive for transient 3D analysis, but are necessary when highly accurate results are required. A typical limitation of CFD analysis is that it is not coupled to the conduction analysis of the solids which thermally interact with the fluid. The CFD boundary conditions at the solids are either constant or prescribed in a time-varying manner rather than coupled to the conduction model. A common technique is to use CFD to find the effective convection coefficient or time-varying thermal boundary condition (BC) for a subsequent thermal analysis of the optical system.

A less expensive alternative to the CFD calculation is to use handbook values for the convection coefficients. For most applications a range of h values is given. By running an analysis at each end of the given range, the sensitivity of the model to the value of h can be determined. If the sensitivity is small, then the handbook value can be used with confidence; otherwise a CFD analysis or an experiment is required to determine a proper convection value.

Radiation

Radiation is a highly nonlinear effect that is especially important in the analysis of spaceborne optical systems. Most surfaces are usually treated as "gray body" such that they absorb a fraction of all incident radiation, and then reflect or re-emit radiation according to Lambert's cosine law independent of wavelength or incident direction. This simplifying assumption allows the use of the radiation equation in Section 9.2 ("Modes of Heat Transfer") to be used with view factors (F) calculated from the geometry. For very simple geometries and small models, simple equations are available for view factors. However, for most geometries found in detailed models, computer-based calculations are required. In NASTRAN, either a finite difference method or a contour integral method is available to calculate view factors between elements. As in most analyses, higher accuracy by using contour integrals or a small finite difference mesh requires more computer time. Typically, the calculation of view factors takes much longer than the solution of the nonlinear thermal problem.

Most optical systems include highly polished mirrors which reflect the incident radiation just as they reflect light. To analyze the heat transfer in systems with highly specular (reflective) surfaces, a ray-trace approach is needed to calculate the effective view factors. To get reasonable accuracy, many rays (>10,000) must be traced through the system, requiring significant computer time. Ray tracing can easily take ten times as long to calculate view factors as the finite difference method for gray bodies.

The radiation properties of real surfaces tend to be wavelength dependent. If high accuracy is required, the analysis should account for the wavelength effects. This complication is usually ignored in most cases, however, some modern programs (i.e., NASTRAN) allow wavelength-dependent effects.

Symmetry

Optical systems often contain high degrees of symmetry in the geometrical design. Just as in structural analysis, symmetry may be used to reduce the size of the analysis model. Symmetry techniques rely on the use of linear superposition to get the final answer. Nonlinearities such as radiation may prevent the use of model reduction in an otherwise symmetric system.

A symmetric body with symmetric thermal boundary conditions but with a generally nonsymmetric thermal load can be solved as a linear combination of two half-models. The half-model is solved once with symmetric BC and loads (Q_S) to give T_S and once with antisymmetric BC and loads (Q_A) to give T_A as in Figure 9.5. Assuming the right side of the body is modeled, then the submodel loads can be found from:

$$Q_S = 0.5 \times \left(Q_R + Q_L\right)$$

$$Q_A = 0.5 \times \left(Q_R - Q_L\right)$$

where Q_R represents the actual load on the modeled (right) half and Q_L is the load on the unmodeled (left) half. If Q_R and Q_L were equal, then there would be no heat flow across the cut boundary, making the symmetric boundary insulated. If Q_R was the negative of Q_L, then there would be no temperature change on the cut, making the antisymmetric boundary fixed. The resulting solution on the full model is T_R on the right side and T_L on the left side, where

$$T_R = T_S + T_A$$

$$T_L = T_S - T_A$$

A common special case is when the applied load is symmetric ($Q_R = Q_L$), in which Q_A is zero, so only the symmetric BC solution needs to be run. Note that in this case, $Q_S = Q_R$, so no new load calculation is required.

Just as in structures (Section 8.3), multiple planes of symmetry can reduce the model even more. This is generally useful only when the loads have the same symmetry, so the antisymmetric solutions are not required. The most common submodels in optical systems are 1/2 and 1/6 models.

Since symmetry can be thought of as a conventional mirror, the radiation view factors can be found for a half-model by placing a reflecting surface at the cut in the ray trace approach. Since the radiation is nonlinear, superposition does not apply. However, a purely symmetric solution is possible without superposition.

Solution Methods

A linear steady-state heat transfer problem has the form:

$$\left(K + H\right)T = Q$$

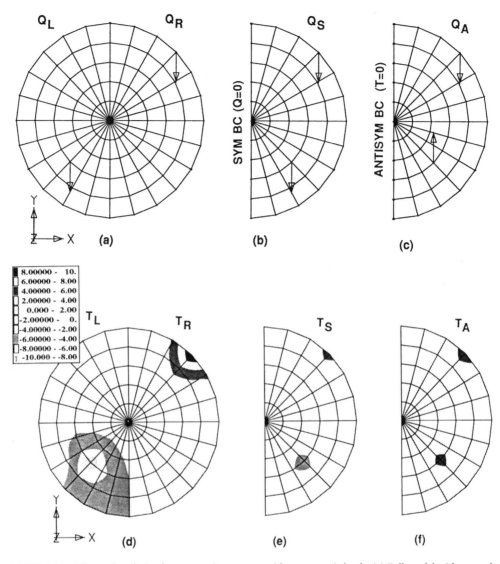

FIGURE 9.5 Thermal analysis of a symmetric structure with asymmetric loads. (a) Full model with general loads (Q_R and Q_L); (b) half-model with symmetric loads (QS) and boundary conditions; (c) half-model with antisymmetric loads (Q_A) and boundary conditions; (d) full model with temperature results (T_R and T_L); (e) half-model with symmetric temperatures (T_S); (f) half-model with antisymmetric temperature (T_A).

where the conduction (K) and convection (H) are temperature independent and the applied load (Q) is constant with time. A variety of solution algorithms are possible for large linear systems. A single pass solution such as Gauss elimination or Cholesky decomposition (NASTRAN), or an iterative scheme such as Jacobi (SINDA) are common techniques. The best method is problem (size, accuracy, conditioning) and computer resource (memory, disk space, speed) dependent.

When the problem becomes nonlinear due to radiation, temperature-dependent properties, or mass flow elements:

$$\left[K(T) + H(T) \right] T + \left[R(T) \right] T^4 = Q(T)$$

then variations of Newton's method are used. Lack of convergence can become a difficulty if the initial starting guess is not accurate enough. Convergence can also be improved by breaking the load into smaller increments so the algorithm can track the solution more closely.

Transient problems require numerical integration to track the solution through time:

$$[C]T' + [K+H]T + [R]T^4 = Q(t) + N(T, T')$$

where T' is the derivative of temperature (T) with respect to time (t) and the coefficient matrix [C] is the thermal capacitance matrix. Nonlinear effects require Newton-like iterations at each time step to stay on a converged solution path. For the best combination of accuracy and efficiency, an automatic time step adjustment algorithm will reduce the step size in periods of rapid change or lengthen the step size in periods of slow thermal change.

The $N(T, T')$ term on the right-hand side represents nonlinear load term which can be used to model thermal control systems. The temperature is sensed at a control node, then, based on its value, a heating or cooling load will be applied to other points in the structure.

9.3 Model Types

The choice of model types depends on the goals of the analysis. A lumped parameter model may be sufficient to determine the net power requirements in an optical system. If thermoelastic distortions are required, then a 2D or 3D continuum model is required.

Lumped Parameter Models

In a lumped model, each optic and major component may be treated as an individual thermal node. Conduction, convection, and radiation links to the other nodes are required. In transient analysis, the total capacitance of each node is also required. The capacitance calculation requires only the net volume or mass of each component which is often available from mass property tables. The thermal links, on the other hand, can be difficult to calculate. Simple geometric calculations or handbook values may be sufficient for many geometries at this level of approximation. Many heat transfer texts use the electrical circuit analogy for lumped parameter models.

One example of a lumped parameter model would be a system level model of an orbiting telescope which is used to determine the net power requirements to maintain the operating temperature when subjected to solar heating while also radiating to deep space. Often a thermal control system must provide heating, cooling, shading, or insulation to maintain the optics at operating temperatures.

A more common example would be a high volume office copy machine. The heat output from several sources in the machine must be controlled for it to function properly. The size and location of fans can be studied using a lumped parameter model.

Even in a more detailed 3D model, small optics may be treated as a single lumped node, creating a mixed model. Physical size is relative and is not always the governing criterion. If a small optic is critical to the overall performance of the system, then a distributed model is warranted.

In a general sense, a lumped model may provide enough information to calculate the despace or tilt in a system, but cannot provide any surface distortion predictions.

Two-Dimensional Models

When thermoelastic distortions are to be studied, the thermal gradient information is required within the optic. A 2D model will provide these results in a plane.

The most common 2D model for optics would be an axisymmetric model of a lens barrel (Figure 8.4). An axisymmetric model assumes that the structure, the boundary conditions (BC), and the thermal loading are axisymmetric.

If thermal gradients through the thickness of the optic are negligible, but the radial or circumferential gradients are not, a 2D shell model of an optic is possible. In Figure 9.6, the in-plane contours are shown for a thin optic with three edge supports subjected to laser heating. Another application would be the nonsymmetric temperature profiles in a thin pelicle due to an off-axis laser beam. Remember, these 2D models cannot account for the temperature gradients in the third direction.

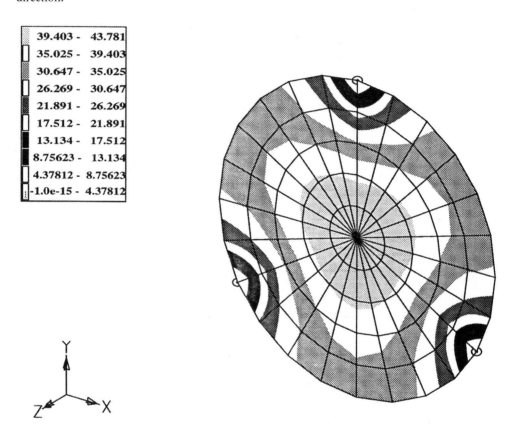

39.403 -	43.781
35.025 -	39.403
30.647 -	35.025
26.269 -	30.647
21.891 -	26.269
17.512 -	21.891
13.134 -	17.512
8.75623 -	13.134
4.37812 -	8.75623
-1.0e-15 -	4.37812

FIGURE 9.6 2D model with in-plane temperature contours.

Three-Dimensional Models

For most thermal problems the temperature varies in all spatial directions requiring a 3D model. If the geometry is very regular, finite difference techniques may be used to create the coefficient matrices. For most real geometries, computer-based modeling schemes such as finite elements are required. The retro-reflector (corner cube) in Figure 9.7 is an example of geometry requiring finite elements to get an accurate heat transfer model. The conduction matrix would be impractical to calculate without a computer-based geometry processor.

In some problems, the models for thermal and structural may be of different order. A thin optic may be represented as a 2D plate model for structural analysis, yet require a 3D thermal model to obtain gradients through the thickness. Most structural FE programs will allow specification of a thermal gradient through the thickness of plate elements. The reason that a 3D thermal model is required is because the temperature on the top and bottom surface are independent variables. A

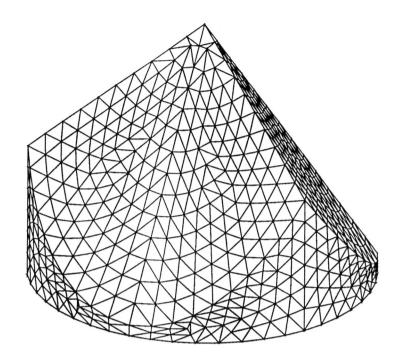

FIGURE 9.7 3D model of a corner cube.

midplane plate model has only a single variable (midplane temperature) to represent the response. In the 2D structural model, the midplane node has both displacement and rotations to represent the structural response. Thus, the out-of-plane bending of plate can be predicted by the equivalent moment loads caused through the thickness thermal gradients. However, the thickness growth is not accounted for by a plate model. If thickness change is important, a 3D solid model is required.

9.4 Interpolation of Temperature Fields

In some analyses, the thermal and structural model are identical, allowing an easy transfer of temperatures to the structural model for thermoelastic response. The models are of the same geometrical order (i.e., 2D or 3D) and the node points have the same numbering and spatial location. In this case, some finite element programs allow for the solution of the thermal and structural problems in a single execution. If the thermal and structural model are solved in different executions, then the temperatures must be passed between the models. If thermal and structural analyses are run in the same program, this is usually automated. If different programs are used for thermal and structural, then the user may have to translate the output format of the thermal results to the input format of the structural program.

A much more complicated interaction occurs when the models do not use a common mesh. This occurs in practical problems because the models are created to study different phenomena. Typically, the thermal model has a coarser mesh within an optic, because the thermal model must also include the exterior surroundings to accurately model convection and radiation effects. The structural model, on the other hand, may include only the optic being studied, but may have much higher refinement around mount points because of the stress gradients. Now the analyst must take temperatures from a thermal model, which has different node locations and numbering, and map those onto the structural model. Three techniques are discussed for the mapping of temperatures on to the structural model.

Nodal Averages

One approach is to search over all thermal nodes to find the N closest nodes to a given structural node. Then average the temperatures (T_j) of the N closest nodes, usually weighting by their inverse distance (d_j) from the structural node.

$$T_{avg} = (1/D)\Sigma(T_j/d_j)$$

$$D = \Sigma(1/dj)$$

This technique is limited to the interior of continuous and uniform structures, because it ignores the gaps, changes of materials, and boundaries. This technique is not accurate at important points like an optic mount location. Also, this technique always underpredicts the temperatures at the curved boundaries since it cannot extrapolate.

Interpolation via Conduction Models

An improvement on the nodal averaging is a conduction analysis in the structural model. In this technique, the analyst maps the nodal temperatures from the thermal model onto the single closest structural nodes. This is usually done 1 for 1, so that there are only NT (NT = number of thermal nodes) specified temperatures in the structural model containing Ns nodes (Ns = number of structural nodes). To obtain the remaining structural node temperatures, the structural model is converted to a thermal conduction model with the specified temperatures as fixed boundary conditions (BC). This model is then run to obtain the temperatures at the remaining structural node points. The output from this model is then input to the structural model with a convenient interface since it has the same node numbers.

The advantage of this conduction interpolation is that no other software is required. Changes of material properties and gaps are accounted for in this approach, so that the mounts can be accurately described.

This technique has the major disadvantage of requiring a point-to-point mapping of thermal node to structural node which is very time consuming and error prone. Interpolation errors are introduced because the thermal BC are not accurately described in the conduction run. The simple model in Figure 9.8 shows the type of error introduced. The bottom edge is fixed to T = 0, the sides are insulated, and a uniform heat input across the top edge causes a temperature of 100. A coarse thermal model (a) shows the resulting temperature contours. In (b), a detailed structural model is converted to a conduction model with the nine common nodes fixed to the temperatures from part (a). The resulting contours show the lack of correlation at the boundaries because the intermediate boundary nodes do not have any boundary effects enforced. In a 2D model the boundaries are lines, but in a 3D model the boundaries are every exterior surface. This is a major flaw in the technique, because the most important structural behavior, such as the highest stress or optical surface distortion, occurs at the boundaries. Similar errors occur around the interior fixed points as well, which can cause unrealistic higher order surface distortions on optical elements.

Interpolation via Shape Functions

This technique[3] was developed to overcome the flaws in the above two methods. In this approach, a finite element representation is required for the thermal model since FE shape functions are used to interpolate temperatures to the structural nodes. The actual temperatures may be computed from any thermal analysis technique, but during interpolation the thermal nodes must be connected with finite elements. A general 3D approach will be described, but this could be specialized to 2D by replacing tetrahedrons with triangles.

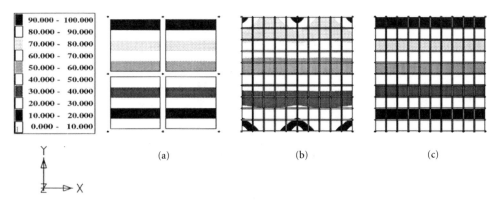

FIGURE 9.8 Interpolation of temperatures from coarse to detailed model. (a) Temperatures determined from coarse model; (b) interpolated temperatures by conduction model; and (c) interpolated temperatures by shape functions.

In step 1, the thermal FE model is converted to a solid model using two primitives, either a parametric tetrahedron or a parametric cylinder. 1D finite elements such as beams or rods are converted to cylinders, whereas all 2D and 3D finite elements are converted to tetrahedra. Only the tetrahedra are described in this section since they are the most general. The subdivision of any solid hexahedron into 5 tetrahedron or the wedge into 3 tetrahedron is obvious. 2D plates are converted to solids by extruding normal to the surface 1/2 of the thickness in each direction. In this manner 4-noded quadrilateral become 8-noded hexahedron and 3-noded triangles become 6-noded wedges, which are then subdivided into tetrahedra. Note that this model is not used for thermal analysis, but only for the postprocessing interpolation step, so the aspect ratios are not important. Only the first-order elements are used because of the math involved. For first-order tetrahedra, the Jacobian matrix is constant throughout the element, making the geometric search for the node points reasonable. Also, most thermal models use linear elements because the temperature tends to behave more smoothly than stress.

In step 2, for any given structural node, all thermal elements are searched to see if the structural node is inside. This requires that the structural node $\{x_p\}$ be converted to an element's parametric coordinates by:

$$\{\xi_p\} = [J]^{-1}\{x_p\} + \{\xi_0\}$$

where $[J]$ is the Jacobian matrix of the element and $\{\xi_0\}$ is the spatial center in element coordinates. The structural node is inside this element if the following conditions are satisfied:

$$0 \le \xi_{pj} \le 1 \quad \text{for} \quad j = 1, 3$$

$$0 \le \xi_{p1} + \xi_{p2} + \xi_{p3} \le 1$$

The search is performed over all elements until the above condition is satisfied.

Once the proper element is found, the corner temperatures are interpolated to the structural node using the appropriate finite element shape functions (N).

$$T(x_p) = \Sigma N_j(\xi_p) \times T_j \quad j = 1, 4$$

This process is performed on all structural points as shown in the flowchart in Figure 9.9.

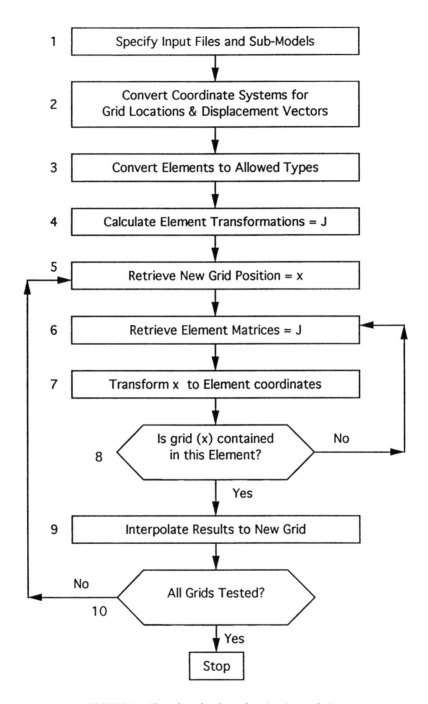

FIGURE 9.9 Flow chart for shape function interpolation.

This is a highly accurate approach, using the same finite element shape functions for interpolation as were used to solve the initial thermal model. The gaps, changes of material properties, and other geometric effects are accounted for automatically.

The first-order tetrahedron used has straight edges. If the real geometry has curved edges, then the boundary nodes on the finer detailed structural model may fall outside of all straight-sided thermal elements. This may be accounted for by taking a second pass through the search routine for any nodes not within an element. On the second pass a user-specified tolerance (ε) is used to modify the search bounds:

$$0 - \varepsilon \le \xi_{pj} \le 1 + \varepsilon \quad \text{for} \quad j = 1, 3$$

$$0 - \varepsilon \le \xi_{p1} + \xi_{p2} + \xi_{p3} \le 1 + \varepsilon$$

This will catch nodes slightly outside an element and the shape functions will extrapolate a nodal temperature.

An example of a circular mirror with a mount ring (Figure 9.10) shows a thermal finite element model with temperature contours from an applied load. The finer detail structural model with its interpolated temperature contours shows the accuracy of this automated technique.

9.5 Thermoelastic Analysis

In many applications the goal of a thermal analysis is to determine the resulting image motion or surface distortion effects in an optical system. Thus, the temperatures obtained must be applied to a thermoelastic structural model. The techniques described in the previous section may be used to apply the thermal loads to the structural model. This section discusses some common analyses for optical structures.

Distortions and Stress

A discussion of proper modeling techniques and model checkout for the structural analysis is given in Chapter 8. When applying a temperature load to the system, a coefficient of thermal expansion (CTE = α) is required for each material. If a strain-free configuration occurs in a simple 1D rod at temperature (T_{ref}), then the equivalent mechanical force (F_T) caused by thermal strain (ε_T) is

$$F_T = AE\varepsilon_T = AE\alpha(\Delta T) = AE\alpha(T - T_{ref})$$

where the cross-sectional area (A) and modulus of elasticity (E) are given. In the general finite element notation of Section 8.2 ("Derivation of Stiffness Matrix"), the equivalent thermal forces are

$$\{F_T\} = \int [B]^T [E] \{\varepsilon_T\} \, dV$$

where B contains the derivatives of the element shape functions. These loads add to any existing mechanical loads and cause distortion and stress in the structure. The calculation of stress requires the correction for stress-free thermal growth:

$$\{\sigma\} = [E]\{\varepsilon - \varepsilon_T\}$$

where ε is the total strain due to all loads. This last effect may require some user action when using a finite element program to obtain stress. For example, in NASTRAN, standard subcases with thermal loads calculate the stress correctly; however, when using subcase combinations (SUB-COM), the thermal load request must be listed again to correctly backout the free thermal growth.

As noted in Chapter 8, accurate stress models generally require more detail than accurate displacement models whether the load is mechanical or thermal.

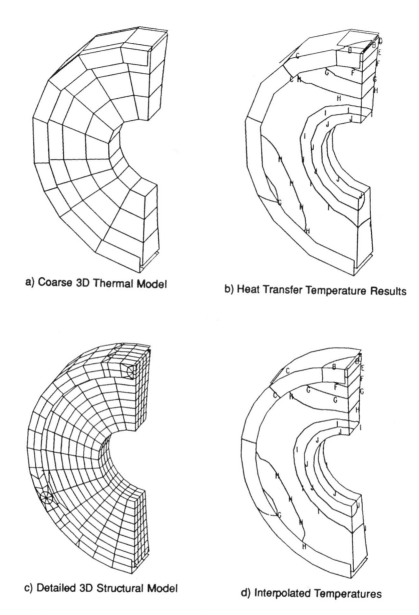

a) Coarse 3D Thermal Model

b) Heat Transfer Temperature Results

c) Detailed 3D Structural Model

d) Interpolated Temperatures

FIGURE 9.10 Shape function interpolation on a mirror and mount ring. (a) Coarse 3D thermal model; (b) heat transfer temperature results; (c) detailed 3D structural model; and (d) interpolated temperatures.

Rigid Element Problems

Rigid elements do not account for thermal growth, thereby introducing potentially large errors into the thermoelastic analysis. Often, rigid elements represent very stiff members which create very large thermal forces and dominate thermoelastic behavior, which is ignored in the rigid element formulation. The magnitude of the error introduced is dependent on the size of the rigid element used. However, even small elements used as offsets can cause large errors, as seen in the shell model with offset stiffeners in Figures 8.47 and 8.48. Offsets on element definition entries are implied rigid links and have the same effect as specifically defined rigid elements. Multipoint constraint equations (MPC) may add errors to the solution in the same manner as rigid elements. After all, the rigid elements are just an automated technique for creating MPC equations inside of the program.

To check for potential rigid body problems, the method described in the next section must be used. An alternative to rigid elements is to use very stiff structural elements which allow for thermal growth. The drawback to very stiff structural elements is that they may create numerical error in the solution. If the structural elements are too soft, the resulting flexibility may cause errors in the solution. Therefore, care must be used in their specification.

Model Check via Thermal Soak

A useful model check for any thermoelastic model is the uniform thermal soak. In this check, all materials used are converted to a common CTE and T_{ref} values. Indeterminate boundary conditions are replaced by a statically determinate BC. A single load case of a uniform temperature change is applied. The magnitude of the temperature change is of the order of the maximum temperature change expected in the problem. The output is then scanned to see if the model predicts a totally stress free condition according to theory. Any nonzero stress is a measure of error created by rigid bodies, offsets, or MPC equations.

Node vs. Element Temperatures

Some finite element programs allow for the specification of either nodal temperatures or element temperatures. In NASTRAN, for example, if element temperatures are specified they take precedence over nodal temperatures. If element temperatures do not exist for a particular element, then the nodal temperatures are used to calculate them. For 3D elements, the nodal temperatures describe both bulk temperature change and gradients within the element. For 2D plate elements, the nodal temperatures can only describe midplane (membrane) thermal effects. The gradients through the plate thickness which cause bending must be specified on element temperature entries. The temperature interpolation schemes described in Section 9.4 only apply to the nodal temperatures, not element temperatures.

CTE Spatial Variation

For large optics, the coefficient of thermal expansion (CTE) may vary throughout the structure. Often an analyst is required to study the effects due to several possible spatial variations of CTE. Since the CTE appears on the material entry, separate models and runs are required to examine different CTE values. Even though CTE appears on the material entry, it has no effect on the stiffness matrix, only on the load calculation as shown in Section 9.5 ("Distortions and Stress"), above. When solving the linear system equation:

$$[K]\{\delta\} = \{F\}$$

for displacements $\{\delta\}$,

$$\{\delta\} = [K]^{-1}\{F\}$$

the major expense is the decomposition or inversion of the stiffness matrix [K]. The multiplication by $\{F\}$ is very cheap computationally. If the vector $\{F\}$ is replaced by a matrix of several columns, there is relatively little additional cost to obtain multiple solution vectors in $\{\delta\}$. Thus multiple load cases are much cheaper than multiple model solutions. An efficient technique is to represent each variation in CTE as a new load case. Using the equations in Section 9.5 ("Distortions and Stress"), the thermal strain caused by a variation in CTE (α^*) with a nominal temperature change (ΔT), is the same as that caused by a nominal CTE (α) with a variation of temperature (ΔT^*):

$$\varepsilon_T = (\alpha \times)(\Delta T) = (\alpha)(\Delta T \times)$$

where

$$\Delta T \times = \Delta T (\alpha \times / \alpha)$$

This method allows several variations in CTE to be studied in a single model run just by creating multiple thermal load vectors.

CTE Thermal Variation

In many materials, the CTE may vary over the temperature range of interest, especially in infrared applications which involve temperatures close to absolute zero. In thermal handbooks, the net thermal strain relative to a reference temperature (usually room temperature) is the quantity plotted as a function of temperature. The instantaneous CTE is the current slope of the curve. However, when studying the net effect of a large isothermal temperature change, the net accumulated strain is the desired quantity from which the net CTE is derived:

$$\varepsilon_T = \alpha (\Delta T)$$

or

$$\alpha = \varepsilon_T / \Delta T$$

If the analysis program allows a temperature-dependent CTE, then this effect can be accounted for by providing a tabular $\alpha(T)$ input. If this feature is not included in the program, the user must input different CTE for each element, or modify the input load as shown in the above section.

Surface Coating Models

Many optical elements have a very thin surface coating applied for various optical reasons (transmission, reflectance, scratch resistance). The coating usually has different structural and thermal properties than the substrate. If the coating is very thin relative to the substrate, its stiffness may be ignored in the model; however, the loads induced by the mismatch in CTE or moisture absorption may be significant. As the relative thickness of the coating increases, it becomes more important to include its structural properties in the model. Several options for models are given below which offer varied complexity.

Model Type 1: Effective Gradient

Model the substrate as a single layer of plate elements with substrate properties, ignoring the stiffness of the coating. Apply an effective temperature gradient through the thickness to approximate the thermal moment effect. In the coating apply an effective temperature which is found by multiplying by the actual temperature (T) in the coating by the ratio of the coating CTE (α_c)-to-substrate CTE (α_s) and applying this temperature variation through the thickness via element temperature input:

$$T^* = T \alpha_c / \alpha_s$$

Model Type 2: Composite Plate

If the analysis program has a composite plate element which is typically used for graphite/epoxy structures, this element may be used for two-layered isotropic materials as well. This accounts for both stiffness and load effects of a coating in a convenient format. Interlaminar shear stress is usually provided which can be used to study layer debonding, as well as laminar stress to study cracking of the coating.

Model Type 3: Offset Plates

Without a standard composite element, the user can create a composite by using two overlapping layers of plate elements, one for the coating and one for the substrate. These must have the proper relative position by using element offsets so the proper moment is created. If element offsets are unavailable, create two planes of grids connected by rigid bars. The coating stress is available, but interlaminar shear is not.

Model Type 4: Solid-Plate

If the substrate is to be modeled with solid elements for other modeling considerations, then the coating can be added as a thin surface layer of membrane (or shell) elements with no additional node points required. Again, the coating stresses are available, but interlaminar shear is not.

Model Type 5: Solid–Solid

When the coating is of comparable thickness to the substrate, then it may be reasonable to model the coating as solid elements as well. This requires additional layers of node points resulting in larger matrices to be solved. In this case, the 3D effects and stresses are available.

Line-of-Action Requirements

Even small isothermal temperature changes can result in large internal forces in an optical system composed of a variety of materials. Using the equations in Section 9.5 ("Distortions and Stress"), a piece of BK-7 will generate a force of 46 lb for every square inch of area per degree F of temperature change. Obviously, internal forces of this magnitude can affect the performance of sensitive optical systems. For a model to accurately predict the behavior of a system, the load paths must have their proper geometric relationships. If two structural members join so that their neutral axes are offset, then internal thermal forces will cause a moment resulting in bending. An example of a mirror supported by a ring with its neutral axis aligned (a) and unaligned (b) with the mirror's neutral axis is compared in Figure 9.11. Obtaining the proper geometric relationship of member locations in a practical model requires additional effort from the analyst, often in the form of offsets, rigid links, or additional model detail.

Note that the amount of distortion caused by these offsets may be small for an automobile or an airplane, but they can be quite large compared to the wavelength of light. Thus, approximations which are valid in other industries may not be valid in optomechanics.

9.6 Analogies

Analogies are useful when an analysis technique developed in one field of engineering can be applied to solve problems in other fields which use the same type of governing equations.[1] One of the older techniques is to solve finite difference thermal problems as an electrical circuit network. Up until very recently, finite element structural or thermal codes were used to solve electrical field problems by analogy. Now several specific electric field finite element programs are available. In this section, some analogies useful in optics will be discussed.

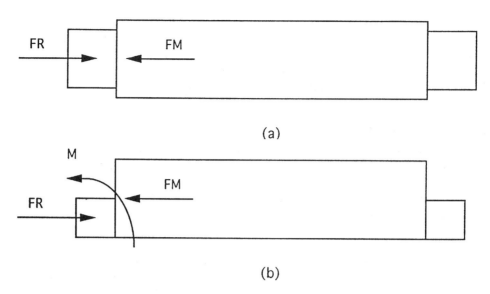

FIGURE 9.11 Line of action comparison on a mirror and mount ring. (a) Ring aligned with neutral axis and (b) ring unaligned with neutral axis.

Structural–Thermal

The analogy between structural and thermal analysis may seem unnecessary because there are many analysis codes available in both fields; however, an analyst trained in one field will appreciate the other topic more if he understands the analogy. Consider the 2D heat conduction problem in the XY plane. This problem can also be solved using a 2D structural analysis capability with the following modifications:

1. Let the X displacement = temperature.
   Set all other displacements = 0.
2. Define the material properties as:
   Young's modulus (E) = thermal conductivity (k)
   Shear modulus (G) = thermal conductivity (k)
   Poisson's ratio (v) = 0
3. Apply fixed temperatures as fixed X displacement BC.
   Insulated boundaries = free displacement = no BC.
4. Apply point heat input as forces in the X direction.
   Apply distributed flux as distributed X traction pressure.
5. Convection can be treated as springs to the ground (ambient point).
6. Structural strain output = thermal gradient output.
   X strain (du/dx) = X gradient (dT/dx)
   Shear strain (du/dy) = Y gradient (dT/dy)
7. Structural stress = negative thermal flux.
   X stress (E × du/dx) = –X flux (k × dT/dx)
   Shear stress (G × du/dy) = –Y flux (k × dT/dx)

This analogy can be extended into 3D solids. Sometimes differences appear between the solutions because structural elements may have used advanced features in their development to improve their structural behavior. If both the structural and thermal elements use the same standard formulations the results will compare exactly.

A complete correlation table between the structural, thermal, and general field problems is given in Figure 9.12. Using this table, an analyst can solve a field problem using either a thermal or a

Field Problem	Heat Transfer	Structures
Variable (Φ)	Temperature (T)	Displacement (U)
X gradient (dΦ/dX)	X gradient (dT/dX)	X normal strain (dU/dX)
Y gradient (dΦ/dY)	Y gradient (dT/dY)	XY shear strain (dU/dY)
X flux (-k dΦ/dX)	X thermal flux (-k dT/dX)	X normal stress(E dU/dX)
Y flux (-k dF/dX)	Y thermal flux (-k dT/dY)	XY shear stress(G dU/dY)
diffusivity	thermal conductivity (k)	modulus (E = G)
1	thermal capacitance (ρc)	mass density (m)
Cauchy BC	convection coefficient (h)	elastic foundation
body force	volumn heat generation	gravity
surface force	surface flux	pressure
point force	point flux	force
Dirichlet BC	fixed temperature	fixed displacement
Neuman BC	insulated BC	free edge
Cauchy BC	surface flux	pressure

FIGURE 9.12 Analogy table for field, thermal, and structural problems.

structural analogy. Generally speaking, the thermal analogy is useful for scalar fields, whereas the structural analogy has the capability to represent vector fields.

Moisture Absorption

Plastic optics may, depending on their composition, absorb moisture and swell, causing a change in shape. The absorption of moisture follows Fick's law which is the same form as transient heat transfer. The following heat transfer analogy can be used to analyze the moisture concentration:

1. Moisture concentration = temperature.
2. Diffusivity = conductivity (capacitance = 1).
3. Moisture gradient = temperature gradient.
4. Moisture flow = thermal flux.

Once the moisture concentration has been determined, the moisture swell is analogous to thermoelastic expansion. Thus the following structural analogy applys:

1. Moisture concentration = temperature.
2. Moisture expansion coefficient = CTE.

The output of the thermoelastic analysis is the deformation due to moisture absorption which may then be added to other deformation effects for surface fitting as in Chapter 8.

Adhesive Curing

Many optics are bonded to their mounts with an adhesive. In the curing of an adhesive, the solvent evaporates according to equations of transient heat transfer. Thus the concentration of solvent is similar to moisture desorption so the above analogy holds. Shrinkage during curing is analogous to thermoelastic distortion. To apply these analogies, the proper coefficients must be obtained from test data since little or no published data are available.

Temperature-Dependent Index of Refraction

In some materials the index of refraction (n) is temperature dependent. As light passes through an optical element with a temperature distribution, the effective pathlength may vary from ray to ray. Thus a planar wave entering the optic may be nonplanar when it exits. For sensitive systems, this can have a bigger effect than the thermoelastic effect. Two modeling techniques are presented here to analyze the pathlength difference.

For windows and nearly plano lenses, the technique given in Genberg[2] is easy and efficient. In this technique, the 3D solid model used to determine the temperature distribution is converted to a modified "thermo-index" model. In this modified model, all displacements except those along the beam path are set to zero. Poisson's ratio is set to zero to eliminate any coupling of the displacements. The CTE is replaced by the index gradient (dn/dT) and the temperature distribution is the applied load. If the first surface is constrained to zero displacement, the second surface's displacement is the shape of the outgoing wave.

$$\Delta L = L\alpha\,\Delta T = L\left(dn/dT\right)\Delta T$$

An example of this application is a window in a test chamber (Figure 9.13) subjected to thermal heating when a laser test beam is passed through. The temperature contours are shown on the 1/12 symmetric solid model. A 1/12 model was the smallest model describing the 6-point mount on the window. The apparent surface distortion from the "thermo-index" analogy model (b) represents the laser beam profile as it leaves the second surface of the window. In this application, the index effect in the window was bigger than the actual surface effects on the test article inside the chamber. With this analysis and a subsequent Zernike fit of surface 2, the window effects could be factored out of the test results to obtain the accurate test article response.

For more complex geometries, a new model is required. First, determine the beam path through the optic of many points distributed across the incoming beam. This may require an optical ray-trace program. Represent each optical ray as a string of truss (rod) elements with enough subdivisions to pick up the variations in temperatures throughout the optic. Use a temperature mapping algorithm to determine the temperatures at each node point along the ray path. Give the truss elements a CTE value of (dn/dT) and apply the temperature distribution as a load on a "thermo-index" model. If the surface 1 nodes are constrained, then the surface 2 nodal displacements represent the exiting wave's profile.

The corner cube in Figure 9.7 shows a complex application of the truss element approach in Figure 9.14. As an individual ray bounces off the multiple surfaces, the index effect must be summed over the ray segments. This is possible by writing multipoint constraint equations (MPC). If the displacement coordinate system is chosen so that δ_x is along each ray path, then at any reflective surface

$$\delta_x\left(\text{outgoing ray}\right) = \delta_x\left(\text{incoming ray}\right)$$

In this example, a plane of symmetry was used requiring an additional internal surface. The exiting beam from surface 6 represented the beam profile effects due to index changes. These were then

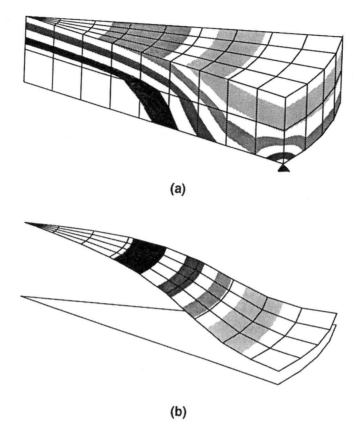

FIGURE 9.13 Optical pathlength effects due to index of refraction thermal sensitivity. (a) Temperatures of a 1/12 model of window subjected to laser heating and (b) resulting beam profile due to index changes.

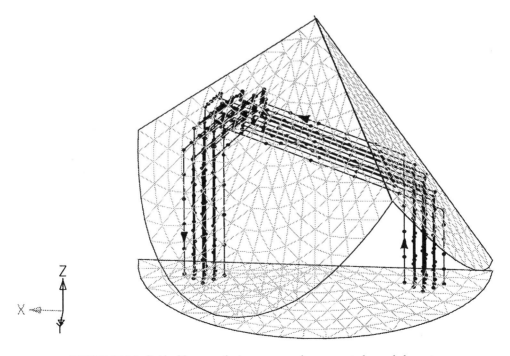

FIGURE 9.14 Individual beam paths in a corner cube represented as rod elements.

fit with appropriate Zernike polynomials and added to other effects such as thermoelastic distortion.

9.7 Summary

The thermal analysis of optical structures requires the accuracy of modern numerical methods such as finite elements to obtain thermal gradients throughout the system. Conduction, convection, and radiation can all be important modes of heat transfer in any system. The resulting temperature profiles are often input into a structural model to obtain the thermoelastic distortion. If the thermal and structural models are not the same, then some form of interpolation is required to apply the temperature results onto the structural model. Other problems, such as moisture absorption and swell, can be solved by analogy to thermal and structural solutions.

References

1. Genberg, V. L. 1986. Solving field problems by structural analogy. In *Proc. Western New York Finite Element User's Conference*. STI, Rochester.
2. Genberg, V. L. 1987. Optical pathlength calculations via finite elements. In *Proc. Structural Mechanics of Optics II*, Vol. 748-14. SPIE, Bellingham.
3. Genberg, V. L. 1993. Shape function interpolation of 2D and 3D finite element results. In *Proc. 1993 MSC World User's Conf.* MSC, Los Angeles.
4. Huebner, K. H. and Thorton, E. A. 1982. *The Finite Element Method for Engineers,* 2nd ed. Wiley, New York.
5. Krieth, F. 1958. *Principles of Heat Transfer,* 2nd ed. International Textbook, Scranton.
6. Segerlind, L. J. 1984. *Applied Finite Element Analysis*, 2nd ed. Wiley, New York.

10

Fabrication Methods

Darell Engelhaupt

10.1 Introduction

Fabrication methods and performance requirements for the production of precise optical and optomechanical support systems are necessarily quite variant. As pointed out in Chapter 4, optical components consist of an almost endless list of forms and requirements. (The reader is referred to Chapter 4 for additional information regarding materials and fabrication requirements). The entire optical or optomechanical device must, therefore, be proposed and subsequently designed as a complete compilation of performance and cost issues. Many optical development efforts have relied on "breadboarding" components to determine the system performance characteristics without regard to subsequent fabrication consequences. Modern integration of computer optical design programs, computer-aided-design (CAD) and thermal, mechanical, and material properties programs using finite element analysis (FEA) now permits a very reasonable prediction of performance. This effort can be used prior to prototyping or breadboarding in the more provincial approaches. By developing the entire system using integrated design approaches, the very data set from the ray-trace can be superimposed on the 3-D CAD drawing package, and critical issues such as interference, vignetting, and component mounting can be resolved prior to material selection or fabrication efforts.

This design data set can next be transferred to a 3-D FEA package and the performance aspects based on conventional machining and materials can be tested with respect to more elaborate and expensive choices. Interactive design efforts then allow for a substantial improvement in the fabrication flow time and the material selections. Also, it is plausible during this design stage to input experienced or intuitive design parameters. As described in Chapter 4, the selection of fabrication methods and materials is dependent on the optical system performance criteria.

Trade-off in cost, materials, and fabrication capabilities must usually be determined in a formal quotation and prepared, without compensation, in a competitive bidding process. This intensifies the need for integrated computer designs beyond the traditional repetitive design, breadboarding, prototyping, testing, and finally fabrication procedures used in the past. With today's demands on optimum performance for the available funding, optical and optomechanical engineers are expected to minimize the steps and the development efforts, yet not incur excessive risk. This, in turn, requires a comprehensive team effort, developing both optical and fabrication design aspects concurrently. Figures 10.1 to 10.3 depict the interactive optical system development, optical design, and optomechanical design processes. Figures 10.4 and 10.5 are typical examples of 3-D models of a UV imager and imaging spectrometer, respectively, which were produced by employing the computer integrated design approach.[1,2]

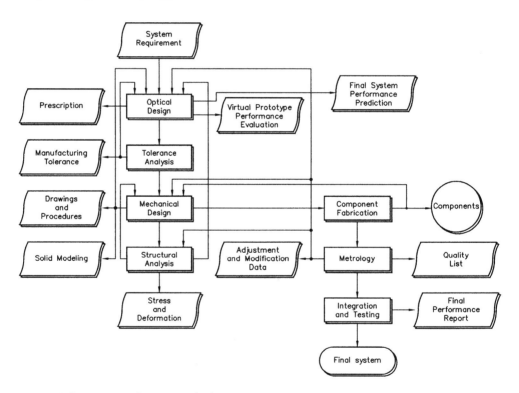

FIGURE 10.1 Major steps involved in an interactive optical system development process.

10.2 Fabrication Method Selection

The selection of a fabrication method for any precision system or component of the system depends primarily on the material selection. The material selection, in turn, is related to the required system reliability, dimensional stability, and thermal and mechanical performance requirements. All of these issues must be balanced by a design which not only functions to meet the design specifications, but one which is also affordable.

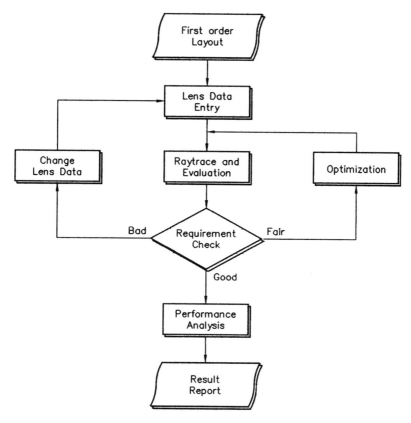

FIGURE 10.2 Major tasks involved in an interactive optical design and analysis process.

Machining Tolerances

Modern optical systems include reflective, refractive, diffractive, fiber optics, acousto-optical, and other disciplines. This includes both imaging and nonimaging systems with possibly reflective and refractive (catadioptric) combinations.

Performance requirements may include functioning for years in deep space for X-ray or XUV (extreme ultraviolet) instruments. Actually these instruments, while operating at very low temperature, usually operate at a reasonably constant temperature. Other systems may need to function over a wide fluctuation in temperature and mechanical shock or vibration with less emphasis on absolute optical finish or figure. The overall system tolerances are determined by the statistical compilation of the individual component requirements.

The selection of modern optical manufacturing methods is dependent on the following information:

- Product requirements
- Material selection
- Personnel capabilities
- Performance vs. cost
- Time to fabricate first system
- Anticipated production quantity

Product Requirements

The utmost in optical form and surface finish is required for the shorter wavelength optics such as soft or medium X-rays and XUV. In this type system, the energy is usually nonimaging and

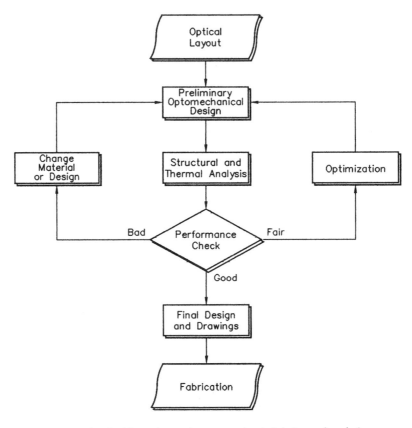

FIGURE 10.3 Major steps involved in an interactive optomechanical design and analysis process.

FIGURE 10.4 A 3-D model of an ultraviolet imaging camera showing the actual ray-trace and optical surfaces.

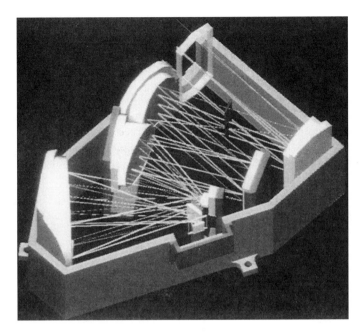

FIGURE 10.5 A solid model of a compact all-reflective imaging spectrometer.

made to collimate or condense only. However, imaging X-ray and ultraviolet systems are also a reality and indeed have been fabricated for about 40 years. It is difficult to think in terms of the scattering and short-period surface finish requirements. Improved and yet affordable systems are eminent which will image or at least guide focused energy in these very short wavelength regimes. These systems include infinite focus instruments for space observation as well as extreme precision clinical devices for advanced medical applications.

Recently reported cancer treatment procedures require photon, gamma, and neutron concentration with focusing of energy, including optical and these other sources to a precision of 1-μm spot size and location. This requires restraining the patient to stabilize the target to the same precision. Such extraordinary optics are saving lives. Also, an intense effort is in progress to focus short wavelength energy to pattern submicron feature microcircuits and nanostructures.

These optical devices may require surface finishes of less than 1 nm rms and surface figure of $\lambda/50$ at 1 μm wavelength. Additionally, this surface must be preserved during operation. This, in turn, requires the use of ceramic coatings to provide an ultrahard optical surface. Such tolerances are in atomic dimension regimes and require the limit of today's capability in manufacturing and performance measurements.

More conventional optical components operating in the ultraviolet to near infrared may still demand optical surfaces of 3 to 10 nm rms to remain diffraction limited and perform with low scatter. Infrared and even millimeter wave devices typically still require precision beyond the best of conventional CNC machining methods. Optical manufacturing procedures may be used to fabricate such devices at a cost savings. Single-point diamond machining and polishing remain as one of the most often used combinations to achieve high surface finish and figure tolerances for metal optics. Further discussion of these requirements will follow in the subsequent sections of this chapter.

Personnel Capabilities

To competitively produce high tolerance optical systems, a team of skilled persons must be available. While this seems obvious, it is not easily accomplished or maintained. For a skilled person to properly perform polishing operations, for example, sophisticated measurements are required on an interim basis. For the production of very high performance systems it becomes mandatory to be able to interpret the measurement data and determine the extent to which a given step has been performed. 2-D and 3-D interferometric analysis, scattering data, BRDF, BTDF, and sophisticated electron beam, X-ray diffraction, and scanning tunneling microscopy (STM) or atomic force microprobe (AFM) analysis are some of the sophisticated inspection tools which might be required to be used.

It is not easy to find a single skilled person to do all of the required tasks. When polishing a surface below 30 or 40 Å rms (3.0 to 4.0 nm) on any given material, the use of frequency spectrum analysis such as Fourier transform frequency or power spectrum methods may be required. For the optician to determine if a given polish medium has extinguished in terms of uniform removal of disparities larger than it creates, the data must be analyzed, not just observed, to conserve time and to avoid losing the overall surface figure of the part.

Optical fabrication may involve experts in the following disciplines:

Optical engineering personnel

- Optical design engineer or scientist
- Ray-trace expert
- Optical materials engineer
- Fiber optic expert
- Diffractive optic expert (depends on system)
- CAD operator
- FEA scientist
- Optomechanical design expert
- Materials engineer
- Purchasing expert
- Manager

Fabrication personnel

- Diamond machining
- Multi-axis work center
- Polishing (automated)
- Coating/plating
- Replication processing
- Material selection and control
- Material processing and heat treatment
- Fabrication supervisor

Inspection

- AFM — nanometer surfaces
- BRDF — reflectance, baffles
- Interferometer — surface, form, figure
- Profilometer — figure, coarse surface
- Scatterometer — surface texture, stray light control
- SEM/STM/AUGER — surface microstructure, chemical composition
- Quality control supervisor

Undoubtedly, many persons will be responsible for multiple tasks during the life of a larger program, particularly as more severe funding limitations are imposed.

When such modern inspection methods are available, it is not unusual to find that the capabilities are personnel rather than equipment limited. With present trends to consolidate cash flow activities and dispense with less profitable capabilities, it is common to lose an important resource in manufacturing capabilities. In the case where this is inevitable, the prudent choice for fabrication of high precision optics is to recruit specific capabilities through subcontractual efforts rather than investing in very expensive equipment and consequent training, which may not be properly utilized to amortize the cost of the personnel or the equipment purchase past a given task.

Perhaps another overlooked issue is the promotional requirements of highly skilled personnel. In order to retain such persons it is absolutely mandatory to develop a professional growth plan which allows personnel with such unique skills to move up as they demonstrate enhanced personal proficiency. This usually means providing an educational and training program and ultimately replacing their previous task assignments with more challenging tasks.

Fabrication Logistics

Information from the CAD and FEA designs, based on the end-product requirements, must be analyzed to determine the type of machining, coating, and inspection capabilities needed. This depends first on the materials and then on the tolerances required.

Next, a schedule of the overall task is prepared including purchased materials and subcontracted efforts and includes input from each responsible person to be involved. When a particular capability does not exist in-house, the technical and cost response of vendors and subcontractors must be verified.

An assessment of common machining methods and tolerances achieved is given in Section 10.3.

Time to Perform

Management decisions on manufacturing methods must be made far in advance of the actual operation. The schedule to which a contract will truly adhere relies on the intelligence of the planning with regards to the fabrication methods. As mentioned both here and in Chapter 4, the material in large part determines the fabrication processes, including machining, heat treatment, etching, annealing (also heat treat), polishing, coating, inspection prior to assembly level testing, and assembly level inspection including the final assembly inspection.

Serious schedule impact is often caused by the selection of the wrong material and manufacturing method to satisfy the performance requirements. This may be the result of either over- or under-specification. It is important to understand appropriate details of the various types of interactions between the materials and fabrication requirements for stabilized optics. This is usually avoided early in the project by insisting on interfacing the design efforts with other tasks to be performed concurrently by a complete project team.

Make Vs. Buy Decisions

The decision to make or buy components is often overlooked from the point of economics. Often a component, which is close to the design requirement, is available commercially and can be suitably redesigned or modified by the original equipment manufacturer (OEM) vendor, thereby resulting in a significant cost saving. Another option is to buy a component which meets most of the requirements and refine it in-house. This may be the case for a lens, since most of the cost of a very high quality lens is in the labor to polish it to a particular aspheric shape. If the material is not correct, however, the lens cannot be produced to high quality. For example, it may be wise to buy a high quality material near-net-shape lens from a production manufacturer and perform the polishing labor in-house.

The reputation of a vendor is paramount in any decision to buy long lead time or extremely critical items. Such is particularly the case with coatings such as electroless nickel and vacuum coatings. If the precision machined parts are not coated properly, expensive rework is in order. Often rework is not possible and the parts must be made a second time. The vendors who coat conventional machine parts may not know the true diligence required for an optical component. The use of modern analytical methods and control should be demonstrated by the selected vendors. The vendors must adhere to performance, material, and selected method specifications.

The most often confused issue between a contractor and a vendor is the controversy over whether a part should be built to an exact method specified by the contractor or whether the contractor should specify an exact tolerance and performance and allow the vendor to pursue his own manufacturing methods. There is no clear-cut answer except that if the contractor insists on an exact method, the vendor is wise to insist on the purchase and acceptance of the part regardless of the outcome. This is not usually the most desirable relationship. Typically, the better choice is to require quality assurance to meet the specifications and allow the vendor to pursue his own course of action in accordance to the specifications set forth. If the item is truly developmental, then the opposite may be true or the contractor may wish to invest in the capability to produce the components in-house so that any engineering changes can be determined experimentally and recorded for future fabrication efforts.

Return on Investment

The bottom line on most projects is the return on investment. Complete courses are taught on this subject but it suffices to say in this chapter that the most significant loss of potential revenue is due to lack of proper communication in almost any unsuccessful optics project. This is especially true if the principals do not understand or seek understanding on the true optical performance, and all of the mechanisms which must go into the manufacturing and inspection to achieve this prior to committing. Tens of thousands of scientists and engineers, thousands of affiliations, millions of man-years experience, and certainly billions of dollars have been committed to the understanding and design of optical systems worldwide. While the research must continue, it is imperative for the staff and management to rely on the almost countless resources through literature searches and council with knowledgeable experts in their fields.

10.3 Manufacturing Methods

Description of Optical Materials

Optical materials are essentially the highest quality available for a given application. The end-product stability requirements are notably more stringent than for most other applications. The material properties must satisfy the reflective, refractive, and perhaps diffractive optical requirements of the complete system. The structural and mounting components must also meet the same stringent stability and thermal performance requirements. Optical device engineers are concerned with not only the most exacting behavior of the material used in the optical path of an optical system, but in the overall performance of the complete system of materials used for the device. A typical example is the need to match the expansion coefficient of a relatively thin coating to the substrate to avoid thermally induced bimetallic deformation. The enhanced performance required of an optical system generally leads to deformation and stability calculations of at least two orders of magnitude finer than most precision mechanical components.

Machining, Finishing, and Coating of Optical Materials

The following guide, shown in Table 10.1, will be used to outline advanced topics covered in this chapter.

Optical Component Machining

Machining, in the more general context of producing shapes from bulk material or from near-net shape billets, is typically not adequate for producing optical components. Final machining on the very best conventional machines is most often referred to as "rough machining" when optical quality components are at stake. This is true in spite of the additional tolerance requirements imposed on the machinist. Experienced machinists and tool and die makers are typically preferred for premachining the optical components.

Principal machining concerns include the precision attained along with temporal and thermal stability of the final optical components. Most important is the concern for the cost in time and

TABLE 10.1[a]

Material	Figure Control Method	Surface Finish Method	Coatings
Al alloys	SPDT, CS, CM, EDM, ECM, IM, PL, IM	PL	MgF, SiO, SiO$_2$, Au, EN, and most others
Al matrix	HIP, CS, SPDT, EDM, ECM, IM, PL, IM, CM (hard)	EN	MgF, SiO, SiO$_2$, Au, EN, AN, PL, most others
Al castings			
A-201	EDM, ECM, IM,	PL or coated	Same as Al
A-356.0	SPDT, CS	EN	Coat Ni
520			Same as Al
Al silicon hypereutectic 393.2	CS, EDM, CE, IM, SPDT, GR, CM	Coated EN, PL	Coat Ni
Beryllium alloys	CM, EDM, ECM, EM, GR, HIP, not SPDT	EN	No coating Coat Ni
Magnesium alloys	SPDT, CS, CM, EDM, ECM, IM	EN	Same as Al
SiC	HIP		
Sintered	CVD		Vacuum
CVD	HIP, added	PL, EN	processes
RB	metal		
Steels	CM, EDM, ECM, GR, not SPDT	PL, EN	EN
Titanium	CM, HIP, ECM, EDM, GR, not SPDT	PL	EN
Glass-quartz most	CS, GR, IM, CE, PL	PL	MgF, Au Al203/Ag
ULE-Zerodur	CS, GR, IM, PL	PL	MgF, Au

[a] Legend: AN = anodize
CE = chemical etching
CM = conventional machining
CS = casting
CVD = chemical vapor deposition
ECM = eclectrochemical etching
EDM = electrode discharge machining
EN = electroless nickel
GL = glazing
GR = grinding
HIP = hot isostatic pressing
IM = ion milling
PL = polishing
RB = reaction bonded
SPDT = single-point diamond machining

materials to achieve proper performance. Dimensional stability of an optical system is related to the distribution of stored energy (internal stress) and the introduction of differential displacements inconsistent with axisymmetric behavior. If a system is truly performing axisymmetrically, then the focus and distortion correction remain accurate independent of thermal changes. A common analogy is that upon thermal expansion or contraction, the axisymmetric unit behaves as a system of a different scale. Unfortunately, the fabrication of a complete system is not easily accomplished with triaxially uniform behavior. The use of different materials is essential and can only be matched within a particular range of precision. Extremely important is the material processing along with machining operations to reduce the induced internal stresses.

The reduction of surface stress has been referred to in Chapter 4. The methods by which this is achieved will be discussed in more detail in this chapter. First, the stress introduced into a surface to be machined may be either compressive or tensile in nature. This stress is (in a complete analysis) always triaxial and will typically diminish rapidly and then reverse from tensile to compressive or vice versa with regards to depth into a surface. Typical stresses are introduced into an optical component by one of several mechanisms. The principle cause is the work put into the machining of a piece of metal. A second reason is the thermal energy in casting of either metal or glass which may account for stored energy (or stress) if the cooling is such that the bulk solidification is not uniform in time. Another common stress malady is the stress induced by a coating or a system of coatings. A systematic discussion of all these causes, effects, and cures for the stress-induced deformation in all optical systems is beyond the scope of this chapter; however, some classic cases will be discussed in later sections.

Finishing and Coating

Analysis of performance for coated parts includes the temperature at which a coating is applied as well as differences in thermal expansion and operating temperature range of the component. A coating applied to one side of an optical component with a different expansion coefficient from the substrate will cause a predictable distortion based on the following factors:

1. Coating physical properties

   Thickness of the coating
   Thickness of the substrate
   Stress in the coating at temperature of application
   Difference in the Young's modulus of elasticity
   Yield strength of coating and substrate
   Subsequent or multiple coating combinations

2. Thermal properties

   Coefficient of expansion difference in materials
   Temperature difference from coating temperature
   Stress relaxation in coating and substrate interface
   Heat treatment process after coating
   Rate of change of expansion coefficient vs. temperature

The system is better stabilized if the coating layer completely surrounds the base component.[6] A commonly applied coating is the electroless nickel–phosphorous alloy applied to single-point diamond-machined aluminum mirrors. The intrinsic or internal stress in the alloy is a direct function of the percentile of phosphorus if all other conditions are the same. The Ni–P alloy in all compositions has significantly lower thermal expansion (or contraction) than nearly all aluminum alloys commonly used in optics such as 6061 or 2024. A commonly desired composition is the ratio of 11% phosphorus and 89% nickel, which corresponds with [Ni_3P] where the brackets indicate the material is more nearly amorphous than the crystalline allotrope. If the nickel is applied at neutral stress at the plating temperature of about 190°F and the coating and substrate subsequently cooled to room temperature, then the aluminum will contract more than the coating, placing the coating in compression. Since the part must come to equilibrium, the substrate interface with the coating is in tension. A flat aluminum plate or strip coated on one side with NiP during plating at elevated temperature will then deflect such that it becomes convex on the plated surface when cooled. If, however, the coating is the same thickness on both sides, the deformation in the coating will be distributed with linear deformation of the substrate interfacial surface and the distortion will be minimal, albeit the stored energy may be higher.

Replication Methods

Many optical components can be manufactured to the needed tolerances by replication methods with significant savings, the most common of which are hot isostatic pressing, casting, electroforming, CVD, and epoxy composite manufacturing processes. Electroforming and CVD processes will be covered in this chapter under Chemical and Vacuum Process in Optics (section 6.0).

A complete description of each is beyond the scope of this book. However, a brief description is in order for selection of the preferred manufacturing method in a given case.

Hot Isostatic Pressing (HIP)

Hipping of powders to form a free-standing shape is commonly used when the material is available as a powder and is not readily cast as a molten liquid. Materials in this category vary widely and for many different reasons. One common example is beryllium, which is often preferred for ultrastiff metal optics for a device that must sustain a high loading in use. Note that this is very different than a device that must sustain the same load but not necessarily while performing an optical function. The beryllium powder is produced as a mixture of Be and BeO. The BeO acts as a second phase for the purpose of pinning the grain boundaries, and also to promote very small grains on the order of the size of the powder used in the HIP operation. The hipping process allows for higher concentrations of BeO to be added to the metal than is possible by casting process.

The anisotropic properties are much improved due to the uniform dispersion of the second phase and small grain size. This also improves the microyield properties, which otherwise mitigates the desirable high elastic modulus of beryllium for use as a stiff optical material, as previously described in Chapter 4.

Another example is the entire family of refractory ceramic materials. The definition of refractory is a high temperature use material. Although few optical systems need this advantage, the refractory materials also commonly share the property of very high elastic modulus and lightweight, thereby imparting them high stiffness.

Some of these materials are of interest in optics such as silicon carbide, with a higher stiffness than beryllium. Since the melt point is very high and the materials often react with air, casting becomes difficult. Usually the HIP or reaction bonding process involves addition of a second lower melt point powder such as glass, which will bond to the ceramic material and yield a high strength free-standing shape. Glass products of lower melting point (and lower elastic modulus) are sometimes fabricated in this fashion using glass powders of different composition to achieve a particular set of properties not otherwise easily obtained. Subsequent heat treatment to partially vitrify one phase in the presence of an amorphous glass can yield a nearly zero or even negative thermal expansion coefficient as in Zerodur™.

The drawback to this process is that unlike most casting operations, the mold must be designed and built to withstand both high temperature and high pressure. A typical complex HIP structure may cost many times that of a cast structure due to first item mold costs. However, this cost can be amortized over a larger quantity of parts to the advantage of the fabricator.

Cast Optical Components

Casting of optical components has usually been limited to aluminum alloys containing silicon, copper, or silver and, of course, materials cast from either glass or plastic. Other materials infrequently cast into optics include titanium and magnesium. Casting methods are currently under development for beryllium aluminum alloys and some composite materials of silicon carbide filled aluminum (DURALCAN™, ALCOA of Canada), and also silicon carbide-filled beryllium. The difficulty in casting composites is in the nonuniform dispersion of the second phase material. While methods have been in place for some time to cast composites, the uniformity has often not satisfied the requirements of an optical system. As pointed out in Chapter 4, the internal stresses may be unacceptable in composite materials unless the dispersion is both very fine and uniform. Additional problems arise in the form of anisotropic properties from preferential alignment or coagulation of particles, particularly fibers. Material experiments in space have shown that in zero gravity, it is possible to achieve very uniform properties for such materials, although these materials are not yet commercially available.

Casting of aluminum optical products is complicated by the issues of porosity. For aluminum alloys, a great deal of information is available on the causes of porosity in castings, but typically not much is done to remedy the porosity to the levels acceptable in optics. Hydrogen formed by dissociation of steam and also air stirred into the aluminum melt are two common causes of porosity. The hydrogen inclusion is reduced by the use of electric heat and a blanket of dry nitrogen or other gas over the casting melt as opposed to gas heat. This is due to the fact that combustion of hydrocarbon fuel produces water vapor, which, at high temperature in contact with many metals, produces hydrogen and oxygen by dissociation. Phosphorus may be added to some alloys as copper phosphate to assist in hydrogen removal. An additional method suitable for some aluminum alloys is the addition of chlorine or another reactive halogen gas to the melt to form hydrogen chloride, which is volatile but difficult to cope with due to corrosion of equipment. The true cure is to first X-ray the casting to assure that a minimal porosity exists, and then to perform a type of autoclaving operation wherein the component is placed in a sealed stainless steel bag and subjected to high pressure at the solution temperature for the alloy. This virtually eliminates the porosity in a normally sound casting. Precision die castings are preferred but are also more expensive. The use of spinning for centrifugal force to remove gas (porosity) and vacuum melt and pour methods are also effective, but increase the cost of fabrication significantly.

The typically preferred cast aluminum alloy is A-201, which is the refined grade of 201 containing a small amount of silver and essentially pure aluminum. The omission of silicon in this alloy allows first surface optics to be fabricated with conventional single-point diamond turning with low wear rates using single crystal diamond tools. This alloy also poses a variety of desirable properties including a somewhat lower thermal expansion coefficient than most wrought alloys, and a low as-cast porosity and inclusion level if filtered while molten through a ceramic filter. Some 500 series aluminum alloys are also used for optics and can be polished better than most wrought alloys for first surface aluminum optics. The 500 series of alloys is not readily heat treated to a high strength or hardness and must be limited to heavier, less stable applications. This material is acceptable for production quantities of small instrument mirrors which are not very lightweight nor subjected to severe shock or vibration.

An interesting cast aluminum alloy is 393.2 combustion engine piston material. This material is a hypereutectic alloy of 23% by weight silicon and other trace elements including vanadium for stabilization. The material has been developed by the automotive industry at great expense for use in high performance automotive and diesel engines. This alloy composition actually is a co-continuous microstructure and will be described in a later section.

This alloy, known as Vanasil, has recently been successfully used in an optical system requiring extreme thermal, shock, and temporal stability, low expansion, and light weight, most all of which are mutually exclusive aluminum properties. The most interesting property of the material is perhaps the microyield resistance, which makes it suitable for millions of mechanical cycles at high loading without any detectable deformation in automotive engines. Also, the coefficient of thermal expansion (CTE) is about half that of most aluminum alloys. This CTE closely matches that of electroless nickel phosphorus at the desired 11% by weight phosphorus commonly preferred for diamond-machined and polished high quality optics. Figure 10.6 shows the stability of an electroless nickel plated 393.2 alloy mirror after thermal cycling.[1]

Most glass components are cast to some form of near-net shape due to the inherent difficulty in machining. When machining is required, ultrasonic methods using diamond powder or water jet machining are often used. One option, as mentioned earlier, is the increased use of HIP processes to form glass components from powders. This is usually different from other refractory materials in the sense that the parts can also be melted or at least sintered completely to an amorphous condition in the mold or die which actually makes the procedures somewhat hybrid.

Composite Material Processing

Matrix Metals/Graphite/Organics

The definition of a composite material is probably best stated as a mixture of two or more very different materials to achieve properties otherwise not achievable in either. This is very different

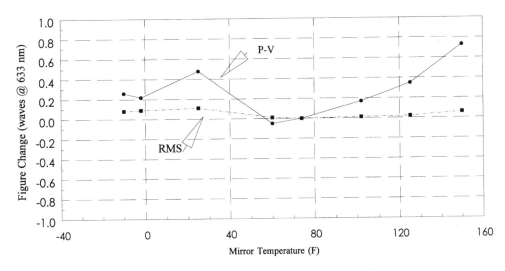

FIGURE 10.6 Figure stability of a Vanasil mirror as a function of temperature.

from alloys or specifically solid solution alloys which are an atomic mixture or a compound as such. On the other hand are the alloys which have precipitated phases or matrices of more than one discernible composition which do qualify as composites rather than alloys. Such is the case for hypereutectic or "oversaturated" alloys which will be described later under lightweight composite materials.

Design Considerations

When two (or more) very different materials are used to prepare a composite structure, the first approximation of the results is obtained by various "Rules of Mixture". It is imperative that the anisotropic (directional) properties be considered. By laws of conservation, the equations for expansion, elasticity, and internal stress must be conserved. Therefore, if the expansion of an epoxy/graphite material is listed as very low, e.g., 1.5 ppm/°C, then a shape and direction must be considered. If low expansion fibers or particles constrain the bulk material from expanding in one direction preferentially, then it will expand much more rapidly in the orthogonal direction.

For a general case, the total forces in the composite are opposite but equal in a given direction. Also, if the temperature is assumed equilibrated, then the strain in the direction of concern for the matrix material and the binding material (epoxy or other resin) is also equal on an averaging basis. Then the axial or unidirectional CTE is

$$\alpha_c = \frac{E_R V_R}{E_F V_F + E_R V_R}(\alpha_R \alpha_F) + \alpha_F$$

where
V_F = volume fraction of fiber
V_R = volume fraction of resin
E_F = elastic modulus of fiber
E_R = elastic modulus of resin

The expansion coefficient in the transverse direction is nearly that of the resin.

When sheets or "plys" of the laminate are laid up, the fiber direction is alternated at 90° to produce a material with low x and y expansion. The diagonal expansion may increase somewhat. The thickness will change in order to compensate for the bulk expansion of the resin. Even though the resin elastic modulus is low, the bulk compression is also low much like liquids. This in turn leads to a large dimensional instability in the unconstrained direction of a uniaxial matrix material.

The thickness variations are not usually as critical since the elastic modulus of the base or resin material is lower than the filament, fiber, or powder filler material. Although the stress in the fibers is often high, the elastic modulus and yield strength of the fibers are extremely high compared to most materials. The net result is such that resin–fiber composites can be fabricated with higher strength and stiffness-to-weight ratios than possible in metals including beryllium alloys. The shock and impact resistance is also higher than for ceramics such as silicon carbide, which makes these composites very attractive for lightweight precision systems.

One drawback is the fact that most resins and some fibers absorb significant moisture. Since the moisture is nearly incompressible, the resin swells upon moisture uptake. The transport of moisture is diffusion limited and obeys Fick's laws for diffusion.

Since the resin is viscoelastic, some stress relaxation is observed upon swelling from the moisture absorbed. This will then cause an opposite stress upon dehydration if the laminate has had ample time to equilibrate. The percentile of change that occurs in a given material due to moisture is[7]

$$G = (M - M_o)/(M_\infty - M_o) = 1 - \exp\left[-7.3\left(Dt/h^2\right)^{0.75}\right]$$

where
D = diffusivity $\approx 1.8 \times 10^{-7}$ mm^2/sec (50 – 150°F)
D = slope of m vs. t$^{0.5}$
t = time in seconds
h = thickness in millimeters
M = present moisture absorbed
M_∞ = end humidity
M_o = starting humidity

Filament Winding Processes

Filament winding is a subset of the above composite material manufacturing process. Filament winding equipment using CNC control permits winding shapes of rotation with a continuously crossed pattern to provide very low expansion components in the circumferential and longitudinal directions. The typical applications include cylindrical or other items of rotation, which can be fabricated with high stability in the direction of a continuous fiber, which in turn is wound with a crossed radial pattern much like a radial tire is wound for the same reasons. Very stable cylindrical mountings for space-borne X-ray telescope optics have been manufactured in this fashion. The primary concern in most optical systems exposed to moisture is the expansion of the material due to moisture absorption, and a very slow release of this moisture over extended periods of time.

Figure 10.7 shows the relative absorption and change in length with time for a 1.4-mm-thick by 76.2-mm-long sample of graphite epoxy composite. The absorption is governed by diffusion and obeys exponential time dependency. The desorption is similar in nature. This in turn places a strong dependency on the thickness of the sample for the time to reach a particular level of the relative humidity present.

The relative change in length is small for a filament wound piece since much of the length change is constrained by the crossed directional winding of the strong fibers. A typical change in length for a properly prepared material is about 50 to 100 ppm over 0 to 100% humidity and long time periods. If, however, the direction of expansion is dominantly governed by epoxy properties, the expansion will be many times this value. This can be serious in a low expansion optical system while probably passing unnoticed in most mechanical designs. The proper development of the material then depends on the application. Since the properties are essentially intermediate to the fiber and the resin, the material should be prepared from a selection of properties which gives a substantial percentile of fibers in order to achieve the desired expansion. This means that the fibers must have lower expansion than the desired end product, but not to the extent that a low volume is required. The graphite fiber is low expansion, although several versions are available with

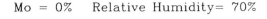

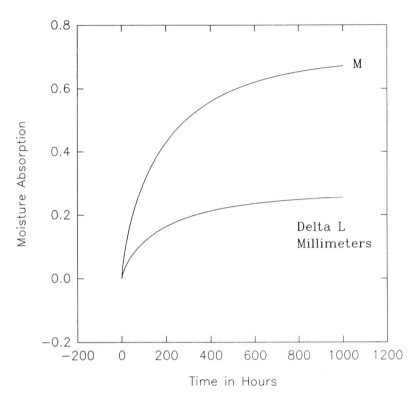

FIGURE 10.7 Moisture absorption and linear dimension change as a function of time for a 1.4-mm-thick sample of graphite epoxy composite.

differing properties. This material will usually be selected to match silicate glass, quartz, or other low expansion optical components.

The substitution of glass fibers can be made to tailor the material CTE upward to match higher expansion materials, such as beryllium, while containing a high percentile of the fiber. Care must be taken if the CTE is to be high such as for aluminum, since the volume of fibers of low expansion will be low resulting in a mechanically poor material.

Assembly Methods and Facilities

Assembly of optical systems usually requires the use of optical alignment equipment such as autocollimators and interferometers for wavefront correction. Automation is usually limited to automated polishing methods for lenses and pick-and-place assembly of the electro-optical sections. The assembly is nearly always performed by the contractor to assure perfection. The use of adjustable components is favored in low volume systems and is described under kinematic mounting, in Chapter 2. Most mirrors are preferably mounted with three-point mounting schemes to avoid torsional stresses set up across any virtual axis of symmetry.

The exception to this procedure is when a mirror system must perform to a specific vibration and shock load which simply precludes a three-point mount due to excessive stress and potential distortion induced. This case can be met with a quasi-three-point mount system with four or more mounting points of which three are match machined. The remaining mounting points may then be carefully shimmed, or a hardenable viscous adhesive such as filled epoxy composites can be used to fill the gaps. These additional mounting points can then actually be used to introduce a correction in the optical performance. This method will provide a very stiff mirror mount with the minimum amount of assembly and alignment effort. The use of two co-aligned autocollimators may be required to align very precision optics such as ultraviolet or X-ray systems. By introducing the error of one into the other it is possible to achieve an arcsecond level of resolution.

By careful design and fabrication using single-point diamond tools or replication from very carefully prepared substrates, it has been possible to fabricate self-aligning optical systems. More of this type of technology is emerging as the precision of the machining and replication processes continues to improve. New diamond turning lathes and fly-cutters are using double-pass interferometers to control the defined position of the tool path.

Piezoelectric or extremely precise DC electrical actuators then resolve the positional errors to submicron resolution over the entire part. Computers capable of real-time translation of more than 1 million data points are in place. The surface finish directly achieved from some of these machines is less than 30 Å rms on certain materials, with figure controlled to less than 1 μm/dm.

The primary difficulty in highly precision alignment methods is the inherent vibration of any building or structure. For this reason the use of an isolated support structure for a complete optics laboratory was employed in the University of Alabama in Huntsville Optics Laboratory. This 110,000-ft^2 facility has the entire center laboratory section of four floors supported by concrete piers 40 ft into the bedrock. A complete vibration absorbing band of resilient rubber then separates the office and laboratory areas. The laboratories are isolated from equipment vibration further by separate support equipment hallways to avoid placing vibrating equipment in the optical laboratories. In this fashion, it is possible to perform optical alignments and assemblies even on the fourth floor of this building.

Inspection

Most optical fabrication requires sophisticated inspection methods and equipment. This can become a very expensive capital investment. The first requirement is to determine the proper methods to assure the quality of the devices under design or fabrication. The typical optical systems contain reflective or refractive components. Certainly more attention is being paid to fiber optical inspection equipment lately. Additionally, diffractive optical components capable of performing tasks of much heavier and more cumbersome conventional optics are being used.

Inspection must address mirror reflective surfaces and refractive lens behavior as well. In addition, nonreflective coatings must be characterized. As the designs of reflective or refractive optics such as collimators, microscopes, and telescopes become refined through better use of computer-aided design, the inspection becomes more difficult. The verification of an aspheric mirror is sometimes best performed by calibration of the machining center rather than development of null correctors, which usually must be as good or better than the mirror being inspected, making them very expensive.

By cutting a cylinder on a diamond turning machine, it is possible to readily inspect the X and Z components of resolved precision. Corrections of repeated error by the machine can usually be made through software adjustments rather than hardware. In extreme cases, such as new equipment setup or after moving such a machine, it may be necessary to align the hardware using a similar approach. Next a spherical shape can usually be cut with near-net dimensions outside the final aspherical part. This piece can then be easily inspected with conventional interferometric methods. Any disparity in the machining can then be adjusted once more through software error subtraction adjustments and the final surface may be cut with very high reliability to proper precision.

Once this procedure has been completed on any given component, the same software corrections apply for additional parts. The final inspection of such parts may be performed at the assembly level. For production runs of an optical system, an inexpensive method is to dedicate a system for use as a test set with various components readily removable and replaced with the component to be tested. The performance of an actual system is then verified. This type of dynamic assembly testing is often preferred over much more elaborate and expensive component testing equipment.

It is important to remember that the surface of a mirror or lens produces a signature characteristic of the fabrication history. The diamond turning marks, for example, have a specific frequency of occurrence on the surface of the turned parts. Polishing of surfaces must start with a larger mesh abrasive compound which fractures and breaks down at different rates under different conditions. Selection of subsequent compounds must be based on the complete, not statistical, background left by the previous operation. For example, if a compound of 1200 mesh is used for an insufficient time and then a mesh of, say, 2400 is used similarly prior to a final polish at 0.1 μm, then a proper Fourier transform analysis of the data will show a frequency characteristic of the 1200 and 2400 mesh superimposed on the final polishing. It will be necessary to return to the 1200-mesh step to correct this without losing the figure in some cases. With too many of these iterations, the dimensions and figure both may be destroyed. The best way to develop a polishing procedure is from the inspection data. Often the best assurance is to perform a frequency analysis on each step during the process development stage. This information is recorded and the repeated performance is compared prior to committing the process. In this way it is possible to sort out a host of problems such as a bad bearing in a machine or polishing table. One must also consider the changes in material such as alpha vs. gamma alumina or synthetic vs. natural diamond. Also, high purity water must be used for aqueous slurries. Sometimes the calcium and iron content of tap water can cause staining. The surface inspection interferrogram may not detect this. For high end optics, such as X-ray mirrors, it may be necessary to use more sophisticated methods such as AFM or SEM instrumentation to achieve the required surface finish.

10.4 Fabrication of Lightweight Components

Material Selection

Fabrication of lightweight components is usually complicated by a number of mutually exclusive material properties and design issues, which are generally mandatory system requirements. This becomes formidable when precision systems such as optics are the end product. In order to develop stable lightweight optical systems it has become necessary to develop materials which can perform to the limit of scientific knowledge of material science. The requirement may be for the material to have very high stiffness or elastic modulus as such, and still absorb shock and vibration without failure. The first objective, therefore, is to determine the actual absolute requirements for the system. This is best accomplished during the virtual computer design stages to allow selection of materials with known properties. In this way, the design can be optimized to take advantage of stiffening ribs or cutout designs. The stiffness of a plate of glass or aluminum, which has about the same elastic modulus and density (specific stiffness = E/ρ), is by experience about 1/8 thickness to span. For a rib design, the face plate should follow this rule for each cell. To make the face very thin requires more closely spaced ribs. At some point the trade-off in weight is lost and should be predetermined.

The most often misunderstood family of properties for lightweight optics relates to the issues of stiffness and yield. First the operating requirements will dictate the material and design selections. The optical system will not remain in alignment during use if the elastic yield of material due to self-weight, vibration, or other forms of loading exceeds the precision optical tolerances. Optical tolerances are far more stringent than those required in most mechanical equipment designs. Furthermore, the system will not return to alignment upon removal of the load if the plastic yield exceeds similar stringent requirements. The optical component yield strength is defined in terms of microyield or changes in a few tens of parts per million at most. The engineering onset of plastic yield strength of a material is usually specified at a stress which causes 0.2% or 2000 ppm offset after the return to zero stress. This is, of course, completely unacceptable in the case of precision optics. Fortunately, the relationship is not typically linear. It is possible to select materials with a yield strength which is nearly the same as the ultimate strength. These materials are known as being completely elastic (or at least nearly so). Materials in this category include glassy materials with the crystalline size below 1 nm typically, and as such these materials are considered for all purposes to be amorphous.

It is important not to confuse the issue of elastic yielding in service with the yielding due to loading, shock, acceleration, or vibration not occurring during use, but perhaps only during transportation. For a system which must image during high vibration or acceleration, the use of very high specific stiffness materials and designs is required and may include beryllium, silicon carbide, graphite epoxy, or Vanasil (Al 393.2) designs. The use of conventional aluminum alloys such as A-201 or 6061-T6 may suffice if the load is temporary and not a service requirement.

Refractive Optical Materials

Most glass used in optical systems is rapidly quenched from liquid in order to accomplish solidification in an amorphous condition. This provides the most clarity by preventing additional transmission loss due to scattering along with the absorption. This also assures from a mechanical point of view that the material is completely elastic. The notion that glass is elastic is not always clear to the inexperienced designer. The elastic modulus is actually quite low and comparable to aluminum at about 70 GPa (10 Mpsi) for most glass. The obvious concern is although the glass is able to recover from any induced strain short of breaking, this amount of strain is very small. Thus, when the glass is anything short of perfectly smooth on the edges and even on the surfaces, the stress will be higher at defect sites, and since the material does not yield significant plastically, the piece prematurely fractures. The area under the stress–strain curve is very low. This value is the strain energy and is very important in the design of lightweight optical components.

The normal design stress limit for most silicate glass including quartz is only 7 to 14 Kpa (1 or 2 Kpsi). This is in spite of the fact that very smooth small fibers in bending modes may exhibit a tensile (and yield) strength of several hundred thousands of psi. This in turn creates a dichotomy with regard to lightweight glass optics. It is possible to improve the resistance to breakage by using resilient mounts at the edges of the glass and also by smoothing the edges of the lens or other component by firing the edges either with a flame or with a laser of proper wavelength for the energy to be absorbed and cause rapid localized heating of the edge. The use of plastic lenses is finding favor in systems of lower precision requirements. The plastic lens is not brittle like the glass and some plastics have higher index of refraction and low density permitting thin and very lightweight designs.

Plastic is not stable over a wide temperature range and also typically exhibits high chromatic dispersion or wavelength-dependent changes. The index of refraction typically varies unacceptably for high end use with both temperature and wide bandpass. For selective wavelength or relatively narrow bandpass, however, one of the better substitutes for brittle glass optics is the use of polycarbonate plastics such as used in millions of eyeglass lenses. These materials can be made into lightweight and shock-resistant optics. The manufacturing methods for various plastic lenses vary, but usually consist of casting or injection molding principles starting with the liquid plastic monomer or dimer and heating to produce the polymeric solid plastic lenses. This can even be performed in an optician's office for a customer on the same day. The molds are typically made by electroforming nickel over a very carefully prepared glass master. The master can be used many times to produce the same lens mold for many vendors and the molds are also used by independent lens fabricators many times over. Certain other plastics may be melted and directly injection molded. In the case of infrared optics, the halogen single-crystal ionic materials are often used for windows or lenses. Strontium fluoride, zinc selenide or sulfide, and some plastics are also selected. Silicon or germanium may also be selected. In this case it is the performance requirement and not the weight which typically becomes the driving motivation for selection, although a great deal of room exists for additional research in this area.

Advanced designs in plastic optics are favoring the use of solid optics. Such designs use plastic or even plastic and glass lenses in contact with each other to eliminate the air space and provide very light compact optical systems with excellent durability. The plastic or glass lens, which is exposed to the environment, must usually be coated to prevent scratching and also for antireflection purposes. This will be discussed later under Section 10.6 ("Chemical and Vacuum Coating Processes in Optics").

Composite Lightweight Optics

The use of composite materials in lightweight optics falls into two general categories, metallic and nonmetallic. Nonmetallic materials include graphite fiber with epoxy or other polymers. Metallic materials typically include ceramic-filled lightweight metals such as Al/SiC or Ti/SiC. An extremely light and stiff material is beryllium stiffened with either beryllium oxide such as in I-250 or I-450 alloys or those filled with silicon carbide. Several manufacturers now offer various aluminum alloys with silicon carbide fillers of either filaments or powder. These materials are usually formed by HIP methods which are generally expensive due to the need for high temperature and pressure molding equipment. Casting efforts have been somewhat successful but typically only produce uniform properties at the lower end of the ceramic material introduced into the melt. A combination of beryllium and aluminum with beryllium oxide is presently under study as a castable material and may soon become available for cast lightweight optics.

The nonmetallic materials with graphite filaments can be produced with stiffness and strength comparable to or even in excess of the filled metals. These materials are listed with the highest of all specific stiffness (elastic modulus/density). Section 10.3 describes the behavior of these materials which, although are extremely stiff, may also suffer from instability due to temperature change-induced internal stress due to the different coefficients of expansion of the materials involved and also due to the moisture absorption as described. For space applications these materials can provide the ultimate in weight savings and performance if the system can be temperature controlled. The atmosphere can be replaced with dry nitrogen or other gas, eliminating moisture issues.

Ceramic Materials

Ceramic materials have been used in numerous high precision applications. One of the most durable and stiff materials is silicon carbide. SiC can be HIP formed with a bonding glass or can be deposited to very thick free-standing shapes from silane and methane gas by hydrogen reduction in a heated chamber. Graphite substrates are prepared and after deposition of the SiC, the graphite is removed by reaction in air or oxygen to form carbon dioxide gas. In the case of nonreentry shapes, a release agent may be used to lower the adhesion and permit separation and reuse of the mandrel.

A common misconception is that this material has "no" elasticity when indeed it possesses almost perfect elasticity. Like glass, however, the material is brittle. The elastic modulus is an astounding five times that for glass and the usable tensile and yield strength are the same number at about 90,000 to 100,000 psi if care in edge preparation and mounting methods is observed. Also, the specific stiffness is second only to the graphite composites. This makes the use of silicon carbide very attractive for lightweight high end optical systems. The cost of the sintered material is lower than for the CVD version, but the mechanical properties are also much lower. Thus, CVD silicon carbide ranks very favorably as the ultimate material for high end optical components. Drawbacks include high cost, relatively long manufacturing lead times, and poor ductility requiring care in handling and mounting of components to avoid formation of nicks or notches.

The silicon carbide is crystalline but is not a two-phase material as are the composites described above. Therefore, this material can be polished directly to extraordinarily fine surface finish in part due to its purity. Samples of less than 1 Å rms surface finish have been prepared for use as measurement standards. Unfortunately, the reflectivity is poor over most of the spectrum for optics, but the material is easily coated with gold or other materials. Reinforced structures and structures with incorporated fibers injected into the CVD gas stream have been fabricated resulting in exceptionally light and stiff components. Mirror substrates of up to 5 ft in diameter have been prepared from CVD SiC for the U.S. Air Force Rapid Optics Fabrication Technology program

(ROFT). Additional reference to actual fabricated SiC mirror blanks is given in Chapter 4, Section 4.4 ("Chemical Vapor Deposited").

Co-Continuous Composites

Co-continuous composites are two or more phase materials which can be molten below vaporization temperatures of either material and quenched sufficiently rapidly to achieve a supersaturated composition of one in the other. One such example is silicon and aluminum. Silicon is soluble in aluminum to about 3.0% by weight. If the melt contains up to about 11% silicon, it may be frozen into the alloy as the eutectic at the melt point. Beyond this, up to about 23%, the excess silicon freezes out as a co-continuous weblike infrastructure in the eutectic. This is known as hypereutectic alloying and requires careful preparation to preserve the maximum properties, but when properly prepared can be cast into low cost components with exceptionally high mechanical properties. One such aluminum alloy is 393.2 piston material. This material has a combination of properties not found in other aluminum alloys. Evidence of the extraordinary properties is found in the most common application, namely, automotive and truck engine components. The use as piston material requires that the CTE be as low as the cast iron cylinder, and also that after even millions of cycles the material must not display any creep or yielding. In this aspect, the material has about one half the CTE of other common aluminum alloys, and does not show any deformation after loading in cyclic fashion for millions of firings in the diesel or gasoline high performance engines or under extreme conditions in racing engines. This material has been successfully implemented into a cast optical system for the U.S. Army.[1]

The casting cost is lower than hipping costs for the powder metal composites and the temporal and thermal stability is outstanding over a very wide range of loading and temperature change. Like silicon carbide, the yield strength is the same as the ultimate strength even on the basis of microyield. This material can be plated for optical mirror performance and can also be relatively easily machined, drilled, and threaded, unlike the ceramic materials and with much less complication than the metal composites. On the other end of the same scale, aluminum can supersaturate silicon forming a high silicon, low aluminum co-continuous composite with potentially interesting applications in optics. This material is presently under study by the Ohio State University.[5]

If investment casting is used, a low cost form of hipping (or autoclaving) is required, wherein the 393.2 aluminum silicon material is placed in a stainless steel bag and subjected to high pressure of about 14 Ksi at the solution heat treatment temperature for about 2 hr to assure the elimination of casting voids.

Although this is often called a hipping operation, it is not to be confused with the powder metal process requiring expensive molds. The cost for the bagging process is about $6/lb of material. This material can be cast into low cost stable optical components of lighter weight than glass or even all but the latest BeO/beryllium composites, silicon carbide or graphite epoxy composite systems. Another feature is the fact that the thermal expansion very nicely matches stainless steel, nickel phosphorus plating, and other common materials used in a complete optical system for athermalization purposes.

Cost Comparisons

A comparison of costs is difficult due to the requirement variations. However, it is always wise to consider the overall needs for the program and use the lowest cost process which satisfies those needs. The use of the graphite epoxy or metal matrices should be carefully judged by the performance drawbacks and the higher machining costs than for the cast hypereutectic aluminum alloy. Likewise, the poor microyield of the low BeO beryllium alloys may require more material to be used than is at first obvious from the engineering stiffness data, as opposed to the microyield properties required as pointed out in Chapter 4.

Added cost of HIP molds and the toxicity issues, which are very real if the beryllium must be machined, polished, or ground in any way, are also issues to be considered. As the high BeO alloys become cost competitive and available commercially, they will be attractive lightweight options. The silicon carbide in either reaction bonded or CVD versions is probably the ultimate high end choice but will likely cost the most. Also, many shapes for optical supports and complex shapes are not readily formed in SiC by any method. If the system must be subjected to wide fluctuations in temperature and humidity, the graphite epoxy composites must be judged very carefully. Likewise, the two-phase powder or filament-filled metal composites may exhibit distortion or even fracture due to internal stresses set up by the different CTE of materials used.

The final design consideration should then be judged based on a complete assessment of system requirements and not limited to simply lightweight or high specific stiffness.

10.5 Dimensional Stability Requirements for Optical Materials

Definition of Requirements

In order to actually perform well in optical manufacturing, often many mechanical engineers have to be reoriented. A typical education in mechanical engineering may not address the issues of tolerances and material requirements needed to understand — or perhaps better put, appreciate — the true requirements to build and test a precision optical system. The units of measurement often look more like something from a physics course than their engineering courses. Indeed optical sciences are generally taught in the physics department of most universities. Some offer independent curricula in optics but omit the precision materials engineering and chemistry issues. This chapter is not intended to replace an education in these subjects but hopefully will make the reader think about the issues prior to under- (or over-) designing an optical system. Above all, it is imperative to appreciate the microscopic dimensional changes due to loading, either self-induced by gravity or by acceleration, vibration, or shock. These changes may be temporary for low loading or permanent for higher loading. The magnitude of either may need to be one or two orders of magnitude lower than for conventionally engineered mechanical hardware.

Additionally, it is also imperative to appreciate the similar changes due to differential temperature, which may include operation over a wide temperature range, or, less stringent, to survive a wide temperature range but recover without displacement hysteresis. The most commonly overlooked factor regarding thermal distortion is the fact that the components may have a temporary thermal gradient due to thermal diffusivity, which ideally is high, but in the case of glass and other common materials may not recover from temperature gradients rapidly due to poor conductivity.

The acceptable stress will be far below the engineering practice for most designs of equipment. For retaining alignment or flatness (curvature control) in precision designs to microstrain units, it is imperative to define both operational (working) loads and nonoperational load conditions such as vibration during transportation or launch of a space-borne system. The operating load is usually much lower, but if the nonoperational loads deform the system, it may not function well at all.

Material Categories for Optical Systems

An optical system designer must obey the same structural rules as any system engineer or scientist. Weight, strength, corrosion, manufacturing methods, and cost are all very important, of course. Greater emphasis must usually be placed on precision system design material choices. In particular, the extended effects of time, temperature, and loading are far more critical for an optical system than for most engineering designs.

Design Criteria for Dimensional Stability

Mechanics of deformable solids involves a knowledge of many engineering and material aspects beyond the scope of this chapter. It must suffice to briefly explain those issues at hand for the typical optical design. Five types of material behaviors regarding stress vs. strain are accepted.

Essentially two of these are to be considered for optics: first, the materials which are completely elastic over a high range of stress required for the system; second, those materials which are not elastic at high stress but behave in a plastic fashion.

All materials respond to mechanical loading by moving. The extent to which an optical component moves and, moreover, the temporal-dependent path through which it moves for a given load both during the application and also after removal of the load is extremely critical to the design performance. This movement must be determined as a function of time-dependent variables.

For the loading regime of a given material, which is completely elastic, the hysteresis or time-lag effects are typically very small compared to normal loading and unloading times. However, for vibrational loading, the time-strain behavior may be important even at very low levels. The slope of the stress/strain over the exact elastic region of a loaded material is the modulus of elasticity or Young's modulus. Yield point is the stress level at which the material becomes plastic. This is a very difficult determination due to the sensitivity with respect to the measurement accuracy requirements.

For most engineering applications, a value of 0.2% plastic offset is used to determine yield strength or the maximum force per unit area at which material should be used. For optics this is totally unacceptable. A strain of no more than 1 to 20 ppm might be tolerated. No exact proportional limit for the stress and strain exists, even for the same materials, due to the sensitivity in processing to be discussed later. This is indeed true in spite of the fact that the modulus of elasticity is very nearly constant regardless of heat treatment or other material history. An example is carbon steel, which may be annealed to a very soft condition with the plastic microstrain yield occurring at a few thousand psi vs. the solution-quenched tool steel and tempered steel with the yield at 100,000 or even 200,000 psi (100 to 200 Kpsi, 700 to 1400 MN/m^2). For many materials used in optics such as Al, Be, or Ti, the microyield vs. the engineering yield is equally disparaging depending on the heat treatment, alloy composition, work hardening, and thermal cycling history to be described later.

Fortunately, the microyield elastic limit and the usual engineering limit are not linearly related. A material such as aluminum 6061-T6 alloy with a 0.2% yield strength of 40 Ksi would only have a usable microyield strength of 400 psi at, say, 2.0 $\mu\epsilon$.

If this material was overaged or completely annealed, this would likely be about correct for the microyield proportional limit. In fact, for most engineering metals, which can be heat-treated to resist yielding, proper conditioning will produce values which are about 1/3 the engineering yield strength for most precision applications. If an optical system is subjected to stress levels beyond this limit, then more serious quantitative results and/or testing would be required to determine the actual microyield strength.

As the design stress level approaches one third the engineering yield strength, then for critical systems a very careful material preparation and history logbook must be maintained. The preferred procedure for producing such machined optical components is to start with a larger material blank not in the solution (or quenched) state. Also, of course, this means that subsequent aging will not be performed. Unfortunately, this generally requires a very careful heat treatment monitoring procedure for proper control. If aluminum alloys containing copper (e.g., 6061 or 2024) are heated only a few degrees Celsius above the proper temperature from the initial melt condition, then the copper migrates to grain boundaries and the material cannot be properly heat-treated at all.

Many materials are also anisotropic in thermal expansion. This is particularly true in polycrystalline materials and fiber-reinforced polymer (epoxy) composites. Low expansion materials are most desirable since the effects of temperature are minimized for that component. In most cases, it is essential to have mounting components or commercial items such as lasers and mounts in the system. In this case, it is far better to achieve a match in the CTE of all the components and the wiser choice may be the selection of a higher CTE material to match a particular optical component to the system. This becomes more difficult when a particular optical item is not replaceable such as a quartz lens, for example. In this case low expansion metal pieces of Invar

(64% Fe/36% Ni by weight) or molybdenum alloys such as TZM or graphite epoxy pieces, which can be engineered to a specific CTE, may be required. The reader is referred to Chapter 4 for additional information.

Hysteresis is the strain from applied mechanical or thermal load which exhibits time dependency upon return to zero strain when the load is removed. Induced distortion is also referred to as hysteresis if the material does not return unless a load is applied in the opposite direction. Atoms of a metal or nonmetal located in an interstitial position (within the lattice structure) of the material may move from one type of lattice position to another when stress is applied. Upon removal of the load, the impurity atom may move back to the original position with a particular time dependency or history. Entire lines or rows of atoms may move similarly and are termed dislocation lines. Randomly oriented solute atoms displaced by stress may be rearranged by time and temperature which will return the strained material to the original position. This is the basis for some of the shape memory alloys such as titanium nickel dental wire alloys.

Grain boundaries can store energy after being strained by excess stress. The lowest energy level is when the grain boundaries have the nearest fit to the solidification configuration. After stressing a material to a small strain value, it may return slowly to the "best fit" of the grain boundaries. This is also true of composites with very small particles or fibers added. If the strain is less than the value required to fracture the fibers or separate the adhesion of the particles, there may be a hysteresis effect upon loading and unloading due to relaxation at the lowest possible level of energy expended. Heat treatment or other methods described later on will often permit additional relaxation.

Materials, which demonstrate a low rate return to the original configuration, are termed anelastic if the return is to the exact same position as at the start. Other forms of hystereses include the return to the starting position by applied compressive loading following a tensile loading. This does not constitute anelastic behavior, however. If a material is stressed such that the dislocations do not return to the original position, the onset of nonreversible plastic strain is observed. If it is forced back into position as above, the disruption of grain boundaries will eventually take place and the material will generally first work harden and then subsequently may be seriously weakened. Continued cyclic stress from tensile to compressive will eventually fracture the material.

Creep is defined as the onset of slow strain changes with loading. Creep or time-dependent strain beyond the elastic (or inelastic limit) occurs when a sustained load allows the slow movement of the dislocations or grain boundaries to proceed irreversibly. In a plastic material, the motion is due to the sliding of rows of whole molecules. Microcreep is defined as the time-dependent strain under a given sustained load which causes 1 ppm change in length. Thus, both the time and load must be considered. Also, this type of deformation is sensitive to the temperature with higher rates at higher loads and temperature.

Creep is more predominant in materials with low melting point and low elastic modulus of elasticity. For example, the use of tin–lead solders in any precision structure subjected to loading is to be discouraged due to the high creep susceptibility.

Stress—Internal to the Material or Intrinsic Stress

A number of issues relate directly to the stored energy in a material or as such within the structure. This stored energy will cause deformation of the structure as it is released. The above discussions relating to anelastic behavior of materials is indicative of the slow relaxation properties. The types of internal stress, sources, and manifestations which may occur will be discussed here.

The internal stresses that exist within a material in the absence of an applied external force contribute to instability when changes are made to the structure such as machining, etching, or during thermal excursions.

Two types of stresses or stored energy are commonly depicted. The first is a long-range stress and the second is termed as short-range. The long-range stress is that which is deeply imbedded in the material typically by processing steps such as welding or early stages of heat treatment involving severe quenches from high to low temperatures to freeze in the desired properties. This stress is on an order much coarser than the grain structure of the material and is usually referred to as the residual stress. The short-range or microstructural stress is related to the types of issues which act as true material properties. The overall sum of stresses in a material must be equal to zero unless the material is moving.

The short-range stress may often be beneficial such as the pinning of dislocations as a result of the two phase alloys or metal matrix materials. The solidification of the material with different expansion rates for the composite additions will cause a local stress at the site of the particle or fiber, and will generally pin the locally surrounding material increasing the elastic modulus and yield strength properties. While the local size of the short-order stress may be small, the magnitude may be very high and even cause cracking of the material in the bulk. Any change in the long-range stress, such as by removing material from one side of a bar with a nonuniform distribution of stress, will certainly cause a distortional change. For the short-order stressed material, removal of the surface by machining or by chemical means is less likely to cause warping than in the long-range stressed material.

However, when the materials with either form of stress are heated and cooled, dimensional changes can occur. Thermal cycling will allow the material to distribute the long-range stress uniformly so long as rapid heating and cooling cycles are avoided. This occurs since the heating cycle will permit the material to flow plastically and conform to the thermal expansion mismatch of the smaller particles or the matrix addition particles. Such materials are generally "age hardenable". This means that an improvement not only in hardness but in the yield and ultimate strength will be realized along with the improvement in stability. Often for an optical system, it is required that the age-hardening steps be carried out past the point of the highest strength in order to fully relieve the internal or long-range stresses. This should be done sequentially to allow rough machining and stresses introduced during machining to be relaxed prior to final precision machining such as single-point diamond turning.

Short-range stresses can sometimes be relieved by low amplitude cyclic stress, which varies from tensile to compressive over many cycles such as by low frequency vibration. A large number of slowly applied thermal cycles including liquid nitrogen immersion with slow cooling permitted, and rapid quench in hot water in reverse of the original sequence of solution temperature (very hot) to boiling water, will also often achieve this, but it can be very cumbersome from the point of view of the manufacturing time. This process reverses the stress direction from the outside (compressive) into the interior of the material (tensile). If performed carefully the total internal stress can be reduced to nearly zero. See Section 10.5 ("Precipitation-Hardened Metals") for additional details.

Strengthening Mechanisms

Stress is typically highest in crystalline materials. This is especially true for small powder HIP and HCP materials. Beryllium is intriguing since loading in it due to acceleration or gravity is low because of the low density. Also, deformation is low due to the high elastic modulus. It is true that pure beryllium and most beryllium alloys close to pure Be have a low microyield stress. However, recent developments in hot isostatic pressing of Be + BeO have lead to more stable alloys. As explained earlier, many materials undergo a stress during cooling or heating cycles. This phenomenon has been studied and it has been shown that for stresses which arise between grains in an HCP material and the bulk material, the stress is due to a mismatch in the CTE of the different axes of the actual individual crystallite. The shear stress τ is given by:

$$\tau_{max} = \int_{Tm}^{T} \frac{\overline{E}(T)\,\Delta\alpha(T)}{4(1+\nu)}\,dT$$

where   E(T) = average elastic modulus at temperature T
        $\Delta\alpha$ = bulk CTE – (c-axis CTE)
        v = Poisson's ratio
        T = temperature considered
        Tm = minimum temperature

For example, when beryllium is cooled from room temperature to
   4K, τ = 7200 psi. Most aluminum alloys are free from this stress.
   The yield strength is also related to the crystalline structure. The grain size is related to the yield strength by the Hall-Petch relationship:

$$\text{Fracture Strength} = \left[K\left(\sigma_y + \sigma_o\right) \times \left(d/r\right) \right]^{1/2}$$

where   K = constant relative to a particular material
        d = grain size
        σ_y = yield stress considered
        σ_o = Peierls stress (friction stress)
        r = distance from onset of crack

   The Peierls-Nabarrow stress or σ_o is actually not constant for a material so much as for the method of forming the material such as heat treatment, impurity additions, and cold work which affect the dislocation density.

Heat Treatment of Optical Materials

As mentioned earlier, it is simply not possible to discuss all of the categories of materials used in optics nor how they are apt to behave to heat treatment in all cases. It is a challenge to convey with enough emphasis the more common material processing steps and perhaps a few which have many applications in optics but are less understood by the optical design community.

Precipitation-Hardened Metals

Precipitation-hardened metals form perhaps the largest category of optical structural and many mirror components. One such metal is 6061 aluminum alloy which can be heat-treated to several conditions and for which stabilization processes are fairly well defined.

   This alloy is characterized by the copper content, which is just below the eutectic value such that when the alloy is cooled from a melt condition rapidly, the solution phase is retained to a degree. The rapid cooling prevents migration of significant copper into the grain boundaries but permits significant pinning and grain size reduction as mentioned above. Chapter 4 shows the effect of cooling the aluminum alloy in different liquids from the solution temperature. This effect is due not only to the ratio of the thermal conductivity of the metal to the solution, but also to the ambient temperature of the solution. The idea is to quench the solution or maximum amount of copper in solid solution, and precipitate the rest including specific addition agents in a fine dispersion, which causes many dislocations and pinning sites in the alloy. If the metal is raised even a few degrees above the solution temperature specified for a given aluminum alloy, then the copper may migrate into the grain boundaries. This will soften and weaken the metal irreversibly. Therefore, the metal is very carefully heated to solution temperature and quenched in a liquid coolant to freeze in the desired structure. The problem is that for thick sections, the cooling rate is not uniform since the temperature of coolant will increase, and of course the cooling effect reaches the interior of the alloy slowly.

   Therefore, as shown in Figure 4.3 (Chapter 4), the bulk piece has different properties including residual stress. This is worse for the case of a very thick piece quenched in a liquid which starts out cool but is rapidly heated. Two causes are known. First, the liquid quickly reaches the boiling temperature in contact with the part, and gives up additional heat due to the latent heat of vaporization and then can only transfer heat at the rate consistent with conversion from steam to vapor or 80 cal/g for water vapor compared to 454 cal/g for water to steam. So at the initial onset of the quench, the outer metal is cooled very rapidly and freezes in the desired structure. Deeper in the metal, the rate of cooling is less as described above and the grain size becomes larger with less pinning. This, in turn, causes a distribution of stress, the average of which must be zero or the material must be in motion as discussed in Chapter 4. For most practical engineering purposes, this is of no concern since the structure as a whole may meet all expectations in performance. For an optical component, however, when this piece of metal is machined removing the preferential stress, the motion that occurs is not acceptable. The next step in metal preparation is the stabilization of this distributed stress to the extent possible. If the aluminum is purchased as T-6, then the entire piece has been through an "aging" process, which will lower the stresses and improve the yield and microyield properties considerably.

If the item to be machined is a mirror, then the trade-off is yield and ultimate strength for stability. Overaging is preferred if a sacrifice in maximum properties can be tolerated. An additional step is to cycle the part after all but the final machining to low and high temperatures by using liquid nitrogen and boiling water cycles. This plays a similar role as the initial quench from solution temperature. The outermost metal first cools faster than the inner metal and the contraction at the surface places the outer metal in compression. This step is usually best done slowly to avoid additional strain in the same direction as before. Then the part is quickly quenched in boiling water or other quenching media of choice such as propylene glycol, and the stresses reverse causing a reversal in the strain and concurrently the stress. This serves to stabilize the metal in the opposite sense from the previous history. Therefore, the last cycle is typically cold to hot for this reason. The final optical machining is then performed.

Work-Hardened Metals

Work-hardened metals include many pure or elemental metals which only form large grain size with poor pinning due to few dislocation or impurity sites from solidification from the melt. Typical examples include copper or solid solution alloys such as brass containing a large percentage of only zinc, which is very soluble in the copper. This category of metals is seldom used for optics unless for some special property which might be desired such as very high temperature use, high thermal conductivity, or some particular corrosion behavior. These metals can be hardened, such as for brass or copper as examples. In this case, the metal is rolled sometimes at elevated temperature, until a particular reduction in cross section is achieved. This breaks down the grain size and increases the strength and yield properties considerably. The trade-off is again internal stress which, in turn, causes motion in the form of the same strain relief as before. Stabilization is similar to the aluminum alloy and some sacrifice in yield strength is generally traded for internal stress relief.

The same effect can be achieved with very low internal stress from electrodeposited copper, which has excellent thermal conductivity and relatively high strength if deposited under certain controlled conditions. Some copper optics have been formed this way, but due to high weight-to-strength ratio the issues of thermal dissipation are usually dealt with in other ways than copper optics.

Stabilization of Selected Advanced Materials

Many methods of forming components to near net shape are used for optical components as mentioned previously. Stabilization of these materials depends on the nature of the material as well as the process. HIP powder metals, CVD, or electrochemically deposited shapes (electroforms) each behave differently for different processes unique to each.

Electroformed Optics

Electroforming is the fabrication of free-standing components by the electrodeposition of a metal. Nickel and copper are the most common metals, however, many others can be utilized such as silver or gold. The requirements for an electroformed optical component are very stringent by most electroplating standards. By proper control of the chemistry and the process, in general, it is possible to deposit stress-free metal shapes with thicknesses of one or more millimeters which replicate a precision master surface. Numerous references are available regarding electroforming. A committee and dedicated symposiums meet to discuss the state of present applications. This is sponsored by the American Electroplaters and Surface Finishers Association located in Orlando, FL.

Electroformed nickel perhaps has the most intense history of stress control of electroformed materials. This is in part due to the utility and relative ease of forming free-standing shapes. Additionally, the elastic modulus of nickel is about 200 GPa (28 Mpsi) with the density at about 8.9 g/cm^3. In order to accomplish fabrication of optical components, the intrinsic or internal stress must be controlled to avoid deformation of the replicate electroform. Stress in an electrodeposit is related to a number of variables. In general, the stress is caused by impurity atoms and by dislocations in the deposit. Hydrogen may cause either compressive or tensile stress and may cause relaxation over time due to the release. The hydrogen enters the metal electrochemically, also. The pressure is related to the pH and the potential at the surface. Also, the absolute rate of deposition of a metal is determined by Faraday's law, which will be defined in the next section. The absorption or uptake of hydrogen can be at pressures greater than the bond strength of the grain boundaries, causing rupture or severe hydrogen embrittlement in some cases. This is, of course, extreme and totally unacceptable in an optical component, although spontaneous in the electrodeposition of chromium, for example.

Typically nickel is plated from a sulfamate solution, which will be described in more detail later. The overall process permits a small amount of sulfur to be codeposited in parts per million, which forms a nickel sulfide striation in the grain boundaries. This in turn causes a minute compressive stress. By controlling the current density to a precise value using real-time stress monitoring, it is possible to continuously deposit zero stress nickel by controlling the diffusion ratio of nickel sulfide-forming compounds to the nickel deposit. Subsequent heat treatment is seldom required or desirable in this case.

Nickel X-ray mirrors have been formed by NASA in the U.S. and by the European team involving Italian and German cooperative efforts.[3,4] These mirrors are of a grazing incidence Wolter I design with extremely fine internal surface finish requirements. The optical surfaces are required to have less than 1 nm rms roughness and circularity of about 1 μm/25 cm diameter.

Cast Hypereutectic Aluminum Silicon

The silicon aluminum hypereutectic mentioned earlier is a multiple-phase material and behaves to heat treatment much like other precipitation-hardenable aluminum alloys. Although the alloy is considered somewhat brittle, it has very good resistance to thermal shock and edge notches. The material can be autoclaved to eliminate casting voids by placing in a sealed stainless steel bag and heating in an autoclave at 510°C (950°F) while applying about 84 Mpa (12 Ksi) pressure. The standard solution and precipitation heat treatments should then follow prior to final machining. The stability is increased by liquid nitrogen to boiling water quenching which will not crack this material.

10.6 Chemical and Vacuum Coating Processes in Optics

The use of both chemical and vacuum coating processes as applied to optics is of paramount importance. Like other material issues, the coating processes are not always well understood by the optical designer. It does not suffice to specify a coating simply because a previous specification was used on a different system.

Vacuum Coating Processes

The most widely used and also variant methods for producing optical coatings is referred to as vacuum coating processes if indeed the parts are placed in a coating chamber from which the air is removed. The methods used depend on the material and thickness to be deposited. For single metals and some of the lower melt point alloys the simple processes are to be preferred in most cases. Common categories include the following:

Physical vapor deposition — evaporation — sputtering

> High vacuum
> Inert gas
> Reactive gas
> Ion assisted
> Bias assisted

Chemical vapor deposition — chemically induced reaction at high temperature

> Hydrogen reduction — $WF6 + 3H2 \rightarrow W + 6HF$
> Organometallic decomposition
> Polymerization from monomeric-evaporated material through dimer chain reactions

Evaporation is one of the more common methods used to coat optics. The material to be deposited is placed in a crucible or "boat" and typically heated either by a resistance heater or by an electron beam. The material is in a hard vacuum of 10^{-6} torr or less. This permits the vapor pressure vs. temperature to be sufficiently low to evaporate the material which then condenses on the colder surroundings forming the coating. This results in a "line of sight" deposition with a cosine-shaped cloud density distribution above the source. Better use of the material occurs when the substrate is covered by the entire "cloud", but this is seldom possible such that surrounding parts of the chamber are also coated. A bias of DC or AC high frequency (RF) may be applied to an inert gas or in some cases a reactive gas to add additional elemental components to the deposit or to enhance the adhesion to the substrate. This also helps to confine the distribution somewhat.

Often more than one boat is used with different materials to form coatings in layers to enhance certain properties such as adhesion. For example, gold does not adhere as well to glass as chromium so the chromium is deposited first followed by the gold layer. In this case the vacuum cannot be interrupted or the chromium spontaneously oxidizes to a thickness of oxide of about 10 to 100 Å, which again is sufficient to prevent adhesion of the gold. The thickness of the deposits can be determined by the use of a quartz crystal oscillator monitor, which changes frequency of resonance with increased thickness of the deposit providing a real-time indication of the thickness. The oscillator portion of this instrument is also placed in the path of the depositing material.

Aqueous Coating Processes

Aqueous coating processes likewise are variant in nature and used for a variety of applications. The processes are by definition those performed in water, but similar processes are possible in organic solvents and even in molten salts, as briefly described in Chapter 4.

Two categories will be briefly described here. First is the issue of producing a coating by an applied current. This is usually referred to as plating when the current is negative on the part permitting transport of the metal to the substrate surface by dissociation from the anion–cation complex. The rate of deposition is governed by the metal valence and the molecular weight. The relationship is defined by Faraday's law as:

$$\text{Grams Deposited} = \frac{\text{MW} \times \text{Time in Seconds} \times \text{Current (Ampere)}}{\text{n} \times 96{,}500}$$

where   MW = molecular weight
           n = valence of ionic metal in solution
     96,500 = Faraday's constant

Often hydrogen is discharged with the metal and must be accounted for in this calculation. The efficiency is the amount of metal actually deposited vs. the predicted amount from this equation.

The coating obtained when the part (or substrate) is positive to an auxiliary electrode is an oxide formed by dissociation of the water and diffusion-controlled metal for combination which is taken from the substrate. This process is referred to as anodizing and is extensively used to produce coatings on aluminum and magnesium. The coating thickness and formation rate are less determinate than above due to the fact that the metal must diffuse through the coating as it forms. In order to maintain a coating with proper density of oxygen and metal, the rate of formation must be limited to that rate for which the metal can equilibrate to the oxide. The coatings are porous, especially when thicker than a few microns, on aluminum. The porosity can be sealed by immersion of the item in boiling water. Dyes can also be added to the process and sealed into the pores in a similar fashion. This is the basis for the black anodize process used to form some nonreflective aluminum parts for optical baffles and other nonreflecting devices.

Electroless Nickel

The second type of aqueous coating commonly applied to optical components is the electroless or autocatalytic nickel–phosphorous alloy applied to single-point diamond machined aluminum mirrors. The term electroless is actually a misnomer in that although no current is externally applied to the plating parts, current is supplied to the surface through electron transfer due to reduction of the nickel from nickel sulfate and phosphorous from hypophosphite. Other mechanisms are possible with additional metals as alloying elements or phosphorous substituted by boron reduced from borohydrates or diethylamineborane. The intrinsic or internal stress in the alloy is a direct function of the percentile of phosphorous if all additional conditions are the same. The Ni-P alloy in all compositions has significantly lower thermal expansion (or contraction) than nearly all aluminum alloys commonly used in optics such as 6061 or 2024. A commonly desired composition is the ratio of 11% phosphorous and 89% nickel which corresponds with [Ni_3P], where the brackets indicate the material is more nearly amorphous than the crystalline allotrope. The plating rate is determined by the temperature and the ratio of nickel to hypophosphite or other "reducing" agent in accordance to an activation or Arrhenius rate law dependent on the specific process. The reader is referred to references on electroless nickel plating for additional chemistry detail.

If the nickel alloy is applied at neutral stress at the plating temperature of about 190°F, and the coating and substrate are subsequently cooled to room temperature, then the aluminum will contract more than the coating, placing the coating in compression. Since the part must come to equilibrium, the substrate interface with the coating is in tension. A flat aluminum plate or strip coated on one side with NiP at low stress during plating at elevated temperature will then deflect such that it becomes convex on the plated surface when cooled. If, however, the coating is the same thickness on both sides, the deformation in the coating will be distributed with linear deformation of the substrate interfacial surface and the distortion will be minimal, albeit the stored energy may be higher. Likewise, if the coating is applied at a sufficient tensile stress, then the relaxation due to lower temperature would allow the coating to come to low or zero stress at room temperature. Heat treatment can be used to relieve some of the stress by exceeding the plating temperature significantly. Permanent distortion occurring at the elevated temperature will relax more nearly to zero at room temperature in some cases. Also, as pointed out in Chapter 4, the lower phosphorous nickel alloys may change from amorphous to crystalline with a reduction in volume resulting in a better match at room temperature. This higher temperature heat treatment is not usually recommended for coated aluminum alloys due to the overaging incurred. Also, the polishing properties of the alloy may be diminished in this way.

The final stress in a plated strip mentioned above can be expressed as a function of the curvature of the strip. This procedure is commonly used to predict the behavior of a coated system. This is done in several ways and depends on whether or not the strip was permitted to bend while plating one side. The typical method of choice is for the piece to be plated on both sides without bending and subsequently stripped on one side to permit the bowing.

The stress due to CTE mismatch can then be accounted for by the ratio of the CTE values for each metal, the delta temperature, the ratio of the elastic moduli, and the overall length of the strip. The stress can be considered biaxial for simplicity with good results. This should be calculated as if the strip was still straight for best results.

Another more direct method is to use a commercially available electronic stress monitor (U.S. Patent 4, 986, 130) to measure the stress in real time. Care must be taken to account for the differential CTE in this case if the optical component is aluminum or another high expansion metal. Since the gauge is made of stainless steel, the CTE is close to the nickel phosphorous at the desired 11% phosphorous alloy such that the reading is very nearly that of the stress in the nickel phosphorus at the deposition temperature. The instrument is also used to measure stress in real time for other coating processes.

Electrolytic Deposits

The most commonly applied electrolytic coatings for optics include gold and nickel. Copper is used as a primary layer in some cases to provide a diamond-machinable surface prior to electroless nickel coating for diamond machining. Pure nickel coatings do not diamond machine well due to formation of carbides, with the very sharp diamond tool edge causing breakage and subsequent dulling of the tool. The phosphorus in the electroless version minimizes this especially in the higher phosphorous alloys. Pure nickel is used, however, for the electroformed optics described earlier and for corrosion-resistant coatings on many major components. Gold can be applied as pure gold or as an alloy. The alloys are unique to specific desired properties. The addition of copper may enhance the hardness and durability of the coating while shifting the best IR performance to longer wavelengths. This also provides for a gold coating which can be applied sufficiently thick and with adequate hardness to allow for polishing or even rework.

The gold alloys are typically plated from cyanide complexes with chelates, such as the versenes of ethylene diamine or triamine tetra- or pentaacetic acids for complexing the copper to maintain some valence two copper instead of the valence one coordination species of cyanide. Also, excess cyanide must be very carefully maintained to prevent formation of AuCN, which is insoluble and can cause roughness in the deposit. Also, cyanide is released as the gold deposits, allowing excess cyanide to accrue. This is offset by oxidation to cyanogen, C_2N_2, which is volatile at the plating temperature and evaporates. If the free cyanide becomes excessive, then the copper is converted to monovalent copper cyanide in one or more of three possible coordination configurations which changes the plating rate and increases the deposit stress. As if this was not enough problems, the copper chelant is also oxidized at the anode, creating up to 17 breakdown products requiring frequent or continuous carbon treatment. It is evident then that both electrolytic and chronological aging of the solution must be dealt with if optical quality components are involved.

Due to these difficulties, only selected plating facilities are willing to coat expensive optical components with gold alloys. Future work in this area should include formulation of noncyanide gold alloy processes such as gold phosphite and copper complexes for improved stability. A study of the reflectance properties of a number of controlled alloy deposits would also be a viable optical undertaking.

Copper may be plated at high efficiency and with excellent quality from acid or alkaline solutions. The material can be plated with controlled stress from acid sulfate processes, with proprietary stress-reducing agents added much like nickel from the sulfamate bath. The material is easily diamond machined and as such is sometimes used for electroformed shapes to be subsequently machined. Copper, however, suffers from lower hardness and elastic modulus than nickel, while maintaining the same density so the specific stiffness is lower. Also, the deposited copper suffers from long-term intergranular migration of the organic additives used in the sulfate processes if the deposit is used at elevated temperature. Even at room temperature, this may be manifested in surface roughness appearing over extended time in the diamond-turned copper surface if the plating process is not controlled to minimize the use of organic additives. Mild heat treatment prior to diamond machining may help stabilize the deposit in some instances.

10.7 Summary

The development of optical snd electro-optical systems must necessarily involve the best design and fabrication recources available. This includes personnel, equipment, and facilities. Often this quality requirement is taken too lightly resulting in confusion or despair. This chapter cannot replace the education and experience required to differentiate between engineering and optical engineering as it applies to the manufacturing of such precision systems. The typical optical system development requires cooperation and coordination of many disciplines of engineering and science in order to establish innovative and functional devices. Modern methods involving computer-aided design and analysis must be incorporated in order to remain competitive. Rapid prototyping and extremely high precision fabrication methods must also be employed. The need to organize a development project is mandatory. The source of personnel, equipment, and materials must be established prior to proceeding in order to avoid delays and difficulties.

While no single aspect of the developmment of an optical system can be listed as most important, it is worth a brief review of those which may be overlooked such that personnel can be assigned to avoid problems which may otherwise arise.

Perhaps foremost is the need to develop a model using a virtual design approach. This can be used to evaluate the design with regards to material selection, mechanical and self-loading, thermal conditions, tolerance analysis, and general issues such as optical path interference or vignetting. Next it is imperative that the subtle manufacturing details be understood. This includes a working knowledge of the ultraprecision machining methods available and the requisite inspection instrumentation and methods to verify the results. Subtle issues such as out-of-plane temporary mounting surfaces can introduce sufficient distortion when an item is bolted to a tooling plate during fabrication to preclude proper operation later as an optical device. Other issues arise from the difficulties in inspection, material stabilization, or perhaps the availability of optically qualified materials. A classic example is the use of recycled materials such as aluminum alloys which may contain large quantities of iron particles and even glass chips and still meet composition specifications. While this may be acceptable for most structural applications, it wreaks havoc on diamond machining of precision optical surfaces.

References

1. Ahmad, A., Engelhaupt, D., Feng, C., Hadaway, J., and Ye Li. 1994. Design and fabrication of low-cost lightweight metal mirrors. In *Proc. Manufacturing Process Development in Photonics Conf.* Redstone Arsenal, AL.

2. Ahmad, A., Feng, C., and Sarapaka, R. 1995. Virtual prototyping: a cost-effective emerging methodology. In *SPIE Proc. Vol. 2537-39.*

3. Citterio, O., Bonelli, G., Conti, G., Mattaini, E., Santambrogio, E., Sacco, B., Lanzara, E., Bruninger, H., and Burkert, W. 1988. Optics for the X-ray imaging concentrators aboard the X-ray astronomy satellite SAX, *Appl. Opt.*, 27, 1470–1475.

4. Citterio, O., Concini, P., Mazzoleni, F., Conti, G., Cusumano, G., Sacco, B., Brauninger, H., and Burkert, W. 1992. Imaging characteristics of the development model of the JET-X X-ray telescope. In *Multilayer and Grazing Incidence X-ray/EUV Optics*, Hoover, R. B., ed., pp. 150–159. SPIE Proc. Vol. 1546.

5. Fraser, H. and Daehn, G. Co-Continuous Ceramic Composite (C^4) Materials, Internal Research Report. Ohio State University, Columbus.

6. Scott, S. 1991. Electroless nickel plating for enhanced gold coating: a lesson in bimetalics. In Proceedings of Metal Platings for Precision Finishing Operations Conf. Tucson, AZ.

7. Wolf, E.G. 1990. Moisture and viscoelastic effects on the dimensional stability of composites, pp. 70–75. In *SPIE Proc. Vol. 1335.*

Index